Springer Series in Advanced Manufacturing

Series Editor

Duc Truong Pham

For further volumes:
http://www.springer.com/series/7113

Y.-S. Ma
Editor

Semantic Modeling and Interoperability in Product and Process Engineering

A Technology for Engineering Informatics

Springer

Editor
Y.-S. Ma
Department of Mechanical Engineering
University of Alberta
Edmonton, AB
Canada

ISSN 1860-5168 ISSN 2196-1735 (electronic)
ISBN 978-1-4471-5072-5 ISBN 978-1-4471-5073-2 (eBook)
DOI 10.1007/978-1-4471-5073-2
Springer London Heidelberg New York Dordrecht

Library of Congress Control Number: 2013940088

Printed on acid-free paper

Springer is part of Springer Science+Business Media (www.springer.com)

Preface

The study of *Semantic Modeling and Interoperability in Product and Process Engineering* deals with an advanced engineering informatics technology in supporting product and process modeling, development, implementation, and management.

This book is a condensed technology handbook for advanced modeling and application of engineering knowledge in product development and process management in industry. Computer-aided tools have been widely used in design and manufacturing activities. However, it is clear that most of these tools are not capable of incorporating engineering semantics into their solutions. Consequently, it is a challenge for the industry to model and apply comprehensive engineering knowledge, constraints, procedures, and concurrent aspects of design and manufacturing in a systematic and sustainable manner. In the past decade, feature-based design and manufacturing applications have gained momentum in the Computer-Aided Design and Manufacturing (CAD/CAM) domains to represent and reuse some design and manufacturing patterns with effective applicability. However, the actual scope of feature application is still very limited.

The editing author of this book intends to expand the scope of feature technology to the more open approach of engineering semantic modeling, and to provide a framework of technological solutions for system integration and information-sharing. This book presents a set of researched methods that can consistently represent and uniformly manage engineering semantics in the ever-dynamic evolvement of modern enterprise. With a proposed common infrastructure of product and process modeling, the interoperability among different computer systems is addressed based on a fine-grain feature-based informatics approach. This book also features some insightful case studies that show promising application prospects at different product stages and in different areas of design and manufacturing. Academics, advanced engineering students, and practicing engineers will benefit from the methodology and techniques proposed in the book, which will serve as useful references, guidelines, and helpful tips for their teaching, research, and development, as well as provide real engineering innovation projects which will increase industrial competency in the field of engineering informatics.

The proposed unified feature scheme, based on the associative feature concept and an important expansion of the well-known feature-based design and manufacturing approach, offers a systematic framework of fine-grain semantic modeling methods for representation and application of engineering knowledge, constraints, and associated engineering procedures. This semantic modeling technology supports uniform, multi-faceted, and multi-level collaborative system engineering with heterogeneous computer-aided tools, such as CAD/CAM, Computer-Aided Engineering (CAE), and Enterprise Resource Planning (ERP) applications. The enabling mechanism, which is essential to enable the proposed technology of implementing associative features, is introduced in detail from concept to implementation, then expanded to real-world applications. Practical case studies are provided to readers with insightful application references that enable readers to use the proposed method with some tested templates.

Semantic Modeling and Interoperability in Product and Process Engineering provides a reference solution for the challenging engineering informatics field, aiming at the enhancement of sustainable knowledge representation, implementation, and reuse in an open and yet practically manageable way. It is the authors' goal that this book becomes a valuable reference for scholars, practitioners, and learners from both academia and the engineering field.

There are 11 chapters in this book. They are organized as follows.

"Introduction to Engineering Informatics" gives a general definition of engineering informatics and reviews the current applications of modern engineering informatics in product development, concurrent and collaborative engineering, and chemical process engineering. A few fundamental technologies for developing engineering informatics solutions are briefly reviewed, including object-oriented software engineering, Unified Modeling Language (UML), multi-faceted data repositories, data mining, and semantic modeling. This chapter describes the context of engineering informatics which is the core theme of this book.

"A Review of Data Representation of Product and Process Models" reviews the state of the art in product and process informatics modeling and implementation. Through a systematic review, it concludes that interoperability among computer systems is a major hurdle. Feature technology is particularly reviewed for its strong versatility in engineering informatics. In this chapter, a special section is devoted to reviewing chemical process engineering informatics modeling systems and processes because this application domain has a unique nature which is different from other typical consumer product engineering practices, and has special local industrial relevance to the authors.

"An Example of Feature Modeling Application: Smart Design for Well-Drilling Systems" provides a typical feature modeling and implementation example, i.e., the development of an oil well drilling design calculation and CAD model generation software tool. Detailed system design modules, such as drilling well casing design, drilling string design, and operation parameter optimization, are described, and case demonstrations for these modules are presented.

"Fundamental Concepts of Generic Features" introduces the concept of *generic feature*, which is a fundamental concept of a common feature class definition that enables representations of different domain features on top of the generic feature with built-in methods to define the most common properties, such as geometric references, attributes, and the related constraints. The geometric and non-geometric consistency-checking methods are also discussed with a multi-view definition approach. Further, a preliminary entire product model with an open framework to accommodate multiple engineering applications is presented; this is the *unified feature* modeling framework.

"Unified Feature Paradigm" is related to "Fundamental Concepts of Generic Features" in that it offers more details about the *unified feature* modeling framework. In this chapter, the modeling scheme, cellular model representation, knowledge-based reasoning, and association and change propagation are presented. Constraint modeling, defined in a broader sense than just geometric constraints, is also illustrated with the help of graph theory. The key mechanism for maintaining feature consistency, i.e., the multiple-view association method in the unified feature system, which enables engineers to access data pertaining to specific engineering domains from the generic feature, is described. A new feature paradigm to support engineering interoperability has been established by "Fundamental Concepts of Generic Features" and "Unified Feature Paradigm".

"Features and Interoperability of Computer Aided Engineering Systems" explores advanced feature-related technologies from the angle of application. It begins with the development of CAx systems and their customization, then reviews the inherent problem of interoperability. After discussing the pros and cons of the most common approach, i.e., STEP-based data exchange and system integration, this chapter explores feature-based solutions with a thorough review of the research field. It provides a well-organized and carefully categorized reference for readers. One section is dedicated to chemical process engineering as well, which can be appreciated for the identification of interoperability gaps. An integrated system architecture is proposed.

"Data Representation and Modeling for Process Planning" presents an insightful research effort in data structure modeling in the domain of computer-aided process planning, which covers manufacturing procedure design with tool and machine alternatives based on a dynamic manufacturing environment. Data search algorithms are suggested with the support of various engineering databases. The system described can generate feasible manufacturing process solutions and estimate cost and time efficiently. With rich data and cases from the real world, this chapter will be a valuable reference for developers and researchers in the field.

"Computation of Offset Curves Using a Distance Function: Addressing a Key Challenge in Cutting Tool Path Generation" illustrates an important aspect of engineering informatics, i.e., the algorithms needed to deal with complex path problems. One of the most commonly used algorithms is curve offsetting; this chapter reports on a recent development of a curve offsetting model with a distance function. The method is different from the traditional approximating methods based on interpolation in that it approximates the progenitor curve with bi-arcs and generates the exact offset curve with direct error control.

"Feature Transformation from Configuration of Open-Loop Mechanisms into Linkages with a Case Study" proposes a feature synthesis method for design manipulators based on a hybrid method of Artificial Neural Network (ANN) and optimization techniques. This approach is useful for solving reverse linkage dimension design problems, such as an excavator design for a predefined reaching envelope profile curve. In fact, the proposed solution to this problem presents a typical example of mapping from the product specification feature to the product configuration feature.

"Feature-Based Mechanism Design" is a continuation of typical mechanism design as shown in "Feature-Based Mechanism Design", but focuses on dimensional synthesis, embodiment, and CAD model generation with minimum designer intervention. Parametric feature-based modeling is successfully applied in mechanism embodiment design. In both "Feature Transformation from Configuration of Open-Loop Mechanisms into Linkages with a Case Study" and "Feature-Based Mechanism Design", features are effectively used for semantic knowledge representation, product modeling, design process interactions, and design intent evaluation. These two chapters demonstrate an advanced feature modeling and engineering approach to embed and evaluate design intent.

"A Smart Knowledge Capturing Method in Enhanced Generative Design Cycles" proposes a new method to capture and reuse engineering knowledge through CAD and CAE interactions by recording journal files and creating reusable source codes for generative CAD and CAE integration. The CAD/CAE feature information and data associations are modeled and implemented in a common data model, which makes data sharing easily attainable. It offers a design automation solution for those products with relatively predictable configurations and constraints.

This book provides a systematic engineering informatics modeling and application methodology that is based on original research carried out over the past two decades. The in-depth descriptions of the new feature paradigm as well as in-depth process planning and product assembly data modeling are the book's primary achievement. Readers will benefit from the systematic theory and the numerous application cases, and are given exposure to the effectiveness and usefulness of the proposed engineering informatics methodology.

The editing author would like to take this opportunity to express his appreciation to all the co-authors of the 11 chapters for their significant contributions to this book. Their excellent research efforts, insightful observations, and valuable commitment have made this book much more solid in theory and rich in case studies. Among them, the editing author would specially thank Professor Qingjin Peng, from University of Manitoba, who contributed "Data Representation and Modeling for Process Planning"; and Dr. C. K. Au, from University of Waikato, who contributed "Computation of Offset Curves Using a Distance Function: Addressing a Key Challenge in Cutting Tool Path Generation". These chapters are substantial contributions complementary to other chapters, and make this book more complete in coverage. The author would also like to extend his sincere thanks to Dr. Rachel Hertz Cobb for her professional editing and patient correction of those numerous typo and grammatical errors; and Miss Katy Moore who diligently formatted the book as per the publisher's requirement in a very short period of time. Their great effort has been instrumental to the timely delivery of this book as planned, and the consistent quality as expected.

Edmonton, AB, Canada, 2012 Y.-S. Ma

Contents

Abbreviations

5 S's	Japanese words: *seiri, seiton, seiso, seiketsu,* and *shitsuke*. English words: sorting, straightening, shining, standardizing, sustaining
AADE	Autonomous agent development environment
ACM	Application cellular model
AF	Application feature
AFM	Application feature model
AFR	Automatic feature recognition
AL	Application layer
ANN	Artificial neural network
ANSI	American national standards institute
AOR	Annual operational requirement
AP	Application protocol
APFM	Assembly planning feature model
API	Application programming interface
ASFM	Analysis feature model
B2B	Business-to-business
BHA	Bottom hole assembly
BIT	Built-in test
BOM	Bill of materials
BS	Bounding sphere
BST	A file extension with an integrated engineering data format
CAA	Component application architecture
CAAD	Computer-aided aesthetic design
CAC	Corrosion allowance constraint
CACD	Computer-aided conceptual design
CAD	Computer-aided design
CAE	Computer-aided engineering
CAI	Computer-aided inspection
CAM	Computer-aided manufacturing
CAPP	Computer-aided process planning
CAS	Computer-aided styling
CATS	Computer-aided tool selection
CSYS	Coordinate system

CDFM	Concept feature model
CDM	Common data model
CE	Concurrent engineering
CEC	Capacity-of-equipment constraint
CEE	Collaborative engineering environment
CIM	Computer-integrated manufacturing
CNC	Computer numerical control
CORBA	Common object request broker architecture
CPCDF	Chemical process conceptual design feature
CPDM	Collaborative product data model
CPE	Chemical process engineering
CRM	Customer relationship management
CSG	Constructive solid geometry
CSM	Component supplier management
DB	Database
DBF	Design by feature
DBMS	Database management system
DC	Design change
DDFM	Detail feature model
DDL	Data definition language
DFA	Design for assembly
DFM	Design for manufacturing
DHT	Design history management tool
DIDE	Distributed intelligent design environment
DL	Data layer
DML	Data manipulation language
DNC	Diameter of nozzle constraint
DOF	Degree of freedom
DP	Diameter of piping
DSS	Decision-support system
DXF	Data exchange file
EAI	External authoring interface
EC	Engineering change
ECM	Engineering change management
ECP	Engineering change propagation
EIDL	End item design life
EPC	Engineering, procurement, and construction
EPM	Entire product model
ERP	Enterprise resource planning
ESA	Engineering server agent
ESDS	Expert slurry-design system
FAT	Fastener-axis action tool
FBD	Free body diagram
FE	Finite element
FEA	Finite element analysis

FEM	Finite element method
FR	Feature recognition
Fr	Flow rate
GA	Genetic algorithm
GT	Group technology
GUI	Graphical user interface
HVAC	Heating, ventilation, and air conditioning
IBIS	Issue-based information system
IA	Interface agent
ICT	Information and computer technology
IEC	International Electrotechnical Commission
IGES	Initial graphics exchange specification
IPDE	Integrated product data-sharing environment
ISO	International Organization for Standardization
IT	Information Technology
JA	Job agent
JSDT	Java shared data toolkit
JTMS	Justification-based truth maintenance system
KA	Kinematic analysis
Kanban	A Japanese term, i.e., "Signboard"—it means a customer requirement driven scheduling system for lean and just-in-time production
KBE	Knowledge-based engineering
KBS	Knowledge-based system
MAD	Mass acceleration diagrams
MDA	Minimum deviation area
MDDF	Mechanical detail design feature
MDO	Multidisciplinary design optimization
ME&D	Mechanical engineering and design
MFC	Microsoft foundation class
MMS	Multi-model structure
MMT	Multi-model technology
MOKA	Methodology for knowledge based engineering applications
MOO	Multiple-objective optimization
MPFM	Machining planning feature model
MSE	Mechanical specific energy
MTBF	Mean time between failures
MTTR	Mean time to repair
MV	Machining volume
NC	Numerical control
NDF	Neutral data format
NMT	Non-manifold topological
NPD	New product development
NURBS	Non-uniform rational B-spline
OBB	Oriented bounding box

OD	Outer diameter
OLAP	Online analytical processing
OO	Object-oriented
OODB	Object-oriented database
OOMF	Object-oriented manufacturing feature
OOP	Object-oriented programming
P&ID	Process and instrumentation diagram
PCM	Product configuration management
PDM	Product data management
PDMS	Product data management system
PFD	Process flow diagram
PLM	Product lifecycle management
PMEX	Performance measurement evaluation matrix
PMPD	Performance measurement for product development
poke yoke	A Japanese term-error proofing
PRP	Product realization process
PVD	Physical vapor deposition
QFD	Quality function deployment
R&D	Research and development
RDB	Relational database
RE	Reverse engineering
ROP	Rate of penetration
RPM	Revolution per minute
RT	Residence time
SMED	"Single minute exchange of dies"
SOO	Single objective optimization
SPM	Supply and planning management
SSL	Semantic schema layer
SQL	Structured query language
STC	Shell thickness constraint
STEP	Standard for the exchange of product data model
TAT	Tool-axis action tool
TPS	Toyota production system
TVD	True vertical depth
UC	Unified cells
UCM	Unified cellular model
UI	User interface
UML	Unified modeling language
VE	Virtual environment
VRML	Virtual reality modeling language
WOB	Weight of bit
X3D	Extensible 3D
XML	Extensible markup language

Symbols

β'	Angular measurement parameter
A	Area
A_{dx}	Area direct stress
A_{tor}	Area torsional
x	Axis x
y	Axis y
σ_{ZX}	Bending stress
β	Boom deflection angle
α_2	Boom lower angular displacement limit
RB_1	Boom rotation matrix #1
RB_2	Boom rotation matrix #2
b	Boom side length
α_1	Boom upper angular displacement limit
α_{bu}	Bucket angle
h	Cross-sectional height
b	Cross-sectional base dimension
t	Cross-sectional plate thickness
dA	Differential area
$dig1$	Digging configuration angle #1
$dig2$	Digging configuration angle #2
σ_{dx}	Direct stress
Q	First moment of area about neutral axis
F_x	Force along x-axis
F_y	Force along y-axis
F_z	Force along z-axis
l_1	Hinge to hinge boom length
l_3	Hinge to tip bucket length
A	Homogeneous transformation matrix
H	Horizontal excavator arm lengths
$J1$–$J11$	Joint 1, … 11
h_J10	Joint J10 extension distance
h_J11	Joint J100 extension distance

h_J2	Joint J2 extension distance
h_J8	Joint J8 extension distance
c	Linear measurement parameter
S_3	Maximum cutting height
S_7	Maximum depth cut at level bottom
S_2	Maximum digging depth
S_1	Maximum reach at ground level
S_4	Maximum loading height
S_6	Maximum vertical wall digging depth
S_5	Minimum loading height
M_x	Moment about x-axis
M_y	Moment about y-axis
M_Z	Moment about z-axis
d_p	Pin diameter
l_p	Pin length
R	Rotation transformation matrix
SD	Scope display output
I	Second moment of area
V	Shear force
l_2	Stick length
RS	Stick rotation matrix
$<JS_3$	Stick-joint J3 interior angle
$<JS_2$	Stick-joint J2 interior angle
$J2_L$	Stick-joint J2 left interior angle
$J2_r$	Stick-joint J2 right interior angle
$J3l$	Stick-joint J3 lower interior angle
$J2u$	Stick-joint J3 upper interior angle
$<JS9$	Stick-joint J9 interior angle
$J9l$	Stick-joint J9 lower interior angle
$J9u$	Stick-joint J9 upper interior angle
τ_{tor}	Torsional shear stress
b_2	Transition four-bar coupler link
b_1	Transition four-bar follower link
b_0	Transition four-bar stationary link
b_3	Transition four-bar driver link
T	Translation transformation matrix
τ_b	Transverse shear stress
σ_u	Ultimate stress
V	Vertical excavator arm lengths
σ_y	Yield stress

Introduction to Engineering Informatics

Narges Sajadfar, Yanan Xie, Hongyi Liu and Y.-S. Ma

1 What is Engineering Informatics?

Engineering informatics is an applied information science sub-domain that is scoped to address information technology (IT) knowledge, methods, models, and algorithms that support engineering and management activities ranging from customer requirements to design and production operations. In fact, this sub-domain overlaps in application scope with the modern product lifecycle management (PLM) concept [46]. Engineering informatics is a growing domain of science for industry applications and is intended to address the basic principles of enhancing the functionality, flexibility, efficiency, and consistency in information and computer technology (ICT) solutions in engineering. Using IT solutions, engineering informatics is aimed at increasing engineering quality of products and the complex associated processes, as well as the management of product lifecycles. The continued success and progress in this field has become critical to the economies of many countries because of IT penetration into almost every imaginable product, project, and work team. The effectiveness and efficiency of IT solutions in a national industry has a direct effect on the country's industrial competitiveness in the global economy. Implementing IT solutions brings changes in almost all aspects of industry, including the intellectual properties of products and production, investment, and human resources. Nowadays, IT has a direct effect on a country's development of the information and communication economy; Internet technology development generates many new jobs.

The implementation of an appropriate engineering informatics framework in a manufactory requires insight into the heavy volume and timely interactions among all aspects of engineering management, found across various disciplines, departments, and geographical boundaries in the current networked information world.

N. Sajadfar · Y. Xie · H. Liu · Y.-S. Ma (✉)
Department of Mechanical Engineering, University of Alberta,
Edmonton, AB T6G 2G8, Canada
e-mail: yongsheng.ma@ualberta.ca

Y.-S. Ma (ed.), *Semantic Modeling and Interoperability in Product and Process Engineering*, Springer Series in Advanced Manufacturing,
DOI: 10.1007/978-1-4471-5073-2_1,

This chapter introduces several key application areas of engineering informatics: product development, measuring product development performance, and concurrent and collaborative engineering. Then, to deepen the reader's appreciation of a special application domain, computer applications in chemical engineering are reviewed in order to illustrate an industry-specific scenario. Finally, the two fundamental technologies of engineering informatics, object-oriented (OO) software engineering and semantic modeling, are briefly introduced.

2 Review of Computer Systems in Engineering Design and Manufacturing

IT tools are increasingly used for engineering activities, from concept development to solving various detailed engineering analysis problems. This trend imposes great challenges for engineering companies to implement advanced ICT tools while managing technological evolution and business efficiency. For example, one of the basic information system implementation challenges is to manage continuously generated, complicated, and tedious engineering information in a consistent data system, such as a unified database [35, 29].

It has been well established that competitive advantages for many enterprises come from timely product development and cost-effective process development, both of which are critical for manufacturing. Product development and process development require a number of complex activities [49]; however, they have to be aligned with the same objective, i.e., making as much profit as possible for the company before other competitors enter the market. Therefore, managing product development and process development with modern information and computer technologies is a useful way to increase quality and decrease cost and time of delivery to market.

2.1 Product Development

Product may be usefully defined as anything that can be offered to a market. The design of a product and its introduction to market is a multi-step process [49]. New product development (NPD), or the product realization process (PRP), is the description of the complete process of producing a new product for the market. The set of activities that makes up NPD begins with the perception of market opportunities and ends with the production, sale, and delivery of the product [48]. Numerous studies have shown that more than 40 % of new products do not have a successful launch [2]. As a result, NPD tries to increase the chance of successful production by effective management of new customer requirements and continual quality improvement. Research on the performance of NPD shows that product strategy, development process, product policy, market, environment, and production activity flow all directly affect success [32]. Therefore, NPD tries to focus on all of these factors. There are many ways to break down NPD into steps. However, the majority of them have six primary phases: envision, plan, design, develop, test, and release [48, 49].

Figure 1 shows the product development processes with five steps [12]. The first step is market analysis, which involves recognizing market requirements, the potential of the market, customer requests, the position of the company, and estimation of the probability of success. This step begins with product definition [49]. Product definition is essentially identifying bright ideas to meet market requirements. Primary questions to be answered during the product definition stage include, what is the target? Who will use this product? How much time is required to develop it? How many competitors does the company have? Once these have been answered, the research and development (R&D) department tries to document the market requirements into a set of well organized and clustered specifications and define a path to the solution by gathering detailed market information and working out the product development process.

The second step is product conceptual design, which considers various aspects of the product domain such as product variety, quality of product, design, features, brand name, performance, packaging, and services. Typically, the R&D department researches production possibilities and the practicalities of converting the bright idea to a reality, and evaluates the feasibility of product development. Innovation is the key responsibility for the R&D department, as an innovative product can increase company sales significantly. Design and simulations are used to determine customer and market demands, budgets, and schedules. The goal of the design phase is finding the best solution to meet all requirements. To prove the feasibility and the superiority of the chosen solution, product design analyses and simulations are commonly conducted from different engineering angles, such as performance, reliability, economics, and so on.

The third step is detailed design, which should include all the details of the product itself, but must also include related information, including the process plans, raw materials, works in progress, tooling, machinery, personnel, production, and delivery schedules. This step may be information dense, but each piece of information is vital to the product's success. The product development process? can be highly iterative. To verify the detailed features of the design, more comprehensive product analyses and simulations are conducted than at the conceptual design stage. If any issue is discovered in the detailed design stage that warrants a

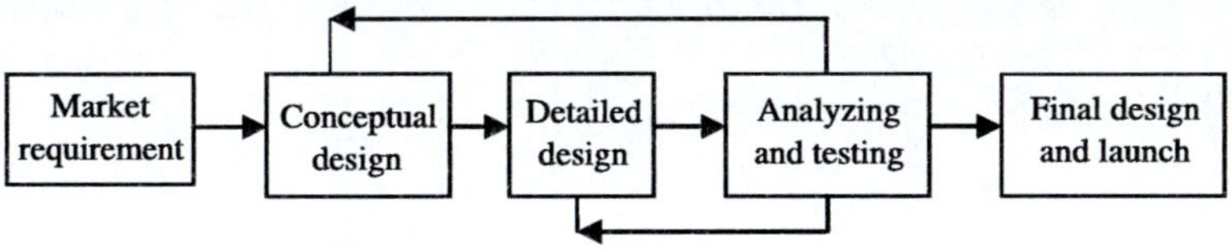

Fig. 1 Product development process

revisit of the conceptual design, then a repeated design loop going back to the conceptual design stage is executed. The evolvement of product development is often a cyclical round of revisions between conceptual and detailed design stages; the contents of product definition are also refined gradually from a macro level, which involves conceptual engineering design, to the micro level, growing progressively more detailed.

The fourth step is the analyzing and testing stage. As the penultimate step, testing is usually conducted to ensure that the product achieves the expected performance and meets quality standards and company expectations [30, 31]. The goal of this step is to produce the first real product according to the product definition. This final product development process? attempts to confirm all aspects of the product definition by verifying all of the previous steps as a prelude to early market entry, and makes every effort to increase quality and performance and further reduce production costs. In summary, managing and controlling the product development process plays an important role in a company's success.

Finally, if the primary product specifications can be fully validated, and the product tests are complete, then the manufacturing process can begin [5, 30].

Computer-aided design (CAD), one of the more important technologies in the computer-aided domain, has been widely used by mechanical industries, such as automotive and aircraft [48]. The predominant commercial CAD systems are listed in Table 1 with their geometrical kernels and applied operating systems.

Current CAD software systems are no longer constrained to solid modeling to assist design activities. Instead, they have become powerful integrated tool packages that support the whole product lifecycle. For example, the latest version of NX, V8, is currently embedded with many capabilities, such as industrial design and styling, mechanical and electromechanical systems design, visual reporting and analytics, mechanical and electromechanical simulation, tooling and fixture design, machining, engineering process management, inspection programming, and mechatronics concept design [41].

Table 1 Key features of popular CAD tools [48]

Company	Product	Geometrical kernel	Operating system	Class
Autodesk	Inventor 2010	Autodesk ShapeManager	Windows	Mid range
Dassault systèmes	CATIA V6	V6	Windows	High range
	SolidWorks 2010	Parasolid	Windows	Mid range
PTC	Pro/E Wildfire 5.0	GRANITE	Windows, Unix	Mid range/high range
SIEMENS	NX 7	Parasolid	Linux, Mac OS, Unix, Windows	High range
	Solid edge	Parasolid	Windows	Mid range

Although software tools are becoming progressively more powerful, interoperability problems still exist among tools from different vendors. The Standard for the Exchange of Product model data (STEP) [44] was developed to facilitate data exchange among various CAD packages. More recently, feature technology has been extended to solve semantic interoperability in the mechanical engineering domain, especially considering the downstream manufacturing phase of a product's lifecycle [8, 11, 21, 26, 29, 50]. However, the interoperability problem is still a barrier for concurrent and collaborative engineering, with the problem even more serious in chemical process engineering projects, in which regulated works are specifically confined to conceptual design and detailed design, and each process takes place in different domains. Unfortunately, little research effort has been expended on systematic collaboration.

2.2 Measuring Product Development Performance

Managing product development is a challenging activity, because the product application environment and the demands of the market are in a constant state of change. Adapting to change is essential for product development; however, these changes can have a negative effect on product development. As a result, product management needs tools to measure product development performance and ensure that the product performs well. Several tools have been implemented in industry to measure product development performance, such as quality function deployment (QFD), balanced scorecards, performance diagnosis, and readiness assessment for concurrent engineering (CE). All of these tools improve decision making during the product development process [13]. Measuring product development performance is a complex activity, but evaluation of the measurement is an even more important task than the measurement itself. Several tools are available for the evaluation of product development; the performance measurement evaluation matrix (PMEX), for example, is a useful tool to help management recognize what is and what is not being measured. This allows management to make adjustments to their measuring tools in order to comprehensively cover all aspects of product development performance.

Current tools for measuring product development have some limitations. These kinds of tools cannot measure all aspects of the product development process. There are also no specific definitions for soft and hard measures of product development performance. The majority of current tools measure internal performances that compare activities and processes to previous operations and targets [13], but some products need more external performance evaluation, such as customer use experience. In addition, current tools ignore the specific nature of a product. Some products need more flexible measuring tools that can be used in all steps of the product lifecycle. Performance measurement for product development (PMPD) is a new methodology that tries to clearly identify the measurement of the effectiveness of the product development process by considering the traditional limitations of nonrationalized indicators in management [13, 43].

2.3 Existing Design Data Exchange and Communication Technologies

Engineering design collaboration in the various engineering phases has been common practice, including product design, analysis, inspection, and so on. However, sharing engineering data generated from CAD/CAE/CAI systems is technically very difficult due to the complexity and size of the data. Furthermore, many CAD/CAE/CAI systems' business barriers constructed for market protection purposes make it very difficult to share data completely for engineering collaboration.

There are numerous efforts underway to settle on a data standard for collaborative design management. So far, most of the research in this area falls into three categories. The first category is the development of a neutral product data model which aims at facilitating product data exchange between CAX systems. The second category is the development of product data models that embody design knowledge and rationales, and intend to promote the share and reuse of product data. The third category is the development of product data models for a particular design domain or with special purposes. The first category typically applies the STEP method, which will be further discussed in later chapters. The second and third categories are introduced in "A Review of Data Representation of Product and Process Models" and "Features and Interoperability of Computer Aided Engineering Systems", with the third category presented in sub-sections on chemical process engineering.

There have been several methods for realizing collaborative design. In this subsection, the STEP method for collaborative design and manufacturing is discussed. This is followed by an introduction to XML-based data processing for collaborative design. Although these methods can solve specific problems in collaborative engineering, the generic data model still needs to be developed and improved. An integrated product information database that is accessible and useful to all the project stakeholders is necessary to support the entire product development lifecycle.

2.3.1 STEP-Based Method for Design and Manufacturing Data Exchange

With the rapid development of industry and manufacturing, the tendency toward economic globalization, cooperation among companies and organizations has become increasingly important. For instance, the outsourcing of contracts and the distribution of bills of materials in the collaboration chain have become common, if not the norm. There is a great deal of product information and data involved in the process of product design, manufacture, utilization, maintenance, and disposal, hence product data access between different companies' computer systems becomes necessary throughout the entire product lifecycle. A common computer-interpretable form is therefore required to ensure that the product model remains complete and consistent when exchanged between different computer systems. STEP was created to meet this need and has been broadly used in recent decades [44].

STEP—ISO 10303, independent of any particular system, is a series of related sub-standards developed by an international network of engineering and IT experts. It provides a neutral computer-interpretable representation of product data throughout the product's lifecycle. STEP is built around an integrated architecture with many domain-specific application protocols (APs) and well defined, generic, and integrated template resources. Each AP specifies the representation of product information for different applications. The APs break STEP into manageable and comprehensible "pieces" that can be more readily implemented. Accurate and complete description of products using a common file format, such as STEP, is essential for fabricating and assembling quality products, because manufacturing is frequently outsourced. Since STEP provides the basic data exchange standards, the market has developed a powerful and robust data exchange technology around it.

STEP also specifies the necessary mechanisms and definitions to exchange the product information into manageable product data. STEP uses a formal specification language, EXPRESS, to specify the product information to be represented. It is expected that several hundred more APs may be developed to support the many industrial applications that STEP is expected to serve. However, STEP is very much hard coded with limited generic capability to support collaborative engineering semantics. It is now recognized that an integrated product information database that is accessible and useful to all the stakeholders is necessary to support a product over its lifecycle [44]. A more detailed discussion follows in "A Review of Data Representation of Product and Process Models".

2.3.2 XML-Based Data Processing for Collaborative Design

Many businesses have used XML-based business-to-consumer (e.g., e-retailing) and business-to-business (e.g., e-marketplace) solutions to reduce transaction costs, open new markets, and better serve their customers. The goal of extensible markup language (XML) was to enable the delivery of self-describing data structures of arbitrary depth and complexity to applications that require such structures [40]. XML provides a specification for business-to-business (B2B) e-commerce data exchange and serves as a common standard for data exchange across various databases and files. In recent years, due to its connatural advantages of easy to design, simple expressions and high flexibility, XML technology is increasingly adopted for product data communication in the engineering fields as well.

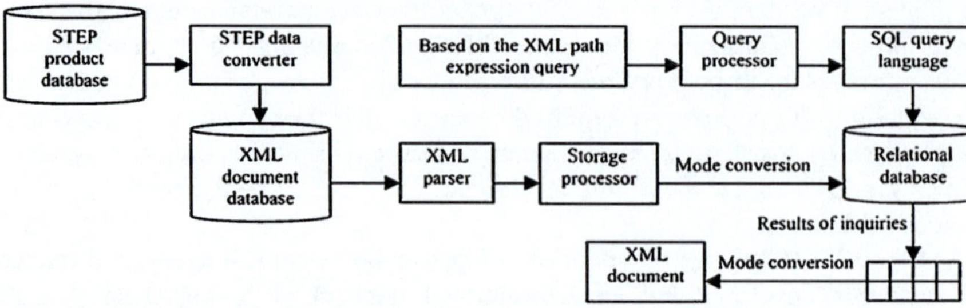

Fig. 2 The XML-based data processing system for collaborative design [53]

Wang and Ren proposed a prototype system based on XML that is used to realize data processing in network-supported collaborative design [53]. First, the STEP data of a product model is converted into XML files. In order to achieve a complete XML document storage and query process in a relational database, four processing procedures need to be followed: XML document parsing, mode conversion, query, and output. A STEP data converter, XML files parser, storage processor, and query processor should be involved in this system. The structure of the system is shown in Fig. 2 [53]. In this system, the researchers considered four major modular functions. The STEP data converter provides a common format and a special format for the XML expression. The XML parser is the basis of information processing for XML data. Extracting the data and structural information from the text form of XML documents, the XML parser can generate the tree structure of the XML file contents. The parser works as an interface for the XML application to instantiate operational data objects with the information processed. The basic function of the storage processor is to store the XML document tree as a two-dimensional table with a mapping scheme. The purpose of the analysis and storage of XML documents is to support XML functions, such as querying functions. The key mechanism of XML query is transforming different forms of path expressions to query sentences, which is the main function of the query processor.

2.4 Concurrent and Collaborative Engineering

Concurrent engineering (CE) is a systematic approach to the integration of all parts of the engineering product and process, such as the parallel production of software and hardware. The term was coined in the United States in 1989 [42]. CE promotes consideration of all aspects of the product and through all engineering processes in order to address increased complexity and reduce the time of delivery to market. Companies have readily practiced CE for the production of simple products. However, CE causes many challenges for complex systems, because dependency relationships between different stages and processes of the development cycles are easier to manage in a simple product; this is not the case for complex systems.

Traditionally, in the early stages of product development of complex systems, the majority of time was spent on design, leaving manufacturing considerations to a much later stage; this causes problems with fulfilling manufacturers' design requirements and delays product release on the market.

The use of CE can decrease the product development cycle time and enhance design quality comprehensively. There are many definitions of CE, and most of them emphasize only limited aspects of it. For instance, Thomas [47] suggests that CE aims for "the simultaneous performance of product design and process design," which typically involves the formation of cross-functional teams; this organizational approach "allows engineers and managers of different disciplines to work together simultaneously in developing product and process design". As per the view of a group of Berkeley researchers, CE is "a business strategy that replaces the traditional product development process so that tasks are done in parallel" [6]. All definitions of CE, however, have the same goals—reducing the total product lifecycle time, lead time, and cost—with a secondary goal of increasing product quality. In addition, globalization and increased product complexity require that more serious attention be paid to CE as a useful approach.

Cost engineering, a basic component of CE, is crucial. Many companies looking for cheaper production solutions will first determine the total profit and then plan and schedule accordingly. Scheduling is quite a complex activity due to the array of conditions that have to be considered. Accurate planning, too, requires determining an enormous number of details in advance. The key challenge is that CE must always consider engineering design changes, such as changing the specifications, complexity evolution, and teamwork mechanisms. To implement CE, the enterprise needs to constantly keep track of engineering changes. The company's production system needs to be more flexible in order to produce a variety of products; this requirement may lead to replacements of specialized machine tools with universal ones.

Collaborative engineering is a systematic business approach further developed from CE, and is designed for collaborative utilization of resources and information among multidisciplinary groups, or even multiple enterprises across the world, in real-time. Collaborative Engineering Environments (CEE) can provide the working platform for ad hoc collaboration and free-flowing processes.

2.5 Lean Manufacturing

Lean manufacturing is a strategy created by Toyota Production System(TPS) in 1980 [37]; the book *The Machine that Changed the World* introduced lean thinking in 1990 [56]. Lean manufacturing is an approach that focuses on the identification of values, finding the existing waste sources in production cycles, analyzing the cycles, and eliminating the waste sources. As a result of lean manufacturing, industry can decrease production time, waiting time, re-work, inventory, and product defects; increase quality, flexibility, and product diversity; save production resources; and create new value [36]. These advantages bring significant financial benefits to companies. Typically, there are several main waste sources in manufacturing processes: transport, inventory, motion, waiting time, over production, over processing, defects, and ineffective production that do not meet customer requirements [57]. To eliminate the waste among these eight sources and create value, lean manufacturing uses tools and techniques such as:

- "Kanban" (a Japanese term, i.e. "Signboard"—it means a customer requirement driven scheduling system for lean and just-in-time production),
- 5 S's (Japanese words: *seiri, seiton, seiso, seiketsu,* and *shitsuke*—English words: sorting, straightening, shining, standardizing, sustaining),
- visual control,
- "poke yoke" (a Japanese term—error proofing), and
- "single minute exchange of dies" (SMED)—a changeover reduction technique [33].

2.6 Review of Informatics Modeling Methods for Concurrent and Collaborative Engineering

To enable concurrent and collaborative engineering, ideally, a semantics-oriented data repository must be used to record the history of a design as a sequence of design decisions. The resulting database records the final specifications, the alternatives that were considered during the design process, and the designers' rationales for choosing the final design parameters. Nagy et al. did some early work in this domain, creating a product data representation scheme based on the issue-based information system (IBIS) method for collaborative mechanical design [35]. This data representation is composed of four data elements (issue, proposals, arguments, and decisions) within a computerized design history management tool (DHT). Figure 3 shows a data representation network.

According to Nagy [35], design *issue* is a set of problems that need to be solved in order to complete the design process. The *proposals* are used to resolve design issues. An *argument* is the basic concept that is used by designers to make decisions in support of or in opposition to a particular proposal. A design argument is a comparison, which can be either absolute or relative. In an absolute comparison, only one proposal is the focus. A *decision* may evaluate only one proposal based uniquely on absolute-type arguments. In a relative comparison, the focus is on a set of proposals, and the ability of each proposal to satisfy the set of requirements is compared relative to the other proposals.

Among many research works in this field, You and Tsou [59] proposed an architecture for a collaborative assembly design, which also adopts STEP-based data representation and CAD feature extraction. Figure 4 shows the architecture and conceptual explanations.

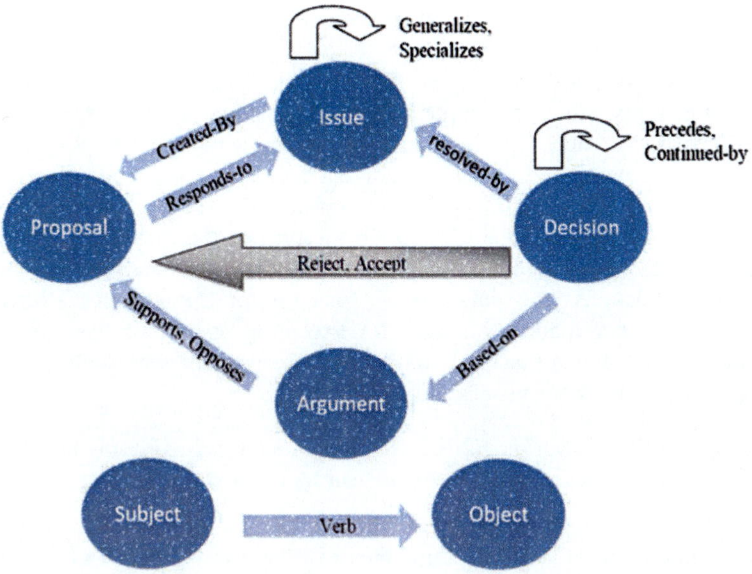

Fig. 3 Conceptual data representation for collaboration [35]

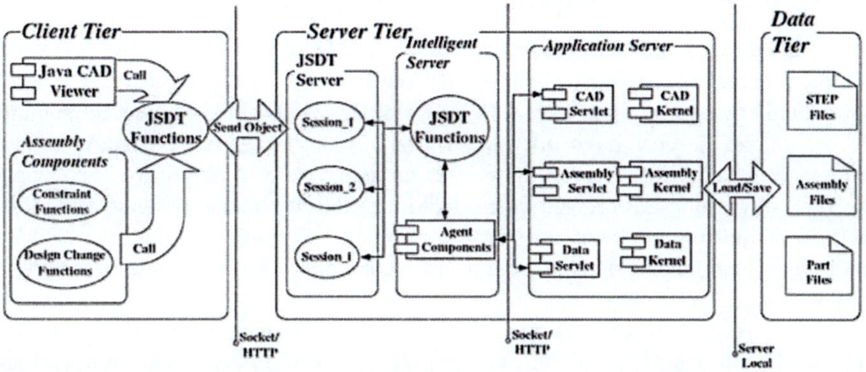

Fig. 4 Network architecture [59]

The client tier provides the exchange for CAD and assembly operations. The CAD module visualizes the solid model. When the model is built in the server tier, it is divided into two parts: the intelligent server and the application server. The application-view model is sent to the client tier using Java shared data toolkit (JSDT) [45] technology.

The server tier is the core of this structure. All relevant data must be computed and stored there. The intelligent server performs three main functions:

1. The session manager maintains all communication and transfers messages between session users.
2. The media between the client and the application server are generated by the JSDT functions, which are supported by the relevant agent components. The intelligent server stores the IP address of the application server. When a client user makes a command and the intelligent server receives a request from the client tier, the corresponding application server will execute the operations. The result is sent back from the application server, and the intelligent server then transforms the solid model for viewing and sends it to the client tier.
3. The session data as "the work model" for a session is stored in the intelligent server. The work model is a basis for executing operations from the client. When a client joins a session, the intelligent server will send the current view model to the client for visualization.

The application server is the computing core of this system, building and operating models. The application server can be divided into three parts:

1. The CAD server reads CAD files and builds the three-dimensional solid model.
2. The assembly server is used to build the order and level of parts in the assembly model and assemble the parts according to the particular conditions.
3. The data server saves or loads files from the data tier, which includes solid model files, part files, and assembly model files. The data server controls user access according to organizational access policies.

In addition to the methods described above, Park and Yang have proposed another integrated engineering data representation scheme [38] centered on an integrated engineering data format (".BST") that was designed to meet three main requirements: it must have all the information that is generated from various engineering activities; it must have a referencing scheme to ensure each engineering information set created by an individual activity is easily relatable to other activities' information sets; and it must maintain the complete data sets used by individual engineering activities. Figure 5 shows the benefits of integrated engineering data.

Engineering collaboration requires sharing engineering data, which can be achieved with a data translation server that translates information from various commercial engineering systems into BST data. This supports not only CAD translation but also CAM/CAE translation. Figure 6 shows the structure of the engineering data translation system. The module can translate a variety of 3D CAD file formats, such as CATIA, PRO/E, SolidWorks, STEP, and so on. However, this approach lacks a specific semantic structure and the necessary details to govern the communication language grammar.

Wang and Tang proposed a collaborative product data model (CPDM) [52] to overcome the shortcomings of the existing product data models. CPDM is established by extending the popular parametric feature-based product data model. Because traditional CAx systems usually employ a feature-based parametric data

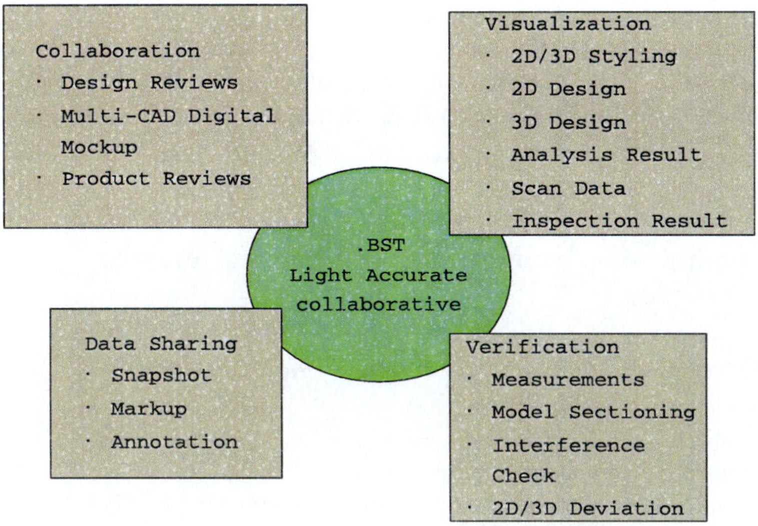

Fig. 5 An integrated engineering data framework for digital engineering [38]

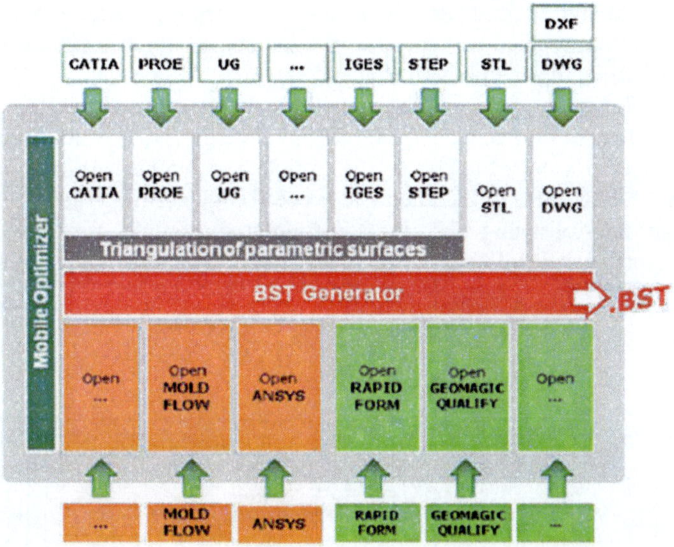

Fig. 6 Engineering data translation system [38]

model, CPDM can take advantage of the mature feature technology. The CPDM consists mainly of three data modules: collaborative design management data, design coordination data, and product data. It contains seven types of information: designer information, design agent information, organization information, project information, version, permission, and alternative. Design coordination data contains the information used to coordinate the collaborative product design process. It contains three types of information: constraint, coordination rule, and history. Product data contains the product's structural and physical information, including assembly and part.

3 Computer Applications Used in Chemical Process Engineering

3.1 Chemical Process Engineering Project Cycles

The complete design and construction of a chemical process plant, which is implemented by engineering, procurement, and construction (EPC) companies, requires collaborative engineering work from different domains [27]. As shown in Fig. 7, the process starts with business planning, followed by conceptualization, conceptual design, and process engineering, which provide the input data for the next phases involving other disciplines, i.e., mechanical design and engineering and electrical design. Together these form the engineering stage, following which are the procurement stage and the construction stage, which includes implementation, commissioning, and maintenance [58].

The complexity lays also in the specific domain, for example, the chemical engineering domain or mechanical engineering domain [12]. Fortunately, software tools have been developed to handle the complexities embedded in each phase or task during the lifecycle of the chemical process engineering project, as shown in Table 2. In the following section, some commonly used software tools are selectively introduced, followed by an analysis of the technological coverage gaps among them.

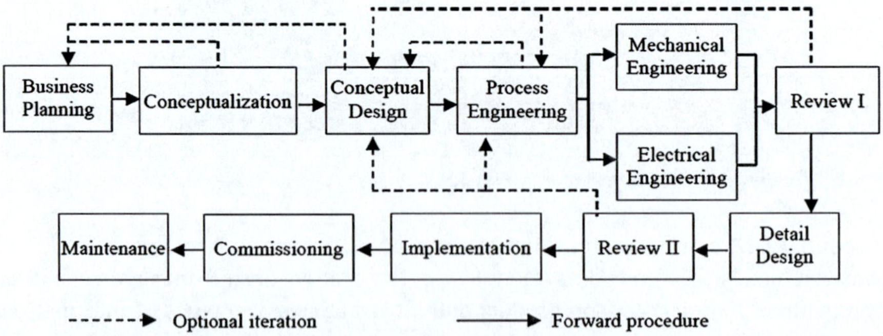

Fig. 7 Lifecycle activities of chemical process engineering

Table 2 Supporting software tools for main activities in process engineering lifecycle

Activities	Supporting software tools
Conceptual design	• SmartPlant P&ID • Microsoft Visio • CAD packages
Process engineering	• Aspen Plus • Aspen Hysys • VMGSIM
Mechanical design	• UG NX/CATIA • Pro/E • SolidWorks • Solidedge

3.2 Domain Software Packages

To appreciate the informatics applications specifically employed in chemical process engineering, and because of their significant market presence, Aspen and Intergraph products are reviewed here.

3.2.1 Aspen Packages

Due to the complexity embedded in chemical process design, a great number of the world's leading process operation and EPC companies have chosen Aspen to facilitate process design or optimization [3]. This is mainly due to the two leading steady-state process simulation software tools offered by this corporation, Aspen Plus and Aspen Hysys, which have already been widely applied in the design workflow of EPC companies. A comparison of the two software packages can be seen in Table 3.

Aspen Plus, a core element in Aspen engineering, has been proven to be an effective process modeling and simulation tool through more than 20 years' industrial use in the field. It is already equipped with the world's largest component database with physical properties and phase equilibrium data, and it still allows users to customize new components as needed. With reliable thermodynamic data and comprehensive equipment models, the process behavior of the system can be predicted by Aspen Plus according to engineering relationships built into the models.

In comparison, Aspen Hysys is an oil- and gas-process-oriented software package running in an integrated environment with a comprehensive thermodynamic foundation and a library of unit operation models. It provides a convenient way to build process flow models with extensive, well-defined component databases. Based on a process flow model accepted by process engineers, Aspen Hysys can provide equipment and other operational parameters with certain

Table 3 Key features of popular tools in Aspen [3]

Solutions	Key features
Aspen Plus	• Streamlined engineering workflows for conceptual design to model deployment
	• Scalability for large and complex processes
	• Column internals calculations for flooding and pressure drop
	• Safety and controllability studies and optimization capability
	• Capability to model batch distillation and reactors
Aspen Hysys	• Physical properties methods and data library
	• Assay management and propagation of refinery molecules across the flow sheet
	• Comprehensive library of unit operation models
	• Pipeline network and pressure drop analysis
	• Integration with Aspen PIMS and Aspen Refinery Scheduler software

optimization to improve production efficiency and productivity through "what-if" and sensitivity analyses.

It should be mentioned that in the latest version of both Aspen Plus and Aspen Hysys (Version 7.3), the interoperability among tools offered by Aspen technology is also enhanced. However, there is still room for further enhancement for both products by a number of add-on applications.

3.2.2 Intergraph Packages

SmartPlant, which was released by Intergraph, has been used by such owner operators as Suncor Energy, Nynas, Neste Oil, SCG Chemical Group, and others, as well as a number of EPC companies, such as Fluor, Bechtel, Foster Wheeler [19], and others.

SmartPlant 3D [19], as the fundamental component of SmartPlant, provides a 3D visualization and modeling tool for chemical process engineers, changing the way that traditional process engineers work. SmartPlant 3D is a forward-looking modeling tool with comprehensive capabilities, customizable design rules, and the capability to store everything built for future reuse. The graphical representation generated by SmartPlant 3D benefits the user during the working session; the perpetual entities created are stored in databases, and the user can easily control the visibility of each object. This greatly facilitates collaboration between global engineers across disciplines, and provides contractors with access to the design process, thereby reducing cost and production time through more efficient project management. The newest version of SmartPlant 3D is equipped with Reference 3D and enhanced model reuse capabilities, and now supports PDS, PDMS, SAT, DGN, DGN V8, DWG, and VUE formats. This means that a number of CAD models are supported, such as VUE and DWG, the file extensions created by Autodesk and AutoCAD, respectively. SAT, the file format of ACIS, is also acceptable; ACIS is a popular kernel used by some mid-range CAD applications.

Table 4 Key features of popular tools in SmartPlant

Solutions	Key features
SmartPlant 3D	• Project setup and reference data tasks • 3D modeling • Drawing and reports
SmartPlant P&ID	• Rule driven design • Engineering standards support • Information sharing support
SmartPlant electrical	• Power distribution network • Interface with ETAP • Interface with SmartPlant 3D
SmartPlant instrumentation	• Lifecycle control system solution • Consistency and quality guarantee • Interface with leading control systems
SmartSketch	• 2D parametrical sketching • Traditional CAD format support

The key features of some popular tools available in the SmartPlant 3D package are listed in Table 4. Each tool is designed for a specific application domain with some unique features, as compared to similar software tools from other corporations. For example, the SmartPlant P&ID module offers an advanced tool to make piping and instrument diagrams (PIDs) with the capability to keep PIDs up to date and accessible to all engineers, owner/operators, and EPC contractors, if needed. With SmartPlant Foundation as their basis, these tools collaborate with each other to offer customers a powerful portfolio, SmartPlant Enterprise Solution, which can effectively improve project execution and operation efficiency.

As is apparent in Table 4, one of the advantages of SmartPlant is the interoperability between the products and solutions within the platform, as it is based on a central data warehouse [55]. This allows the importation and distribution of the data created by application tools, which makes it possible to view combined 3D effects and perform interference checks between models created with different tools. For example, SmartPlant 3D models for piping design can be created with reference to the output of SmartPlant PID module, or electrical cable routing from SmartPlant Electrical module. Furthermore, numerous interfaces have been developed to further enhance interoperability within SmartPlant or with another computer system, such as between SmartPlant Instrumentation and a control system.

Although there are few efforts underway to integrate Aspen with SmartPlant, these two systems have commonly been adopted concurrently in EPC companies. They both provide capability for information sharing and design reuse, and hence facilitate collaborative work between engineers around the world. However, interoperability between such packages within the chemical process design domain still needs further enhancement.

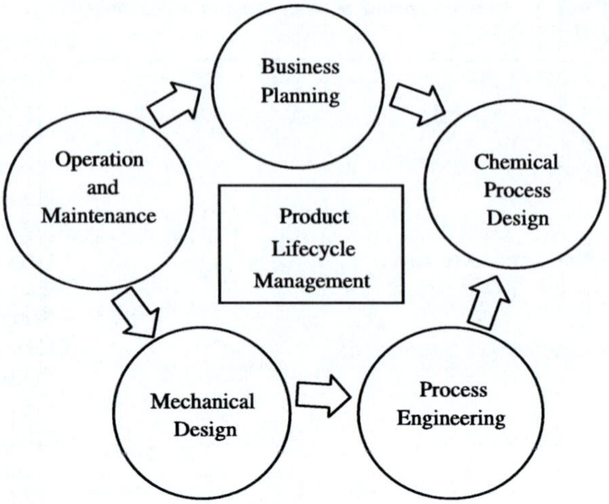

Fig. 8 Work flow of a chemical process engineering project

3.3 Integration Gaps between Chemical Process Design and Mechanical Design

In a chemical process engineering project, the traditional workflow includes a project proposal, chemical process design, engineering design, mechanical design, operation, and maintenance (see Fig. 8). As an upstream process, the output of chemical process design will first pass through engineering evaluation, then provide the design requirements for downstream engineering activities, such as mechanical design, which determine equipment size, material, and pressure requirements.

The common challenge in a typical chemical process engineering project is that the activities involved are interdependent, which leads to the biggest disadvantage of the traditional sequential working practice: the long lead time and the necessity for iterative design cycles [20, 56]. To shorten the lifecycle of chemical process

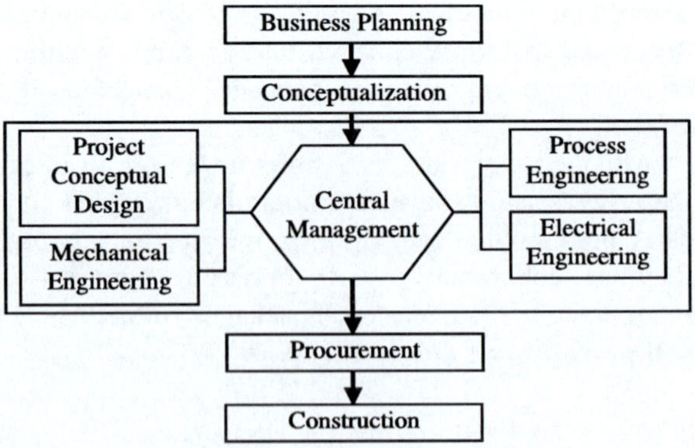

Fig. 9 Collaborative engineering processes for EPC

engineering, the authors propose a collaborative working process for EPC companies based on collaborative engineering concepts (see Fig. 9).

This approach requires a great deal of interoperability among the software tools involved, as the process requires a massive number of interactions. Without interoperability, the impact of each process design decision cannot be evaluated accurately or efficiently. For example, one parameter change, such as flow rate, could have a significant impact on downstream mechanical design, which may lead to high increases in cost or a delay in schedule. Unfortunately, interoperability currently exists only among tools provided by the same vendor. The computer system modules listed above need to run in parallel independently, and hence cause isolated information islands. There is currently no computer system that can act as a bridge to connect those isolated islands. Moreover, little research attention has been focused on this topic. There is therefore an urgent need for an architecture or platform that can provide efficient information sharing to facilitate elements of collaborative work such as timely communication and information sharing between chemical and mechanical engineers.

The close connection between chemical process design and mechanical design requires great interoperability between computer systems across disciplines. Therefore, to make a chemical process engineering project run smoothly, the tools listed above must work collaboratively, and the problem of interoperability among them needs to be addressed. In "A Review of Data Representation of Product and Process Models", existing technologies to facilitate interoperability are elaborated upon and a semantic modeling method is proposed. A semantic integration framework for chemical process engineering is then proposed. Under this framework, engineers from different disciplines and using a variety of software tools should be able to work collaboratively in a more efficient way.

4 Fundamental Technologies for Engineering Informatics Solutions

This section briefly reviews the three most important aspects of implementing engineering informatics solutions: OO software engineering methodology, unified modeling language (UML)-based informatics conceptualization, and some potential effective informatics solution drivers, including multifaceted data repositories, data mining for decision making, and semantic modeling.

4.1 Object-Oriented Software Engineering Methodology

To support CEEs and to work more efficiently and flexibly, we need to change our software development approach, from the past function-oriented and procedure-based implementations such as C programming, to an OO approach, such as C++

programming. The OO approach focuses on both data consistency and procedure encapsulation. It has numerous useful features, including data abstraction, inheritance, polymorphism, and dynamic binding. This approach is also useful for streamlining software development and the reuse of codes and software modules.

4.2 Informatics Modeling Conceptualization with UML

UML is a visual software modeling language used to formulate a model of a system using all available information and relations [14]. UML is an effective industry standard at the system engineering level for illustrating software requirements and complicated interfacing and integration relationships. UML includes several modeling diagrams, such as use-case diagrams, class diagrams, and sequence diagrams as demonstrated by Thimm et al. [46]. In their work, UML diagrams are uniformly used to explain the PLM system modules using a vacuum cleaner as an example. The next chapter will further discuss the use of UML as an effective tool for PLM modeling.

4.3 Promising Engineering Informatics Drivers

4.3.1 Multifaceted Data Repository

Organizations have traditionally used data files and documents to store their information. This kind of data storage is simple and fast, but it creates data redundancy, lack of flexibility, and poor security; it also impedes data sharing. The competitive marketplace has forced industries and organizations to find more effective ways to store, reuse, and analyze data. Organizations and manufacturers have thus turned to database management systems that can control data and information more effectively.

A data repository is a logical and physical approach that essentially gathers together several databases that have the same set of objectives and tasks into one vast unified database. Many computer systems have data repositories for the collection and organization of data, but among them, goals and definitions vary. However, in recent years, industry and the research community have driven a trend: data repositories are now designed to create a unified information platform that includes everything from business services and processes to interfaces between databases. In the latter case, a data repository works to increase the level of effectiveness in a system. Data repositories need to provide a strong and comprehensive relation graph among all the data entities to assist in engineering change updating, constraint evaluation, and decision-making validation by the supporting systems. In a manufacturing enterprise, having one collaborative data repository that reflects all the available knowledge and information is also useful for managing the process of updating data in a short period of time [23].

Developing a database management system (DBMS) involves two activities: conceptual or logical design according to business, market, and organizational strategies, and physical design. Logical design results in the development of three main components: a data definition language, a data manipulation language, and a data dictionary.

A data structure definition language is a formal programming language that defines the database layout, ideally using an OO language. A data manipulation language is used to define computer-interpretable queries for interactions with the database; examples include SQL and Oracle. A data dictionary is a tool for storing and managing data. In addition to the languages used to run it, physical design can also focus on creating a united database or a distributed database. Distributed databases are useful for integrating various resources from collaborators, but can create redundancy for data contents and relationships among data entities. Furthermore, distributed databases are less secure than single-location databases. Data mart and data mining are useful methods for searching specific data or finding the relationships between entities. This kind of information manipulation function is useful for reusing data and for data synthesis with certain analyses [4].

Although database management systems are useful methods for massive and persistent data storage, manufacturers need additional tools to support diverse, distributed, agile, and associated engineering activities with heterogeneous informatics systems. One such tool is a workflow management system, which can systematically define final outcomes for different types of jobs and processes [1]. Workflow management can support the business process modeling and business processes reengineering that are necessary for enterprise [16].

4.3.2 Data Mining to Support Engineering Decision-Making

If an enterprise wants to lead in the market, or wants to pursue high-value business activities, it has to be adaptable to market changes. Decision tree techniques can help determine what exactly the customer wants from a new product. Decision trees offer a shortcut through the product lifecycle that offer a company a number of advantages. When a company begins the process of producing a new product, its first step is market research, gathering information on markets and customers; data mining techniques are then used to distinguish unknown information from known information.

Westphal and Blaxton introduced four primary functions for data mining: classification, estimation, segmentation, and description [54]. *Classification* tries to detect previously unseen record patterns. There are several techniques for classification data mining, including decision tree methods, rule-based methods, memory-based reasoning, neural networks, and support vector machines. *Estimation* is the phase that attempts to fill in missing data. This is sometimes the result of market research: the information collected is not always cohesive, nor does it always display strong or obvious relationships among the data. Estimating information can serve to fill the gaps between existing pieces of information. The *segmentation* step further divides the market, clustering customers according to various demands, preferences, behaviors, or other demographic data; segmentation can also be used to cluster products for better control and monitoring of the database. Finally, *description* defines the relationships among the data for a more cohesive database. These four steps in data mining aim to extract vital and useful information from a database to improve product development.

An example of the use of data mining to support NPD can be found in the work of Bae and Kim, who researched the use of using data mining techniques in the production of a new digital camera in 2011 [5]. They used three main steps in their research framework. The first step was data collection through questionnaires and surveys. This step aimed to answer the following questions: What are the customers' "needs" and "wants" for digital camera products? What features are customers looking for in digital cameras? How much money can or will the customer pay? What kind of strategy must be used in designing a new camera? At the end of the first step, the researchers determined that there are nine Attributes that can have a significant effect on customer satisfaction regarding new digital camera design: price, size, resolution, functions, colors, weight, LED screen size, battery category, and ease of use.

The second step was model construction by associating rules with the decision tree. In this step, the researchers developed a decision table with the numerous statuses for each attribute. The researchers then defined rules for the decision tree, such as, *if "color = green" and "function = digital zoom or does not matter" and "weight = less than 125* g *or 125–320* g*," then "style A is selected"*. According to this kind of rule, the company could design a digital camera with more tailored details in the initial design [5]. They could then select one model whose final design would be the most popular on the market.

The third step was the development of a new product design strategy and a product variation profile. At this stage, a series of products were selected to form the product portfolio of a company [5].

4.3.3 Semantic Modeling

In general, engineering activities and manufacturing rely on two types of information support: data and algorithms. Careful observation of modern industrial engineering practice illustrates that many advanced, challenging, knowledge intensive, and system-oriented activities (such as information sharing, exchanging data, data mining, and expertise modeling) are carried out on a regular basis. Manufacturing information services are usually based on separate computer systems, and cannot provide a consistent, systematic semantic representation of the continuous manufacturing cycles. Modeling real-world concepts in a virtual knowledge domain can create an efficient and effective system [9].

Semantic modeling is being developed for knowledge classification and processing based on modular, OO software engineering methods that are found to be versatile and useful for engineering informatics modeling. This approach is a kind of formal knowledge modeling that describes the meaning of data from the user's point of view. The objectives of semantic modeling are abstracting knowledge and understanding the implications of data and information, and then processing them accordingly. One additional objective of semantic modeling is implementing highly cohesive informatics models [17]. This section provides a review of semantic modeling in applications used for product development and process modeling, and its advantages and disadvantages.

Semantic modeling is a way of mapping the real world onto the virtual world. There are several forms of information in real-world models, including knowledge, semantics, features, geometries, and data. Semantic models are similar to graph models in their ability to bind the semantic objects to physical models. There are several types of data models that can be used to represent the intricate relations and dynamic processes of the real world, such as relational, hierarchical, and graph models coupled with functional algorithms. Each of these models shows the relationships among entities for software development, runtime queries, and database-maintenance purposes. Semantic modeling maps such different kinds of information onto the virtual world.

Figure 10 shows the concept of semantic modeling applied in the product development domain and illustrates relationships across the real world and the virtual world. For example, a mechanical component part design has a number of features. Semantic modeling has to implement cases and express the embedded design intent of product design via associated attributes. Then it has to define some classes and class diagrams to create the systematic structural data models for

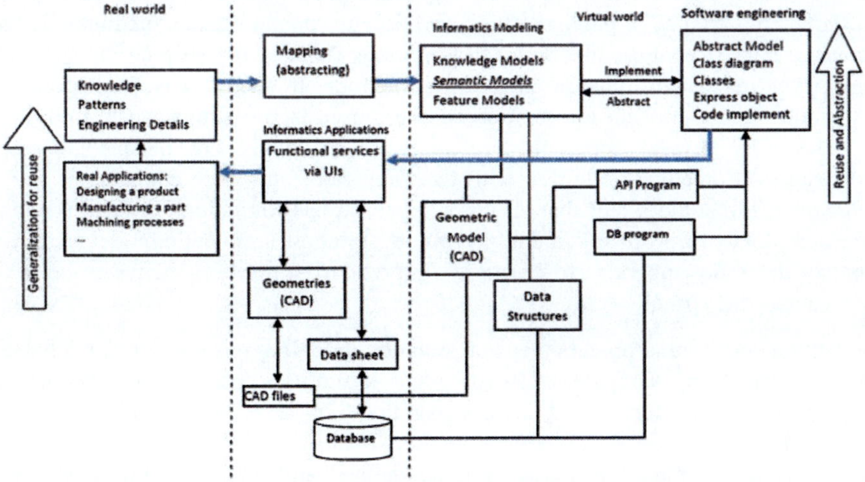

Fig. 10 Semantic modeling in engineering informatics cycles

storing parts and assemblies into a product database. Definitions and the semantic relations of functions, features, data blocks, classes, attributes, and their patterns must be done clearly [50]. Identifying the representative classes and objects is the most important part of semantic modeling.

For example, in manufacturing process planning, complicated reasoning is necessary for the selection of adequate resources, tooling, and the sequencing of operations for the execution of a part's production. In such typical engineering activities, a resource database, a technical machining database, and a CAD database are all necessary for effective process planning. Semantic modeling is required to facilitate and manage the interfaces between the real world and the virtual world. One of the important roles that can be played by semantic modeling is managing the correspondence between different process plans and the details of part design, including the downstream packaging and labeling of the final product. Semantic modeling can also be useful as the driving technology for the generalization of different products' design feature justifications as well as individual parts' process plans. The feature justifications are useful for evaluating and making decisions in product change management; individual process plans have to be effectively synthesized and built into a production plan. A production plan has to achieve equipment reusability and process coverage for all the required parts. These two capabilities are vital for connecting the real manufacturing world to the virtual world.

Semantic modeling can be used to generalize the part designs for reuse in the real world and to represent the associated reusability in the virtual world. The virtual world also needs to store data and information in a database; this can also be effectively carried out through semantic engineering. For example, the software routines must be reusable. Achieving a high level of code reusability in software engineering is a complex task. Object-oriented software engineering methodology is a well-proven method that can support reuse of templates, dynamic linking of libraries, inheritance of parent entities, and polymorphism. Hence, the method can manage data structures, functions, and scenario changes by defining classes and implementing object dynamic processing methods. In such a way, both product Attributes and algorithm information can be reused in the virtual world [39].

There are several approaches to semantic modeling. For instance, Zhang introduced a practical definition with two predicates and three connectives [60]. However, the authors feel that Zhang's approach has limitations in dealing with the complexity of problems in engineering informatics; a semantic model needs to satisfy the following rules to create a comprehensive mapping between the real world and the virtual world:

- All the entities and process steps of semantic modeling must be clearly defined.
- There must be a consistent and reproducible mutual corresponding scheme in designing an abstract model that can map between the real world and the virtual world.
- The semantic modeling items must be defined and directly mapped with an object, which in turn is defined by a class that is part of the domain-related class

diagram. We strongly believe a class diagram is a useful tool for abstract modeling, and the relationships built in are implemented into related class definitions.

- A semantic model must specify all the relationships and interactions among the modeled objects to describe the logics of a semantic model [10].

Even if a semantic model follows all of the above rules, it still may encounter some problems. A semantic model is usually defined from a specific point of view, such as that of the product designer. Semantic models of a single product can therefore differ from each other due to their different purposes, which can lead to different models of the same product. Certain mutual transformation mechanisms need to be developed to manage the associations among semantic models. The other potential problem is that constructing a working semantic model requires an engineering expert who understands the Object-oriented modeling concept. Fortunately, this expert's contribution does not necessarily require detailed information about process and product development.

There are several approaches to constructing a semantic product model [24]. The first is to use a tree structure of the top assembly, sub-assemblies, and components. This model divides the product entities into the above-mentioned classes. Top-level assembly is a total combination of the product with two or more sub-assemblies, in which each sub-assembly must be defined with specific functions or purposes. The assembly or sub-assembly level can also include additional components; an assembly and a component must be clearly distinguished.

An alternative semantic modeling structure is a dual representation of conceptual-detail product entities. This model has two associative levels: the conceptual model and the data structure model. The conceptual model represents the relationships between Attributes and entities in Object-oriented class/object diagrams. This model is a useful tool for developing and understanding informatics modeling concepts. The second level consists of a more detailed implementable data structure model. For example, to develop a semantic model for a manufacturing management system, a conceptual model is usually required to illustrate the

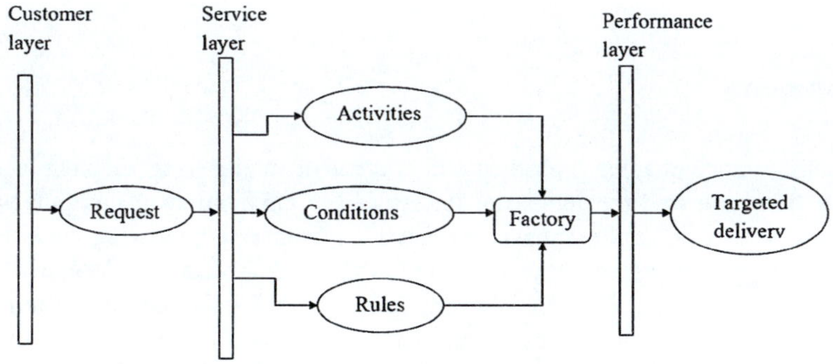

Fig. 11 Conceptual process model of handling customer orders

definitions and relationships of an organization's departments or units, while the detailed model represents the database design table schemas. Figure 11 shows a simplified example of a conceptual process model of handling customer orders.

A third semantic modeling structure for products is based on a component tree structure model, which distinguishes in-house-manufactured from outsourced parts. When manufacturers want to design a part in-house, they use a specific system of modeling that incorporates the company's standards and conditions. However, for outsourced parts, the purchasing company only needs to select or preliminarily specify the outsourced parts with a set of attributes that describe the quality and properties of the part according to the demands of the market or their business strategy. In this situation, it is the supplier's responsibility to design the outsourced part in detail according to the prescribed specifications.

The fourth structure for semantic modeling is created by leveraging relationships among the conceptual, detailed, or manufacturing features [7]. There are numerous ways to define features [15, 23, 29]; here, *feature* is defined as a specific associative geometric pattern or configuration formed to reflect a certain meaningful semantic entity for certain engineering reasoning. In keeping with this definition, there are several ways to model a feature. In general, feature modeling is a useful technique for analyzing a product line [14]. A conceptual feature model is a set of simplified or abstracted modeling features that are expressed by referenced member elements and their interdependencies. According to Vranic [50], a feature-based semantic model involves two domains: application and solution domains. CAD programs have a number of predefined features, but they are not intended to provide rich engineering semantics, such as manufacturability attributes and evaluation methods; rather, most CAD features are meant only for geometry construction. Any manufacturing company that wants to capture and reuse its practical "know-how" needs to achieve a high level of semantic modeling competency; the company needs to define and develop a set of features according to the available manufacturing technologies and resources. The company has to incorporate many decision-making attributes, such as the justification attributes for any selected process, into the company-wide semantic model for reasoning purposes [25].

5 Summary

This introductory chapter presented the *context* of engineering informatics. The chapter first reviewed the industrial reality of the application of computer-aided solutions, with a special emphasis on chemical process engineering as a sample sector. The challenges and research directions for engineering informatics have been clearly identified. Based on the increasing use of information technologies, such as multifaceted data repositories, data mining, and semantic modeling, the potential leveraging applications that could significantly increase the effectiveness of engineering informatics applications have been projected. While only minimal space was devoted here to engineering informatics itself, the remaining chapters will address the topic in greater detail.

References

1. Abbott KR, Sarin SK (1994) Experiences with workflow management: issues for the next generation. In: Proceedings of ACM CSCW 84, Chapel Hill, North Carolina, USA
2. Adams M (2004) Comparative performance assessment study. In: Proceedings of the presentation at the comparative performance assessment conference, PDMA Foundation
3. Aspentech (2012) http://www.aspentech.com/
4. Babcock B, Babu S, Datar M, Motwani R, Widom J (2002) Models and issues in data stream systems. In: Proceedings of ACM PODS, Madison, Wisconsin, USA
5. Bae J, Kim J (2011) Product development with data mining techniques: a case on design of digital camera. Expert Syst with Appl 38:9274–9280
6. Berkeley (2012) http://best.berkeley.edu/~pps/pps/concurrent.html
7. Bidarra R, Bronsvoort WF (2000) Semantic feature modeling. Comput Aided Des 32:201–225
8. Bronsvoort WF, Noort A (2004) Multiple-view feature modeling for integral product development. Comput Aided Des 36:929–946
9. Cai M, Zhang WY, Chen G, Zhang K, Li ST (2010) SWMRD: a semantic web-based manufacturing resource discovery system for cross-enterprise collaboration. Int J Prod Res 48:3445–3460
10. Chan FTS, Zhang J (2001) Modeling for agile manufacturing systems. Int J Prod Res 39:2313–2332
11. Chen G, Ma YS, Thimm G et al (2006) Associations in a unified feature modeling scheme. ASME Trans J Comput Inf Sci Eng 6:114–126
12. Contea E, Gania R, Malikb TI (2011) The virtual product-process design laboratory to manage the complexity in the verification of formulated products. Fluid Phase Equilib 302:294–304
13. Driva H, Pawar KS, Menon U (2000) Measuring product development performance in manufacturing organizations. Int J Prod Econ 63:147–159
14. D'Souza DF, Wills AC (1999) Objects, components, and frameworks with UML—the catalysis^{SM} approach. Addison Wesley Longman, Reading, Massachusetts
15. Geelink R, Salomons OW, van Slooten F, van Houten F, Kals HJJ (1995) Unified feature definition for feature based design and feature based manufacturing. In Busnaina A (ed) Comput in Eng, ASME Conf., 517–533
16. Georgakopoulos D, Hornick M, Sheth A (1995) An overview of workflow management: from process modeling to workflow automation infrastructure. Distrib Parallel Databases 3:119–153
17. Hanis T, Noller D (2012) The role of semantic models in smarter industrial operations. IBM DeveloperWorks. http://www.ibm.com/developerworks/industry/library/ind-semanticmodels/ind-semanticmodels-pdf.pdf. Accessed 18 Nov 2012
18. Herder PM, Weijnen MPC (2000) A concurrent engineering approach to chemical process design. Int J Prod Econ 64:311–318
19. Intergraph (2012) http://www.intergraph.com/
20. Jouini BM (2004) Time-to-market vs. time-to-delivery: managing speed in engineering, procurement and construction projects. Int J Proj Manag 25:359–367
21. Kima J, Prattb M, Iyer RG et al (2008) Standardized data exchange of CAD models with design intent. Comput Aided Des 40:760–777

22. Körtgen A, Nagl M (2011) Tools for consistency management between design products. Comput Chem Eng 35:724–735
23. Lee SH (2005) A CAD–CAE integration approach using feature-based multi-resolution and multi-abstraction modeling techniques. Comput Aided Des 37:941–955
24. Leibrecht S, van Pham T, Anderl R (2004) Techniques for the integration of expert knowledge into the development of environmentally sound products. J Eng Des 15:353–366
25. Lohtander M, Varis J (2007) Manufacturing features in cutting shapes and punching holes in sheet metal. In: Proceedings of the 19th international conference on production research, Valparaiso, Chile
26. Ma YS, Bong CH (2010) Fine grain associative feature reasoning in collaborative engineering. Int J Comp Appl Technol 37:210–216
27. Ma YS, Hadi Q (2012) Unified feature based approach for process system design. Int J Comp Integr Manuf 25:263–279
28. Ma YS, Tong T (2003) Associative feature modeling for concurrent engineering integration. Comput Ind 51:51–71
29. Ma YS, Chen G, Thimm G (2008) Paradigm shift: unified and associative feature-based concurrent and collaborative engineering. J Intell Manuf 19:626–641
30. Marion TJ, Simpson T (2009) New product development practice application to an early-stage firm. Des Stud 30:256–587
31. Marion TJ, Friar JH, Simpson TW (2012) New product development practices and early-stage firms: two in-depth case studies. J Prod Innov Manag 29:639–654
32. McCarthy IP, Tsinopoulos C, Allen P, Rose-Anderssen C (2006) New product development as a complex adaptive system of decisions. J Prod Innov Manag 23:437–456
33. Melton T (2005) The benefits of lean manufacturing—what lean thinking has to offer the process industries. Trans IChemE Part A, Chem Eng Res Des 83:662–673
34. Morbach J, Yang A, Marquardt W (2007) OntoCAPE: a large-scale ontology for chemical process engineering. Eng Appl Artif Intell 20:147–161
35. Nagy RL, Ullman DG, Dietterich TG (1992) A data representation for collaborative mechanical design. Res Eng Des 3:233–242
36. Nasab HH, Bioki TA, Zare HK (2012) Finding a probabilistic approach to analyze lean manufacturing. J Clean Prod 29–30:73–81
37. Ohno T (1988) Toyota production system. Productivity Press, Portland
38. Park J, Yang S (2008) Collaborative engineering and product quality assurance based on integrated engineering information management. In: Proceedings of international conference on smart manufacturing application
39. Ramaraj E, Duraisamy S (2007) Design optimization metrics for UML based object-oriented systems. Int J Softw Eng Knowl Eng 17:423–448
40. Renner A (2001) XML data and object databases: a perfect couple. In: Proceedings of the 17th international conference on data engineering, IEEE, Heidelberg, Germany
41. Siemens (2012) http://www.plm.automation.siemens.com/
42. Sohlenius G (1992) Concurrent engineering. CIRP Ann Manuf Technol 41:645–655
43. Stage-Gate (2012) Measuring and improving product development performance and practices. http://www.stage-gate.eu/article-how-to-measure-innovation.asp. Accessed 18 Nov 2012
44. STEP application handbook ISO 10303 Version 3. 30 Jun 2006
45. Sun Microsystem Inc. (2012) Java shared data toolkit API 2.0. http://java.sun.com/products/java-media/jsdt/. Accessed 28 Aug 2012
46. Thimm G, Lee SG, Ma YS (2006) Towards unified modeling of product life-cycles. Comput Ind 57:331–341
47. Thomas S (2001) Managing quality: an integrative approach. Prentice Hall, New Jersey
48. Tornincasa S, Di Monaco F (2010) The future and the evolution of CAD. In: Proceedings of 14th international research/expert conference: trends in the development of machinery and associated technology

49. Ulrich K, Eppinger S (2011) Product design and development. http://www.ulrich-eppinger.net/
50. Vranic V (2004) Reconciling feature modeling: a feature modeling metamodel. In: Proceedings of net object days, pp. 122–137
51. Wang H, Xiang D, Duan G et al (2007) Assembly planning based on semantic modeling approach. Comput Ind 58:227–239
52. Wang J, Tang MX (2007) Product data modeling for design collaboration. In: Proceedings of 11th international conference on computer supported cooperative work in design
53. Wang Q, Ren Z (2010) XML-based data processing in network supported collaborative design. Int J Autom Comput 7:330–335
54. Westphal CR, Blaxton T (1998) Data mining solutions: methods and tools for solving real-world problems. Wiley, New York
55. Wiesner A, Morbach J, Marquardt W (2011) Information integration in chemical process engineering based on semantic technologies. Comput Chem Eng 35:692–708
56. Womack JP, Jones DT, Roos D (1990) The machine that changed the world. Free Press, New York
57. Womack JP, Daniel TJ (2003) Lean thinking. Free Press, New York
58. Yeo KT, Ning JH (2002) Integrating supply chain and critical chain concepts in engineer-procure-construct (EPC) projects. Int J Proj Manag 20:253–262
59. You CF, Tsou PJ, Yeh SC (2007) Collaborative design for an assembly via the internet. Int J Adv Manuf Technol 31:1217–1222
60. Zhang L (2009) Modelling process platforms based on an object-oriented visual diagrammatic modeling language. Int J Prod Res 47:4413–4435

A Review of Data Representation of Product and Process Models

Narges Sajadfar, Yanan Xie, Hongyi Liu and Y.-S. Ma

1 Product Modeling Methods

As a type of semantic data model, a product data model is a set of data in a consistent data structure that ideally represents product information efficiently, effectively, and concisely [8, 20]. In the current information era, a product data model should satisfy the technical and quality requirements of the whole product lifecycle. Engineers and manufacturers need a product data model that has a common, unified, and global definitions of the product information resources and also can be interpreted by various computer programs. In recent decades, a variety of product modeling methods and incidental software has been created, developed, improved, and used effectively. The main methodologies can be categorized into four classes: solid product modeling, feature-based product modeling, knowledge-based product modeling, and integrated product modeling methodologies [42].

1.1 Solid Product Modeling

Solid product modeling was created as a technology to precisely embody 3D product geometry information. The most common solid product modeling methods are boundary representation (B-Rep) and constructive solid geometry (CSG) [42]. The B-Rep modeling method uses a model to bound the edge and vertices of the solid object in order to clearly store and speedily display geometric information, including the faces, edges, and vertices in the representation. In contrast, the CSG modeling method is based on primitive solids (e.g., cubes, cylinders, and spheres). Boolean operators are used to define a set of operations to put together a complex

N. Sajadfar · Y. Xie · H. Liu · Y.-S. Ma (✉)
Department of Mechanical Engineering, University of Alberta,
Edmonton, AB T6G 2G8, Canada
e-mail: yongsheng.ma@ualberta.ca

Y.-S. Ma (ed.), *Semantic Modeling and Interoperability in Product and Process Engineering*, Springer Series in Advanced Manufacturing,
DOI: 10.1007/978-1-4471-5073-2_2,

product solid model by adding or subtracting volumes. The common operations are *union*, *subtraction*, *supplementary set*, and *intersection*. The CSG modeling method can define objects with a relatively concise and simple data structure.

Solid modeling methods are now very mature, and are widely utilized in various product development phases. However, because solid modeling concentrates on geometric details, it lacks the functionality to present other information indispensable to the entire product development lifecycle. To solve this problem, the feature concept was created and has since been utilized in many computer-aided product modeling processes.

1.2 Feature-Based Product Modeling

1.2.1 Definition of Feature Concept

A feature "represents engineering meaning or significance of the geometry of a part or assembly" [32]. Features can be understood as information sets that refer to aspects of form or other attributes of a part. A feature can be thought of as a representation of an engineering pattern that contains the associations of the relevant geometry data with other kinds of data, such as manufacturing data, in order to provide sufficiently rich and versatile information and to speed up product engineering processes. It therefore represents a great improvement over B-Rep and CSG techniques [5, 17].

1.2.2 Definition of Feature Model

A feature model is a data structure that is comprised of a variety of types of features. These features are all recognizable entities that have specific representations. The choice of features depends on what function the feature model is intended to support. Additional features can be added into the feature model according to new requirements. For example, manufacturing companies can choose a specific range of machining features that include the geometry to be produced and the related nongeometric technical information. Such a manufacturing feature model is not only related to the manufacturing requirement for customizability but also makes feature technology more influential in related industrial applications [32].

1.3 Knowledge-Based Product Modeling

Knowledge-based product modeling uses Artificial intelligence technologies to model product development expertise and rules and to automate the many logical reasoning and optimization processes of engineering design and manufacturing.

Systems developed using this approach can also store a large amount of information and knowledge about previous designs, which can help avoid unnecessary time spent on planning and redesign. This approach can be used to simplify the modeling tasks and enhance modeling quality. Although capturing, representing, and using knowledge is both costly and risky, knowledge engineering plays an important role in business globalization and product development.

1.4 Integration-Based Product Modeling

The integrated product modeling methodology can be considered a functional combination of several modeling methods by associating data and processes of engineering in a systematic approach. All kinds of product data, including feature information, geometric data, and product knowledge, can be modeled and stored in a comprehensive and thoroughly integrated product model. This is an active research domain; the modeling methodology is not well established and still needs to be explored systematically.

1.5 Data Requirement in Product Lifecycle Management

Product lifecycle management (PLM) is the business activity of managing production effectively, from the initialization of a product to its withdrawal from the market. Full PLM involves various applications and several complex processes. It has to support cyclic engineering process modeling in order to represent, exchange, reuse, and store knowledge, and track decision-making processes in all application domains. To achieve these goals, PLM needs information technology (IT) to support its connections and a central data repository for gathering all the information during data exchange at all stages of manufacturing and production. IT services have to connect PLM to the product design and analysis processes. PLM also needs to have a relationship with the supply chain process, which includes processes such as enterprise resource planning (ERP), customer relationship management (CRM), supply and planning management (SPM)), and component supplier management (CSM) [28]. Information and communications technology (ICT) can support PLM to cover product process such as holding, retrieving, manipulating, sending, or receiving knowledge. In a new PLM paradigm, PLM is defined as an integrated business model that employs ICT technologies and implements an integrated cooperative and collaborative management system for product-related information throughout the entire product lifecycle, from a product's conceptualization until its dismissal. In this paradigm, ICT solutions are expected to bring many advantages for PLM, including customer satisfaction requirements, reduced time-to-market for new products, and decreased environmental issues in product manufacturing [13].

2 Data Repository

Every computer system must have a data repository for the collection and organization of data [20]. The advantage of storing information in a data repository is that it creates a supporting module in a demanding information system with a high level of data organization and reusability. Usually, the database involves several tables where each row stores the same sort of information. The tables can reference each other by building in data relationships. Each table holds the information about a particular entity, which is a virtual data representation of an existing artifact that is important to the application system and on which the database wants to keep information. To achieve the advantage of efficient data access and storage, a data repository needs data schema to efficiently manage information that can support reusability.

A modern networked data repository is a logical and physical facility that can be a collection of databases with objectives and tasks combined into one vast unified database. It can also be defined and developed as a unified and global business solution from customer services to manufacturing and shipping, which are essentially managed by accurate information interaction processes interfacing many database entities. A data repository needs to provide a strong and comprehensive relationship among all the data to help with decision-making and decision-support systems (DSS). Having one data repository that reflects all the information and knowledge in a system is useful for updating data quickly and efficiently [25].

A data repository can also bring competitive advantages for businesses. When several companies or organizations want to improve their business collaboration relations, they can use the semantic repository approach to integrate with each other via their databases. A semantics-oriented engineering data repository can support the complex and dynamic engineering and business information flows among associated and networked databases in the modern economy. For instance, the ICT project aiming to decrease the distance between government and business is using the semantic data repository approach. Figure 1 shows a few key components of a semantic data repository [24].

A data repository consists of data records and interrelated tables [45]; therefore, a data repository first needs to collect data. Although modern informatics tools enable active data collection, such as Web-based data crawling from numerous sources, the majority of engineering systems use passive data collection, such as designing a specific application programming interface (API) function to extract information from existing data sheets or files. The data gathered through active or passive data collection is initially stored in a local storage system, but needs to be integrated into a networked data repository to support accessibility expectations. At the end of the collection process, the data is accumulated and ready for use. In addition, the type of data source determines the structure and the module of data storage. Because a data repository has to keep data current and consistent, it will be designed to trace data change impact and schedule data updating procedures

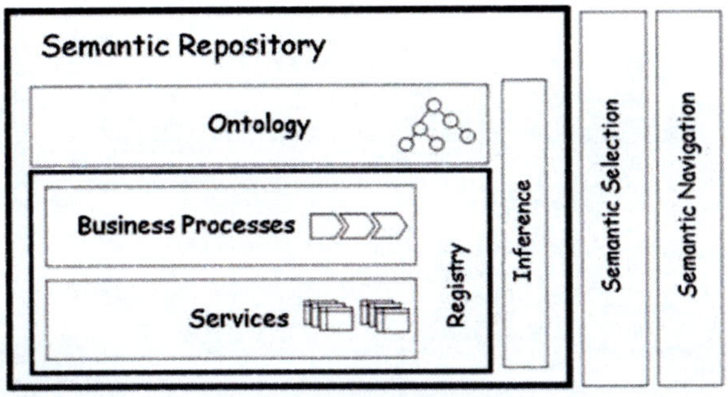

Fig. 1 Layers of a semantic repository of services [24]

constantly and dynamically with a generic "road-map". Scheduling frequent data access and updates requires the prioritizing of data operations [45].

Operating on a larger scale than can be managed by a database, data warehouses are a kind of optimal data repository system that include historical and static data. "Data warehouses are built in the interests of business decision support and contain historical data, summarized and consolidated from detailed individual records" [31]. A data warehouse can also be defined as an abundant database that combines distributed operational data in one place, linking a collection of subject-oriented data with the original sources. The objective of warehouse design is to provide a database that can support different kinds of applications such as decision making, online analytical processing (OLAP), data mining, and (DSS) [6].

2.1 Engineering Database Technology Status

Implementing an engineering data repository system requires designing data integration schemes and interfacing information sources that can be in a variety of forms, such as artifacts, functions, failure signals, physical objects, performance indicators, sensory input, and media related records. The scheme designer has to select the best type of category to store and extract data. For instance, in a product database [3], *function type* is useful for describing product features or finding an existing product. After the type of category has been selected and the schema designed, the database tables must then be designed. A design repository can store two types of data: artifacts and taxonomies. An artifact contains the field name and data type, and a serial-based ID as in a typical database. A serial ID makes a unique number for each artifact stored in the database, and data can be extracted according to this ID. Taxonomies, by contrast, make data interpretable by associating more information together such as a product's color, parameter, functions, material, and/or sensors. In addition, a design repository can support inheritance relationships to organize the inherited Attributes. Figure 2 shows a representation diagram of data repository tables. Usually, a repository model needs to be implemented into a structured query language (SQL) compatible database or a set of networked databases [3].

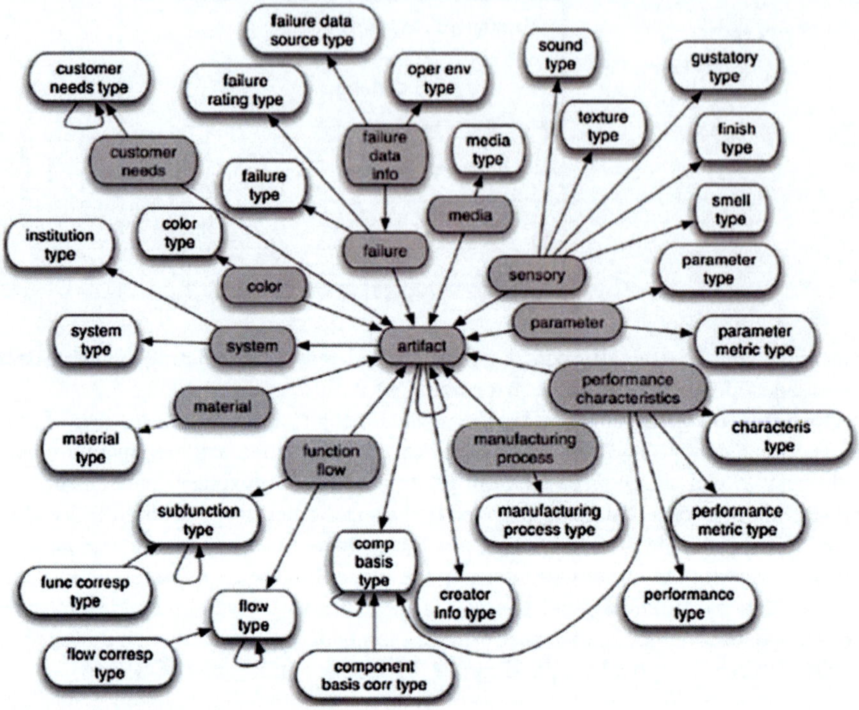

Fig. 2 Graphical view of repository database tables [3]

2.2 Data Repositories in Integrated Systems

Collaborative engineering is a systematic business approach that facilitates the exchange of useful sources and information-sharing for multi-disciplinary groups in real time [23]. A data repository supporting collaborative engineering must be designed to support dynamic data interactions, which requires integration and collaboration tools [18]. Information integration for a product model can have different layers of data granularity, i.e., functional applications, dynamic and persistent information entities, structural representations, and physical storage records, as suggested by Tang [35]. The application layer (AL) is the top layer, which contains various feature-based functional applications. The information layer consists of a feature-oriented meta-product model, unified feature components, application feature components, and STEP EXPRESS-based [14] specifications. The representation layer keeps the relation between EXPRESS-defined and database schema. The physical records layer forms the basis of a data repository, which includes feature properties and geometrical entities. The data repository can exchange and share information with various levels of granularity. For instance, the data repository can call on interactive system class methods

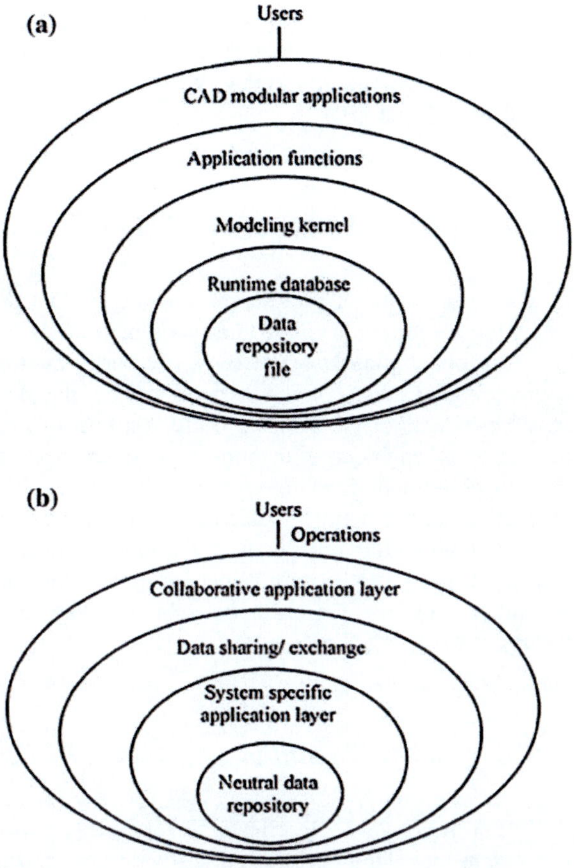

Fig. 3 Comparison of traditional and proposed CAx structure [18] **a** Traditional CAD structure. **b** Proposed CAx structure

(functions) to synchronize data transactions between computer-aided design (CAD) systems and their databases. The goal of the integrated engineering data repository is to support concurrent engineering activities, such as design phases and CAx analyzes, and to reduce the cost and time involved in data management. In general, this kind of repository can use an Internet browser as a front end, and can interface with application servers. Figure 3 compares the traditional CAD structure and proposed CAx structure [18].

Efficient information sharing and data interchange can create competitive advantages for companies and organizations. One of the most notable companies that focuses on collaboration in the supply chain is Dell. Dell has a unique supply chain management system, which has created outstanding sales for Dell. It uses the direct sales system to build exactly the product that the client wants. The computer industry grows significantly every year, but rapid technology changes are to the disadvantage of this industry. Dell must therefore keep little inventory and introduce new products to the market quickly. Dell fosters a close relationship with its suppliers and uses five key strategies to create a unique supply chain: rapid time to accommodate low-volume orders, customized products built to order, elimination of the retailer, superior services, and support, and low inventory and low capital investment [10].

3 Informatics Modeling in Chemical Process Engineering

In a typical chemical process engineering (CPE) development project, such as designing a refinery facility, thousands of process flow diagrams (PFDs), piping and instrument diagrams (P&IDs), electrical circuit diagrams, and mechanical engineering drawings are generated during the design, engineering, and construction phases of the project. There is a host of information embedded in these diagrams and engineering documents which also serves in the downstream phases, such as equipment procurement, construction engineering, operation, and maintenance, as either input or reference. Currently, most engineering documents are generated with domain-specific software applications, such as SmartPlant and Aspen. In such projects, it is necessary that the effective flow of engineering semantics, not just the data, be achieved via an integrated computer system throughout each CPE project.

3.1 Embedded Semantics and Issues in CPE Documents

A CPE project involves a series of development phases across various departments and disciplines. Domain-specific software applications are used to complete the activities in each phase but also to generate heterogeneous data sources [40]. The typical engineering documents generated in a chemical engineering project are summarized in Table 1, from a very general block diagram to a detailed flow diagram. Some of these documents have semantic information beyond one specific domain, i.e., chemical engineering alone. For example, a P&ID specifies many relevant semantic constraints for mechanical engineering and electrical engineering as well. Data files, such as spreadsheets or a 3D process model, provide even more detailed specifications to downstream engineering activities. A variety of types of flow sheet can be found in the work of Ludwig [15].

To generate more and more detailed engineering designs, engineering activities involved in the chemical process project often use information from those data files resulting from earlier engineering phases. Associations between activities involved and files generated are shown in Fig. 4, with three critical activities (conceptual design, process engineering, and mechanical engineering and design) as an illustration.

One major problem faced by engineers is the conflict among intense associations and the heterogeneity of data sources. As illustrated in Table 1, the data files

Table 1 Semantics embedded in a typical data sources in chemical process engineering

Engineering documents	Software tools	Semantics
Input/output flow diagram (IOFD)	Any flow chart packages, e.g., MS Visio	Raw materials Reaction stoichiometry Products
Block flow diagram (BFD)	Any flow chart packages, e.g., MS Visio	Everything above, plus: Materials balances Major process units Process unit performance specification
Process flow diagram (PFD)	Any flow chart packages, e.g., MS Visio CAPE packages, e.g., Aspen	Everything above, plus: Energy balances
	CAD packages	Process conditions (T & P) Major process equipment specification
Process & instrument diagram (P&ID)	Any flow chart packages, e.g., MS Visio	Key piping and instrument details Process control schema
	SmartPlant P&ID CAD packages	Symbol representation of all equipment and components involved
Mechanical flow diagram (MFD)	Any flow chart packages, e.g., MS Visio CAD packages	Pipe specification All valves (sizes and types) Operation condition specification
Spreadsheets	Any spreadsheet packages, e.g., MS EXCEL	Engineering calculations Materials balances Process conditions Detailed equipment specification
3D process model	SmartPlant 3D	Piping routing and specification Process conditions Process equipment and component topological and geometrical features

generated by CPE projects come in a variety of formats, including unstructured data, semi-structured data, and structured data; this situation leads to structural heterogeneity [15, 22, 36]. For example, the requirement analysis document used at the start of a chemical process project is an unstructured data source, while the specification in the spreadsheets is a kind of semi-structured data source, and the data stored in the databases of, for example, SmartPlant 3D belongs to a structured data source.

To make the engineering processes even more complicated, differences also exist in the interpretation of the semantics of the data, which leads to another level of heterogeneity, called semantic heterogeneity [11]. For example, process engineering requires PFDs, and the information embedded in them is used as the input for process simulations; the simulation results will in turn influence the conceptual

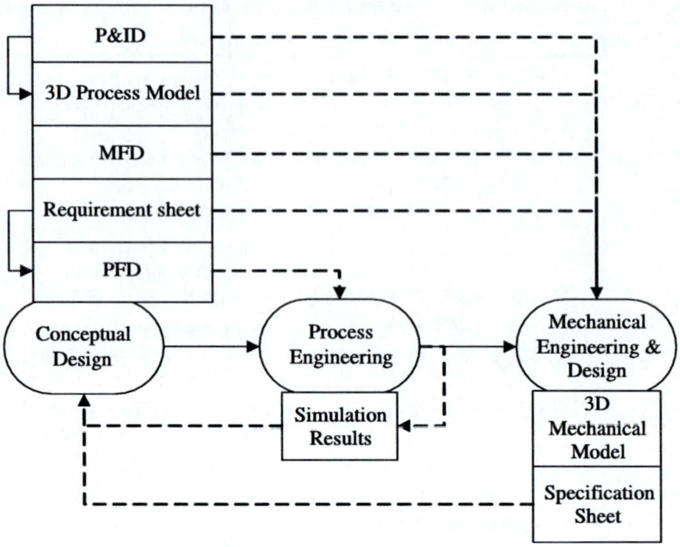

Fig. 4 Associations between activities involved and files generated

design activity. To implement the downstream mechanical engineering design (ME&D) with the accurate interpretation of the constraints embedded in data, i.e., using the semantic information, consistency-checking with the data sources in the conceptual design phase has to be achieved.

Meanwhile, the output of the ME&D, i.e., the 3D models as well as the specification sheets, need to be interpreted face-to-face by engineers in order to provide information feedback for conceptual design.

As discussed in chapter "Introduction to Engineering Informatics", some neutral data formats have been developed to deal with geometry heterogeneity, such as the STEP standard, but they can do nothing with semantic heterogeneity. Existing feature technology has improved interoperability between heterogeneous applications but still has limited capability to handle semantic heterogeneity. This is due to the fact that, currently, feature semantics has not been well-defined or maintained [2]. The lack of semantic interoperability may lead to severe information loss and, further, economic loss due to potential operation breakdown and maintenance.

To achieve interoperability on the semantic level, a specification of the interpretation of terminology used in different computer-aided systems has to be formalized [40]. However, there is little, if any, representation of semantics embedded in the data files generated with the existing software packages. Fortunately, the increasing research trend in semantic modeling, which covers semantic conceptual schema with embedded semantic information, provides reason for optimism in overcoming the semantic interoperability problem.

3.2 The Current State of Informatics Modeling Research in CPE

The emerging informatics methodologies for the integration of CPE activities, including mechanical engineering activities, lead in two different directions.

On the one hand, semantic modeling (which comes from the domain of software engineering, as does the development of database design theory) is now applied in many other disciplines, including CPE and mechanical engineering. In the chemical engineering domain, ontology has been applied to facilitate semantic technology [21, 40]. A large research project, IMPROVE, is being conducted in Germany, with a goal of developing a collaborative environment for CPE based on ontology [12, 21, 40]. A flexible and extensible data structure is also proposed to apply to heterogeneous and distributed data. With the knowledge representation capability embedded, it provides a basis for knowledge engineering to incorporate knowledge to guide the activities. Semantic modeling offers a common ground to enable interoperability for both disciplinary domains. However, there is currently no formal definition of semantic modeling. Semantic modeling is understood in the engineering domain to be the information-modeling activities that develop a high-level representation of semantic schema, which provides specifications for the interpretation of data and relations that are then used to capture comprehensive information from different entities [29, 39]. The essence of semantic modeling is to represent relationships between data elements in an explicit way, which helps maintain the consistency of semantic information. Ontology, defined as "an explicit specification of a conceptualization, typically involving classes, their relations and axioms for clarifying the intended semantics," is perfectly suitable for use in implementing semantic modeling in CPE [38].

On the other hand, feature technology, which is believed to be versatile enough to support the encapsulation of tedious mechanical engineering parameters, Attributes, and constraints, is also flexible enough to be associated with semantic modeling entities in an abstracted and declarative form. Feature is still the main technology in the mechanical engineering domain, but is equipped with more capability to represent semantics, such as design intent. Hence, a hybrid semantic modeling that bridges ontology modeling and feature modeling is believed to be a practical way to realize interoperability between multidisciplinary systems. In the authors' recent the research [16], with reference to the feature models proposed by Bidarra and Bronsvoort et al. [2, 4], a declarative feature modeling method called semantic feature modeling was proposed. In the proposed framework, chronological order dependence is removed and the semantics of all the features are well-defined and maintained through the product lifecycle by means of a detection and consistency-checking mechanism. Although some modeling freedom may be sacrificed, this is acceptable considering the improvement to semantic representation capability.

3.3 Semantic Modeling Methods in CPE

There are two approaches to semantic modeling in CPE: integration and mapping.

3.3.1 Semantic Integration Model

The architecture of the proposed semantic integration model is shown in Fig. 5. It consists of three layers: a data layer (DL), a semantic schema layer (SSL), and an AL. The SSL works as a mediator to generate a semantic view for a particular user/engineer from the original distributed, heterogeneous data embedded in various formats. The model proposed here forms the core of the integration framework, which will be discussed in detail in "Features and Interoperability of Computer-Aided Engineering Systems".

Data Layer (*DL*). The data sources, such as those listed in Table 1, lie in the DL in various formats.

Semantic Schema Layer (*SSL*). The central unified semantic mediator lying in this layer includes three basic modules: (1) information extraction, which extracts necessary information from the data sources according to local ontology; (2) semantic mapping, which maps the information extracted into consensual and formal specifications according to the semantic schema; (3) view generation and management, which generates the views with only the necessary semantic information. The knowledge library, holding the domain-specific knowledge, exists to

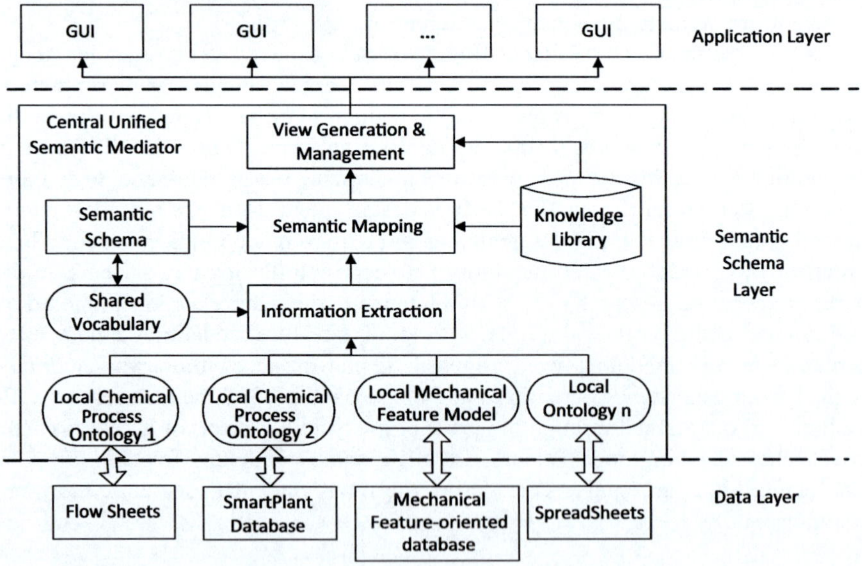

Fig. 5 Architecture of semantic integration model

facilitate the mapping process as well as view generation and management. A hybrid approach for content explication is applied here, as it is scalable and supports heterogeneous views with reasonable implementation cost, as compared to single- and multiple-ontology approaches [34].

Application Layer (AL). Graphical user interfaces (GUIs) for different potential users, based on the views generated in the SSL, are provided in this layer.

3.3.2 Semantic Mapping

The information retrieved from distributed data sources can be first mapped based on the semantic schema defined, and then mapped to any other engineering disciplinary view as required by referencing. The partial semantic schema of pressure vessels is given as an example in Fig. 6. To generate a requirement for a vessel, the material as well as its temperature and flow rate will be grouped together to form the input. Pressure, capacity, and temperature retrieved from P&ID will be grouped into the operating conditions. Design pressure, material, and thickness retrieved from the specification sheet supplied by the vendor will be grouped into vessel specifications.

However, to collaborate with different systems, there are still two potential causes of semantic inconsistency: different meanings attached to the same terminology, and the same meaning represented by different terminologies [34]. For example, as shown in Fig. 6, vessel specification has two pressures specified, design pressure and operating pressure, but only one pressure is required in the operating conditions. During ontology mapping, it should be therefore explicitly

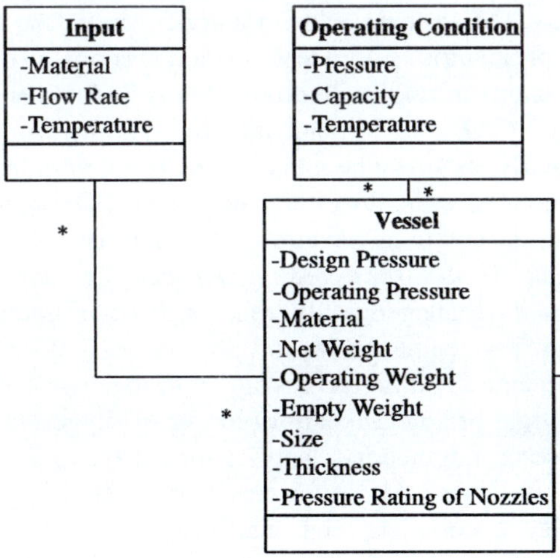

Fig. 6 A semantic schema of vessels

reflected in the shared vocabulary that the *pressure* in the *operating conditions* should be interpreted as the *operating pressure* in the *vessel* vocabulary. In the same example, the *material* in the *input* and the one in *vessel* specification sets do not mean the same thing: one is the material of input and the latter means the material of the vessel. Within the shared vocabulary, these two instances of the term *material* have to be interpreted into different meanings automatically.

Therefore, in semantic modeling, a well-structured semantic schema, along with the shared vocabulary that serves the function of "dictionary" to accurately interpret the meaning of the information, keeps the semantics consistent among the instances of different entity types, and provides the schema to map the data elements accordingly.

4 New Development in Product and Process Modeling with Engineering Informatics

Storing and extracting geometric and nongeometric feature information for engineering design in an integrated and dynamic data repository is an important step in managing and improving product and process engineering lifecycles. Currently, the state of industrial practices is still based on holding engineering design models in file storage [1] and then managing data through various data controllers. This situation constrains the integration of applications, and implementing concurrent design development processes also demands more complicated data controllers. Further, the situation leads to redundancy in storing consistent product and process information. An unfortunate fact is that a lack of integration among CAD and CAM systems leads to a deficiency in the computer numerical control (CNC) programming process, which seriously limits the implementation of digital manufacturing technology. Many industries find that they need to convert various CAD file formats into one another for different engineering applications. Clearly, as many people have realized, an interoperable engineering platform will help industries to effectively share their digital assets among various computer systems and to achieve full return on their investment in digital intellectual properties. If the projected vision can be achieved, the centralized data repository can be managed at different levels of abstraction based on the root-level geometric or nongeometric data.

To leverage the widely-accepted feature technology with database technology, a unified feature-based fine-grain product repository has been suggested with the aim of interoperability among engineering applications [37]. For instance, Ma et al. [18] provided a fine-grain and feature-based product data repository design. They suggested using a complete SQL database to accommodate the complex neutral features and extracted feature information via database API functions. However, this database has limitations, as the table of databases must be created manually, and the database cannot perform validation of feature changes.

Global research in the area of integrated feature-based systems can be divided into two approaches: developing the architecture of an integrated feature-based system, and implementing feature-mapping functions between systems. The majority of the current research focuses on the latter approach and explores the concept of reusing CAD models and converting design results from one feature model to another. Much of the previous research also focused on the integration of CAD and CAE processes, but was limited to operating geometric entities or storing layer-base information in a database. The most common problem is that CAD designs have much more complex structures than CAE geometric meshes. Consequently, full-scale implementation of CAD and CAE integration has yet to be attempted.

5 Engineering Change Management in Design: The Propagation Method

Engineering change management (ECM) is an essential aspect of concurrent engineering, and comprises all related activities during the design and manufacturing stages. To reduce product development time, many companies adopt the concurrent engineering approach by stressing a parallel and collaborative engineering procedure. ECM is typically time-consuming, as it frequently involves disparate information systems issued frequently during the product lifecycle. Due to its complexity, building a system that enables seamless design change propagation is likely to be a very demanding task. This section introduces a preliminary propagation-based methodology for design change management in collaborative design and manufacturing.

5.1 The Design Change Process Framework

Design changes (DCs) are known as engineering changes (ECs), which is an important phase in the computer-integrated manufacturing system [19]. Reducing the time for ECs can greatly shorten the product lifecycle and improve the productivity of an enterprise. Figure 7 shows a framework for the process of DC.

Due to the application of modular technology, companies can choose a particular supplier who is focused on a series of components, and build a collaborative unit in order to maximize the benefit margin and shorten the product lifecycle. In this collaborative product development process?, speeding up EC requires that EC information should be represented precisely and clearly in a standard format and be shared among participating companies. However, part suppliers do not always use the same CAD system and are often unwilling to share their CAD data with other cooperating companies, in order to honor policies of protecting corporate intellectual property. These circumstances make it difficult for collaborating

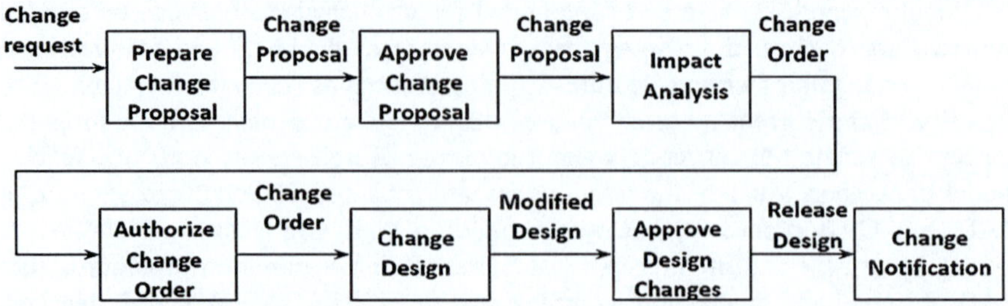

Fig. 7 Framework for the process of design change

companies to conduct efficient EC, since a part supplier who is responsible for one part of a product needs other CAD part model data designed by other companies for the ECs in the typical CAD product assembly modeling process. While much research still needs to be done to address this issue, there are a few methods that have been both published and practiced in the industry domain in recent years.

5.2 *Recent Research and Implementation of ECM*

ECM (especially design change management) occupies an irreplaceable position throughout the product lifecycle. Much effort has been put toward using a propagation-based approach. With the development of product data management (PDM) systems, conceptual design change management based on product structure has been studied by Peng and Trappey [26] and Do and Choi [7, 8]. Upon improvement in the parametric modeling capabilities of commercial CAD systems, a parameter-based approach was suggested for ECM [41]. Recently, common platform specifications and implementation guidelines have been developed by standardization research organizations, such as ProSTEP, toward the development of an ECM system [27, 30]. Two methods in particular, an engineering change propagation (ECP) system based on the STEP neutral data format and a neutral reference model based on parameter referencing, are discussed below for representation and propagation of ECs in collaborative product development.

5.2.1 Engineering Change Propagation with STEP Data Structures

You and Yeh [44] proposed an ECP system based on STEP. This system for modeling ECs in models of engineering data, geometry, and features using the STEP standard provides a flexible, virtually integrated framework to enable EC. Figure 8 shows the basic conceptual structure of the approach.

Figure 8 illustrates the overall structure of the ECP system. The ECP, CAD, and PDM system databases are organized as a three-tier architecture of individual

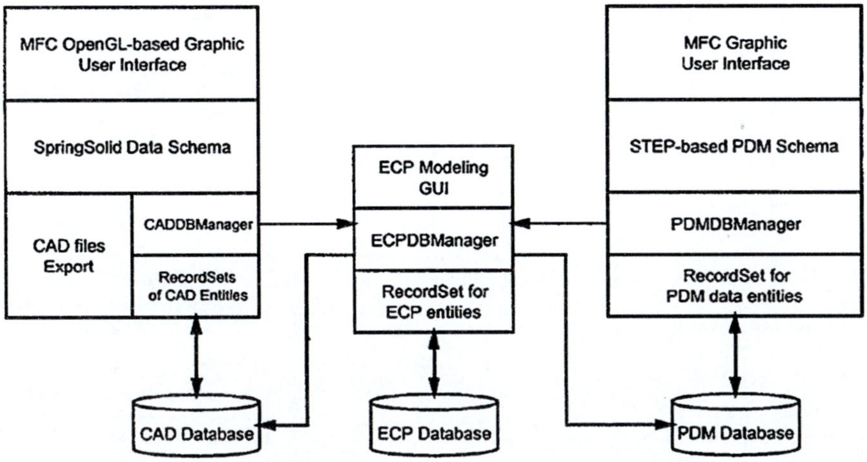

Fig. 8 The ECP system architecture [44]

databases, which stores the original data. The open database connectivity (ODBC) protocol and the record sets class of Microsoft foundation classes (MFCs) are applied to the database operation modules of these systems. Two additional modules, *CADDBManager* and *PDMDBManager*, are used to handle the EC transactions when a change is issued. When a DC is issued in the CAD system, the *CADDBManager* searches and triggers the *ECPDBManager*, which manages the database of the ECP system. *ECPDBManager* obtains the ECP network of the affected data section and gains all the changed items in the CAD and PDM systems. The ECP system then propagates the change from CAD to PDM. The CAD feature editing functions and PDM system commands are applied via implemented COM technology to assist interoperation. Functions defined by these systems are made available to the public and compiled as COM automation documents. The client can request, through a public interface, access to the functions defined in the server. The automation server inherits *IUnkown* and *IDispatch* interfaces in the MFC class and provides clients with public methods to call automation objects. The operation of the ECP network in the study [44] relies on the COM-based change propagation mechanism in the ECP system. The ECP triggering module, CAD feature editing, and the PDM system are all COM object servers that can call and be called by other COM automation objects.

A DC may be propagated through the related features, if they are affected. Collaborative product design can be classified into two categories: collaborative component design and collaborative assembly design [33, 43]. Changes in the design of a part often involve the shape modification of other parts in the assembly model, especially in places where there is a tight connection between parts. When the feature of a part is changed, the tight connection feature with this part should thus also be changed. The process shown in Fig. 9 facilitates the propagation of changing information throughout the whole model.

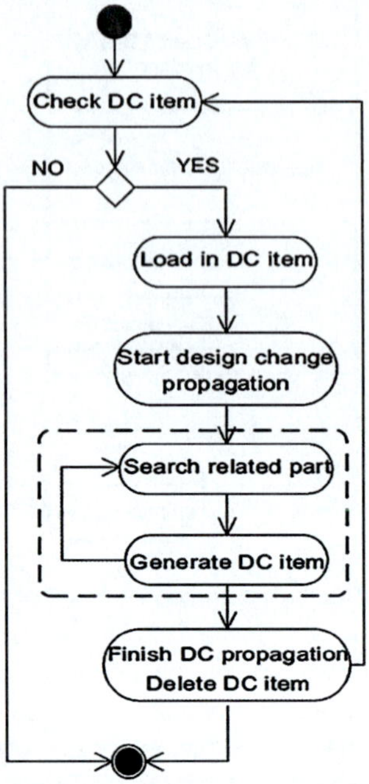

Fig. 9 The propagation process [43]

Searching the propagated parts and establishing the items impacted by the DC is the critical algorithm of the propagation process. Therefore, the related parts must satisfy the following conditions:

1. A relational part must have the characteristics of features.
2. It has to be one of the mating couples with the parts feature that has the DC data.
3. The mating condition has to be a tight connection.

After the above conditions are satisfied, when a related feature is changed, the corresponding features should also change shape in order to keep the same mating conditions of the assembly relationship. In order to accelerate the process, before propagation is begun, it is better to initialize the information in the assembly model for higher efficiency and shorter search time.

5.2.2 Engineering Change Propagation with a Feature Reference Model

Hwang and Mun [9] proposed a neutral reference model for the representation and propagation of EC information in collaborative product development. This neutral reference model consists of a neutral skeleton model and an external reference model, which is implemented on the parametric referencing functions supported by most available CAD systems. If the referenced geometric entity is changed, the consequent parameter value changes are automatically propagated to the referencing geometry entities, which trigger an automatic change of the referencing model. Employing this mechanism, an external referencing model was used by Hwang and Mun [9] with the aim of managing the relations between the original skeleton and the NSM CAD files. Do and Choi [7] also proposed a comprehensive procedure for ECP in order to maintain consistency between various product data views. Their procedure used the history of product structure changes based on an integrated product data model. The effectiveness of these methods is, however, still to be proved.

Clearly, much research has yet to be done to set up an integrated product information database with a common standard for ECM that is accessible and useful for supporting the entire product lifecycle.

6 Summary

This chapter provided a review of the state of the art in product and process informatics modeling and implementation. It is clear that currently there are numerous computer solutions that are addressing engineering application support requirements piece by piece. However, there are many gaps among these piecemeal solutions in communicating associated information effectively. Such integration and data sharing difficulties have been the main hurdles in realizing the potential economic benefits of engineering informatics. A systematic study of interoperability among computer systems is in high demand, which happens also to justify the purpose of this book, i.e., exploring a theoretical framework for interdisciplinary and multifaceted informatics engineering. To do so, the authors believe extended feature technology will play a pivotal role in future technology development.

References

1. Anderson RA, Ingram DS, Zanier AM (1973) Determining fracture pressure gradients from well logs. J Pet Technol 25:1259–1268
2. Bidarra R, Bronsvoort WF (2000) Semantic feature modeling. Comput Aided Des 32:201–225
3. Bohm MR, Stone RB, Simpson TW, Steva ED (2008) Introduction of a data schema to support a design repository. Comput Aided Des 40:801–811
4. Bronsvoort WF, Bidarra R, van der Meiden HA et al (2010) The increasing role of semantics in object modeling. Comput Aided Des Appl 7:431–440
5. Chen YM, Wei CL (1997) Computer-aided feature-based design for net shape manufacturing. Comput Integr Manuf Syst 10:147–164
6. Dey D (2003) Record matching in data warehouses: a decision model for data consolidation. Oper Res 51:240–254

7. Do N, Choi IJ (2008) Propagation of engineering changes to multiple product data views using history of product structure changes. Int J Comput Integr Manuf 21:19–32
8. Do N, Choi IJ, Jang MK (2002) A structure-oriented product data representation of engineering changes for supporting integrity constraints. Int J Adv Manuf Technol 20:564–570
9. Hwang J, Mun D (2009) Representation and propagation of engineering change information in collaborative product development using a neutral reference model. Concurrent Eng 17:147
10. Kapuscinski R, Zhang RQ, Carbonneau P, Moore R, Reeves B (2004) Inventory decisions in Dell's supply chain. Interfaces 34:191–205
11. Kim W, Seo J (1991) Classifying schematic and data heterogeneity in multidatabase systems. IEEE Comput 24:2–18
12. Körtgen A, Nagl M (2011) Tools for consistency management between design products. Comput Chem Eng 35:724–735
13. Kovacs G, Kopacsi S, Haidegger G, Michelini R (2006) Ambient intelligence in product lifecycle management. Eng Appl Artif Intell 19:953–965
14. Liang J, Shah JJ, D'Souza R et al (1999) Synthesis of consolidated data schema for engineering analysis from multiple STEP application protocols. Comput Aided Des 31:429–447
15. Ludwig EE (1964) Applied process design for chemical and petrochemical plants, vol 1. Gulf Publishing Company, Houston
16. Ma YS (2009) Towards semantic interoperability of collaborative engineering in oil production industry. Concurrent Eng 17:111–119
17. Ma YS, Tong T (2003) Associative feature modeling for concurrent engineering integration. Comput in Ind 51:51–71
18. Ma YS, Tang SH, Au CK, Chen JY (2009) Collaborative feature-based design via operations with a fine-grain product database. Comput in Ind 60:381–391
19. Marri HB, Gunasekaran A, Grieve RJ (1998) An investigation into the implementation of computer integrated manufacturing in small and medium enterprises. Int J Adv Manuf Technol 14:935–942
20. Marsh GL, Smith JR (1984) Exploratory well design for 5,000- to 7,500-Ft water depths U.S. East Coast. Offshore Technology Conference, Houston. ISBN 978-1-61399-077-3
21. Morbach J, Yang A, Marquardt W (2007) OntoCAPE: a large-scale ontology for chemical process engineering. Eng Appl Artif Intell 20:147–161
22. Murphy RM (2007) Introduction to chemical process: principles, analysis, synthesis. McGraw-Hill, New York
23. Olsen GR, Cutkosky M, Tenenbaum JM, Gruber TR (1995) Collaborative engineering based on knowledge sharing agreements. Concurrent Eng 3:145–159
24. Palmonari M, Viscusi G, Batini C (2008) A semantic repository approach to improve the government to business relationship. Data Knowl Eng 65:485–511
25. Patent Application Publication (2007) United States, Pub No: US2007/0022081 A1, Pub Date 25 Jan 2007
26. Peng TK, Trappey A (1998) A step toward STEP compatible engineering data management: the data models of product structure and engineering changes. Robot Comput Integr manuf 14:89–109
27. ProSTEP iViP ECM Implementor Forum (2012). http://www.prostep.org/fileadmin/freie_downloads/Empfehlungen-Standards/ProSTEP_iViP/ProSTEP-iViP_Recommendation_Engineering-Change-Management_Engineering-Change-Order_3-2_0.9.pdf. Accessed 23 Aug 2012
28. Rachuri S, Subrahmanian E, Bouras A, Fenves SJ, Foufou S, Sriram RD (2008) Information sharing and exchange in the context of product lifecycle management: role of standards. Comput Aided Des 40:789–800
29. Rishe N (1992) Database design: semantic modeling approach. McGraw-Hill, New York
30. SASIG ECM group (2006). http://www.sasig.com/site. Accessed 23 August 2012

31. Sen A, Jacob VS (1998) Industrial-strength data warehousing. Commun ACM 41:28–31
32. Shah JJ, Mantyla M (1995) Parametric and feature-based CAD/CAM concepts, techniques and applications. Wiley-Interscience, New York
33. Shyamsundar N, Gadh R (2002) Collaborative virtual prototyping of product assemblies over the internet. Comput Aided Des 34:755–768
34. Stuckenschmidt H, Harmelen FV (2005) Information sharing on the semantic web. Springer, New York
35. Tang SH (2007) The investigation for a feature-oriented product database. PhD thesis, School of Mechanical and Aerospace Engineering, Nanyang Technological University, Singapore
36. Thomas CE (2007) Process technology equipment and systems, 2nd edn. Cengage Learning, Delmar
37. Uddin MM, Ma YS (2011) Towards a feature-based and fine-grain product repository for heterogeneous computer-aided systems. In: 4th international conference on changeable, agile, reconfigurable and virtual production (CARV2011), Montreal, Canada
38. Uschold M, Gruninger M (1996) Ontologies: principles, methods and applications. Knowl Eng Rev 11:93–155
39. Wang H, Xiang D, Duan G et al (2007) Assembly planning based on semantic modeling approach. Comput Ind, pp 227–239
40. Wiesner A, Morbach J, Marquardt W (2011) Information integration in chemical process engineering based on semantic technologies. Comput Chem Eng 35:692–708
41. Yang J, Goltz M, Han S (2004) Parameter-based engineering changes for a distributed engineering environment. Concurrent Eng: Res Appl 12:275–286
42. Yang WZ, Xie SQ, Ai QS, Zhou ZD (2008) Recent development on product modelling: a review. Int J Prod Res 46:6055–6085
43. You CF, Tsou PJ (2007) Collaborative design for an assembly via the internet. Adv Manuf Technol 31:1217–1222
44. You CF, Yeh SC (2002) Engineering change propagation system using STEP. Concurrent Eng 10:349
45. Zamite J, Silva F, Couto F, Silva MJ (2010) MEDCollector: multisource epidemic data collector. In: Proceedings of the 1st International Conference on Information Technology in Bio- and Medical Informatics

An Example of Feature Modeling Application: Smart Design for Well-Drilling Systems

Rajiur S. M. Rahman and Y.-S. Ma

Abstract The reported effort is intended to develop a semi-automated, knowledge-based, and integrated petroleum well-drilling engineering design system, considering various aspects such as drill-string [40] and casing design models. The goal was to significantly increase the dynamic drilling engineering responsiveness to real field changes through the automation of conceptual design and 3D modeling processes. Built-in rules and knowledge are used to develop the conceptual design; the system then automatically generates the assembly configuration and retrieves part specifications from a data sheet to generate the CAD parameter files. These parameter files are used to further generate the full CAD model. The conceptual design and CAD models are integrated in such a way that any changes in the design can be reflected automatically throughout the system. Hopefully, this chapter serves not only as an example application for feature-based design, but also as a research reference for the energy industry to leverage modern informatics advancement for its efficiency and cost effectiveness.

1 Introduction

Traditional feature technology has been reviewed thoroughly in previous two chapters, i.e. "Introduction to Engineering Informatics" and "A Review of Data Representation of Product and Process Models". The effective use of feature technology varies considerably from industry to industry. This is due to both the nature of an industry's engineering process and the extent of new technology

R. S. M. Rahman · Y.-S. Ma (✉)
Department of Mechanical Engineering, University of Alberta,
Alberta, Canada
e-mail: yongsheng.ma@ualberta.ca

R. S. M. Rahman
e-mail: rrahman@mccoyglobal.com

Y.-S. Ma (ed.), *Semantic Modeling and Interoperability in Product and Process Engineering*, Springer Series in Advanced Manufacturing,
DOI: 10.1007/978-1-4471-5073-2_3,

penetration into industrial practice. Kasravi [21] has pointed out that human expertise and knowledge are scarce, and that the need for knowledge embodiment within the geometric model is obvious. However, most 3D CAD software offer only simple geometric modeling functions and fail to provide users with sufficient design knowledge. Design knowledge embedded in a computer system is of great help in most engineering tasks. Therefore, the design of automatic and knowledge-based systems has been an active research topic for quite some time [25]. Zha et al. [48] have developed a knowledge-based expert design system for assembly-oriented design. Koo et al. [22] have constructed an expert paper-feeding mechanism design system, where the physical part of the paper-feeding mechanisms are represented as objects, and the design knowledge and constraints are represented by rules and object methods. The researchers did not extend the program to CAD application, however. Myung et al. [37] have proposed a design expert system to redesign assemblies of machine tools in a CAD environment. Roh and Lee [43] have created a hull structural modeling system for ship design, which was developed using C++ and was built on top of 3D CAD software.

Transferring knowledge-based engineering (KBE) intelligence to a CAD system presents a challenge because there is no readily available mechanism to enable such information flow, as identified by Ma et al. [27]. As introduced by Kasravi [21] preliminarily, parametric engineering uses the design requirements as the input data, and the parameters of the key features of the constituent components as the output. More recently, numerous industrial applications have been developed with feature technology; for example, Chu et al. [11] have constructed a parametric design system for 3D tire mold production. Lee et al. [24] have developed a parametric tool design system for cold forging using Autolisp. Researchers have commonly used parametric part templates to generate new 3D designs; changes are realized by setting values to the driving parameters [46]. Ma et al. [26] have considered the topological and configuration changes of parts.

Feature technology has been a cornerstone of engineering informatics since the 1990s. The abundant literature on feature technology shows its versatility in capturing, modeling, and deploying the best engineering practices in a number of industries. This chapter showcases an example of a typical feature technology application in a less common CAD application area: well-drilling system design for the oil and gas industry.

The research effort at the University of Alberta is intended to develop a semi-automated, knowledge-based, and integrated petroleum well-drilling engineering design system, considering various aspects such as drill-string [40] and casing design models. This proposed comprehensive and intelligent drilling design package is named "DrillSoft." The goal is to significantly increase the dynamic drilling engineering responsiveness to real field changes through the automation of conceptual design and 3D modeling processes. Built-in rules and knowledge are used to develop the conceptual design; the system then automatically generates the assembly configuration and retrieves part specifications from a data sheet to generate the CAD parameter files. These parameter files are used to further generate the full CAD model. The conceptual design and CAD models are integrated in such a way that any changes in the design can be reflected automatically throughout the system. Such intelligent CAD design practice is new in the drilling industry.

It is the authors' intention that this chapter serves not only as an example for feature-based engineering design, but also as a research reference for the energy industry to leverage modern informatics advancement for its efficiency and cost-effectiveness.

Naturally, intelligent computer tools supporting petroleum well-drilling design have attracted a lot of attention. As pointed out by Mattiello et al. [32], an accurate well and casing design can significantly reduce drilling costs and risks. A computer-aided support system for casing design and shoe depth selection has been reported [33] and was intended to improve the reliability of solutions, reduce total project time, and help reduce costs. However, their well casing design is still manual. Casing design is rigorous, time-consuming, and obsolescent; it is also error-prone due to the difference in the analytical computation of stresses and their graphical representation against depth [1]. The authors strongly believe that modern CAD technology can play a significant role in enhancing the consistency and efficiency of well-drilling design.

As evidenced by this body of research, the petroleum industry has been using knowledge bases and expert systems for decades; Hayes-Roth identified as early as 1987 that expert systems would play a dramatic role in the success of the outstanding performers in the petroleum industry [18]. Mabile et al. [28] developed an expert system that helps in formation recognition. Martinez et al. [31] constructed a directional drilling expert system for the use of advisory tools, which recommended changes in the bottom hole assembly (BHA). Kulakofsky et al. [23] proposed an expert slurry-design system (ESDS) in order to guide users in the selection process of cement slurry. Chiu et al. [9] implemented an expert system that can be used efficiently as a tool to advise engineers of the proper base fluids and additives to be selected for a given set of well conditions. Fear et al. [16] created an expert system for drill bit selection; their system uses a knowledge base of bit selection rules to produce a generic description of the most suitable bit for a particular set of drilling and geologic conditions. Their approach has several limitations; for example, the bit selection cannot demonstrate best use of past experience and relies too heavily on data that is conveniently available rather than the best fit for the purpose. Several case-based systems have also been reported [35, 36, 45]. Mendes et al. [35] developed a petroleum well design system capable of reusing previous designs which included considerations for potential failure in new designs. Shokouhi et al. [45] integrated real-time data with past experience to reduce operational problems. Al-Yami et al. [2] developed a software tool to guide drilling engineers in formulating effective cement slurries for entire well sections.

This work studies well-drilling planning and design applications with an automatic and generative approach. The first step in well-drilling is to plan the well. Current well-planning practice is usually done section by section with limited help from computer-based tools. Well-drilling planning has to follow a systematic approach. It involves several stages, as shown in Fig. 1. The planning tasks at different stages depend on one or more other stages; for example, casing selection

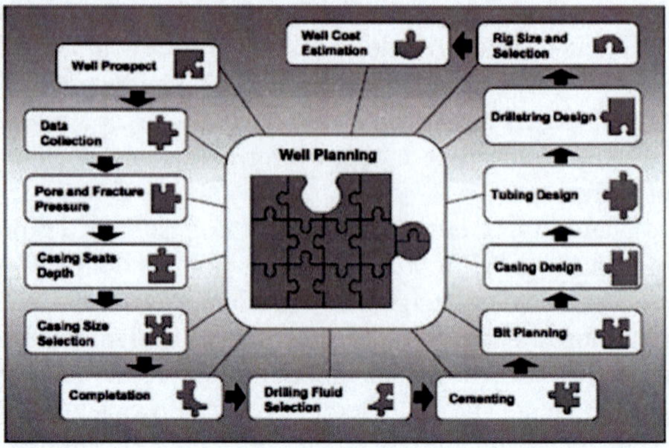

Fig. 1 Well-drilling planning stages [36]

requires input of casing setting depth and casing size. Therefore, the planning stages are commonly developed concurrently and interactively by a team of experts. A great deal of specialized knowledge is required to achieve a safe and economical design. Although many drilling software tools are available on the market to assist the planning team, most of them are standalone and support only one or two stages of the planning process.

One critical application of expert systems in the drilling industry is the design of casing strings, as identified by Heinze [19], who constructed an expert system to design the casing and hydraulic system of a drilling well. Jellison et al. [20] proposed a rule-based expert system for casing design, but casing setting depth was not included in their model. Wajtanowicz and Maidla [47] proposed an optimization program for minimum-cost casing design. This method was, however, unable to handle complex load conditions. Roque et al. [44] have developed an optimized methodology and an algorithm to minimize the cost of combined casing design for complex loading conditions. Unfortunately, the researchers separated the load calculation from the optimization model, and as a result sacrificed the efficiency of the overall design process. In addition, their system required many hours to complete a single string design. A solution to this problem was attempted by Halal et al. [17], who proposed a minimum-cost casing design technique that employed a recursive branch-and-bound search method together with a streamlined load generator for complex loading conditions. Their technique designed the casing string quickly, but their system did not include casing setting depth. Rabia [39] pointed out that not every company has unlimited access to all grades and types of available casings; he argued that cost calculations come into play after the grades and weights are selected. Akpan [1] created a computer program for selecting casing design using a graphical method, but his program does not automatically select the casing; instead, each casing has to be chosen by the user and fed to the system manually for evaluation.

As can be seen in the literature, to date most of the existing drilling software tools are standalone. Those reviewed computer systems suffer from a common shortcoming, i.e. the software packages require a tedious setup process to run for each analysis or generation cycle, and the solution development process is geared toward experienced users, hence the limited popularity of such applications. From the author's point of view, the knowledge modules implemented in such packages are not adaptive enough to new input from real situations in the field. The need for a comprehensive casing design program still exists. A properly developed well-drilling program can improve design cycle time and reliability while reducing the total cost.

With the advancement of information technologies, an integrated and comprehensive well-drilling engineering system is now feasible, because data sharing and constraint management can be carried out in a coherent manner with sophisticated new methodologies [26, 27]. As will be described below, an integrated computer software package, "DrillSoft," has been designed at the University of Alberta to meet this objective. The package currently consists of three modules: casing design, drill-string design, and operational parameter optimization.

By using a feature-based approach, the system under research is also integrated with CAD software to parametrically model the 3D well structure and drill-string. To the best of the author's knowledge, there is no such software solution that integrates well planning with 3D CAD modeling. The overall objective of this project is the integration of well-drilling planning with automatic model generation.

2 Research Approach

Knowledge-based engineering is commonly applied in capturing and structuring reusable engineering cases to create and enhance solutions for a product during its entire life cycle [11]. Knowledge bases can exist in many forms, such as spreadsheets, handbooks, engineering formulas, drawings, and documents. The current research at the University of Alberta is aimed toward developing a generic and parametric drilling system driven by knowledge-based rules and constraints, which can be reused repeatedly. At the current stage of the research, this system can now produce the conceptual design and further generate the parametric 3D models of the well casings and the drill-string. Unlike the efforts using CAD templates, which require part libraries and are difficult to manage, this research uses a generative approach to program generic drill-string models. There are many advantages of using a generative approach instead of template files, including: geometry and features can be easily created and edited, parameters can be created and manipulated in a more controlled manner, geometry analysis and part standardization can be easily achieved, files can be managed more efficiently, and finally, data access and family parts creation are more convenient.

To develop the target knowledge-driven system for well-drilling system design, feature technology has been employed, and the software has been prototyped. It considers the geological input, such as pore pressure or overburden, to generate a step-by-step interactive drilling plan. The implemented well-planning stages include the casing setting depth, casing and hole size determination, casing selection, and finally drill-string design and modeling. The system is integrated with a feature-based CAD system for generating 3D parametric models. An operational parameter module is developed in the system to predict the drilling coefficients and to minimize the drilling cost per foot by using offset well data, determining the optimum WOB, and optimizing the drill-string rotation speed for a single bit run. Based on this approach, an integrated well-planning system can be fully developed and will be very useful for the decision making of drilling companies.

3 Well-Drilling System Design Principles and Processes

Well-drilling planning is a systematic and team-based process. Achieving a safe and economical design requires a great deal of specialized knowledge. Although drilling software is widely available on the market to assist planning teams, data integration and information exchange still cause serious inefficiencies in response to application conditions and low-quality output models.

3.1 Well Casing Design

A casing is a collection of steel tubes that becomes a permanent part of an oil or gas well. A well consists of several sections of holes of different diameters, and a string of casing is run after each section of hole has been drilled. Casing serves many important functions during the life of a well. The major functions of the casing are as follows [8, 32]:

- Maintaining the structural integrity of the bore hole;
- Serving as a high-strength flow conduit to surface for both drilling and production fluids;
- Providing support for wellhead equipment and blowout preventers;
- Preventing contamination of nearby fresh water zones;
- Facilitating the running of wireline equipment up and down for testing;
- Allowing isolated communication with selectively perforated formation(s) of interest.

At nearly 20 % of overall well cost, casing design engineering for well-drilling represents a significant amount of well expenditure [44]. A small reduction in cost will therefore result in huge savings. But just as importantly, the casing design solution should satisfy all the constraint and loading requirements. The casing design module starts with casing setting depth determination, which is the most critical step in casing design. Many parameters must be considered, including pore pressure, fracture pressure, geophysical conditions in the area, bore hole stability problems, corrosive zones, environmental considerations, regulations, and company policies.

Among the input parameters, pore pressure and fracture pressure are the most widely used to determine the setting depth. The rocks inside the earth contain pore spaces, which are filled with fluids in the form of either gas or liquids. These trapped fluids cause the rock wall to experience a pressure known as *formation pore pressure*.

There are two different methods used to determine the formation pore pressure: the geophysical method and the logging method. The geophysical method helps to predict the formation pore pressure before the well is drilled, while the logging method is applicable after the well has been drilled. In this study, *pore pressure* is considered to be an input, while *fracture pressure* is the pressure at which a formation matrix opens to admit hole liquid through an actual crack in the matrix of the rock, as opposed to invasion through the natural porosity of the rock [8].

Two methods are used to determine the fracture pressure: direct and indirect. There is only one direct method: fracture pressure required to fracture the rock or to propagate the resulting fracture can be determined directly by stress analysis so as to predict the fracture gradient. On the other hand, there are three different indirect methods: Hubber and Willis's method [30], Matthews and Kelly's [4], and Eaton's [14]. As a modified version of the Hubber and Willis method, the Eaton method has gained wide acceptance. Eaton suggested that Poisson's ratio for a given field should be fairly constant and can be determined from the data obtained from the nearby well. In this prototyped software tool, Poisson's ratio is considered to be a user input, and the Eaton method is used to determine the fracture gradient of the prospective well.

$$FG = (v/1 - v)(\sigma_v - P_f)/D + P_f/D \tag{1}$$

here,

FG	Fracture gradient;
D	Depth, ft;
υ	Poisson's ratio;
σ_v	Over-burden, psi/ft; and
P_f	Formation pore pressure, psi/ft.

In the process of determining the well casing settings, two more safety factors related to downhole pressures have to be considered: trip margin and kick margin. They are considered during mud density determination according to the following two rules: the mud density should be slightly higher than the formation pressure (trip margin); and the mud density should be lower than the fracture pressure (kick margin).

Trip margin allows the mud density to be slightly higher than the formation pore pressure and eliminates the negative surge effect. A negative surge pressure is produced during tripping of the pipe. When making a trip the pipe is pulled upward, and due to this pulling action a negative pressure is created inside the hole; this results in a reduction of hydrostatic pressure. This phenomenon is known as the negative surge effect. Conversely, in order to eliminate the positive surge effect, another safety factor, the kick margin, is considered. When the drill-string is put back into the hole, a positive surge pressure is produced. If the pressure is more than the fracture pressure of the well, the stability of the well will be compromised.

After calculating the casing setting depth, the second step of the casing design cycle is casing size determination. Usually a well consists of several sections; it is an important task to determine the bore hole and casing size in each section. The following rules should be considered during casing size determination [8]:

- The bore hole must be large enough for the casing to pass freely with little chance of getting stuck;
- There should be enough clearance around the casing to allow for a good cement job; and
- The hole should be minimized because the bigger the bore hole, the more costly it is to drill.

The third step of the casing design cycle is casing selection. Devereux [12] identified two important aspects of casing selection: the casing strength to resist the forces that are imposed on it during drilling and its reliability throughout the life of the well without requiring a workover. Three basic loads are considered: collapse load, burst load, and axial load. *Collapse load* can be defined as differential pressure load between the external and internal pressures when the external pressure exceeds the internal pressure and causes the casing to collapse. During the design process, worst-case scenarios are considered. When the collapse load is calculated, for example, the minimum internal pressure and the maximum external pressure are both considered. A safety margin is also included. *Burst load* is defined by the difference between the internal and external pressure when the internal pressure exceeds the external pressure and causes the casing to rupture or burst. *Axial load* is the cumulative tension or compression load caused by gravitational and frictional forces on the pipe. As mentioned earlier, a well has several sections and casing design is required for each of them.

During casing selection, the engineer first calculates collapse and burst loads and sets their values as constraints. Depending on the type of casing section, the engineer selects the appropriate rules from the rule base for load calculation. For example, the collapse load calculation for surface and intermediate casings are not the same, so different procedures have to be followed. Next, the engineer checks whether the available casings are capable of meeting the total depth. If not, the engineer asks for more casings. The engineer then determines the allowable length for each available casing based on collapse and burst ratings. Once the casing selection has been made, the next step is to check whether the selected casing is capable of sustaining the axial load. After the conceptual design of the casing is completed, a formal report has to be developed.

3.2 Drill-String Design

Oil well drill-string design is another major task in drilling engineering that requires geological information input and, informed by knowledge and experience, is carried out upfront. A drill-string is a collection of drill pipes, drill collars, heavyweight drill pipe, crossover sub, and bit sub that transmits drilling fluid and rotational power to the drill bit. Drilling pipes connected by drilling collars and crossover subs form hollow shafts through which drilling fluid can be pumped down, and the fluid and cutting (such as the produced drilling mud with rock chips) can be brought back to the surface through the annulus.

According to Chuna [10], drill-string design is the most important part of operations in drilling engineering. It is the responsibility of the drilling engineer to design a system suitable for varying field conditions. The success of a drilling job is very much dependent on the design of the drill-string. In order to reduce the risk of drill-string failure, the design should be justified beforehand by simulation or finite element analysis. Austin [5] has argued that a 3D model can provide closer links between geoscientists and reservoir engineers, while promoting the integration as well as the interaction of the two. Although it has long been clear that 3D well-design models, especially the drill-string model, will be quite useful in simulation and finite element analysis (FEA) [34] to predict the behavior of the well, it is cumbersome to develop repetitive 3D models for each section of the well in order to perform such analyses. At present, analytical models, or rough FEA for the whole drill-string, are used to compute torque and drag.

3.3 Drilling Optimization via Operational Parameters

The design of a well program must satisfy all the technical considerations. Considering only the normal design aspect of drilling tools may not result in the most economical design; operational parameters must also be considered. Drilling optimization can be carried out by selecting the best combination of drilling parameters. Drilling parameters are divided into two groups: alterable and unalterable. Alterable drilling parameters, or *variables*, are related to mud, hydraulics, bit type, WOB, and rotary speed. Unalterable parameters, or *conditional parameters*, are weather, location, depth, rig condition, and so on. A comprehensive drilling optimization program has been one of the most important research areas in drilling engineering since the 1960s.

Several researchers have developed algorithms, models, and programs to optimize drilling performance by maximizing the rate of penetration (ROP) and minimizing the cost per foot. Bourgoyne et al. [7] constructed a comprehensive drilling model to calculate formation pore pressure, optimum weight of bit (WOB), rotary speed, and jet bit hydraulics, and also provided a multiple regression approach to determine the drilling coefficients in order to calibrate the drilling model with different field conditions. Maidla et al. [29] used a computer program to select drill bit, WOB, and rotary speed. Bjornsson et al. [6] proposed a rule-based bit selection expert system by employing the mechanical specific energy (MSE) concept. Their system aimed to increase the ROP and bit life and significantly reduce drilling time. Dupriest and Koederitz [13] effectively used the MSE concept in evaluating the drilling efficiency of bits in real time. Rashidi et al. [42] used both MSE and inverted ROP models to develop a method for evaluating real-time bit wear; the method can be useful in assisting the field engineer in deciding when to pull the bit. Eren et al. [15] constructed a real-time optimization model for drilling parameters.

4 Proposed Software System Structure

This section investigates an integrated and comprehensive system for well design: the computer program DrillSoft. The program is intended and prototyped to address the difficulty of manual management of a complete drilling plan. Figure 2 shows DrillSoft's various modules. The program consists of two functional modules: casing design and drill-string design. Another parameter determination module, the drilling coefficient calculator, is designed to work out those commonly used engineering coefficients related to real application conditions. This module serves as a central engineering data sharing block among various functional modules.

For implementing engineering rules and calculations, the Microsoft Visual Basic application in Excel was used. Siemens NX6 was used for CAD modeling with its "Open C" application programming interfaces (APIs). The concept is to make full use of the capability of CAD API functions that can be called within an object-oriented system environment to generate standard component and assembly models. The CAD API functions can also integrate CAD functions with Excel application programs. Figure 3 shows the main user interface.

4.1 Casing Design Module

As shown in Fig. 3, DrillSoft first takes input from the user. The following inputs are required: type of well, unit system, pore pressures at various depths, kick margin, and trip margin. Pore pressure and fracture pressure are the main determinants for casing setting depth and size calculation, as shown in Fig. 4.

The design software tool contains a rule base to calculate collapse, burst, and axial load. Every company has its own set of standards to calculate these loads during casing design, using flexible software that can be adapted according to need. This flexibility can be achieved by adding new rules to the Excel model used in this work. A casing selection process flowchart is shown in Fig. 5.

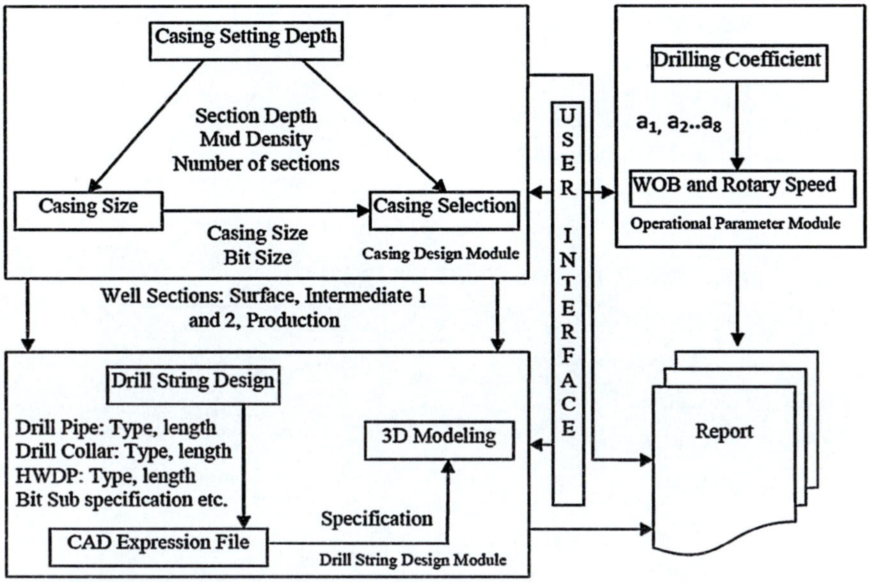

Fig. 2 DrillSoft modules

Depending on the type of casing section, the software selects the appropriate rules from the rule base for load calculation. The system identifies which rules are to be applied by considering the type of section. After determining the collapse and burst rating, DrillSoft checks whether the available casings are capable of meeting the total depth. If the available casings do not meet whole depth, then the software module asks to provide more casings. Once it finds that the total depth is achievable with these casings, the program then determines the allowable length for each available casing based on collapse and burst rating. The allowable length is used as break points for the algorithm.

The potential candidates are those types of casings that can be used safely in the concerned depth interval. It is assumed that the available casing input is provided sequentially, moving from those with the lowest cost to highest costs. DrillSoft selects the lowest-cost casing type (and thus the most economical) from the potential candidates and adds a length equal to the minimum casing section. The minimum casing section is a key factor in this algorithm, as it limits the number of different types of casing used in a combined casing string. Byrom [8] suggests that this minimum casing section length should not be less than 500 ft. The system selects the first casing type and adds a length equal to the minimum casing section length. The next step is to check whether the casing string achieves the total depth. If the total depth has not yet been achieved, the program chooses again the lowest-cost casing type from the available candidates. If the candidate selected in that stage is similar to the previous casing type, the system works out the remaining

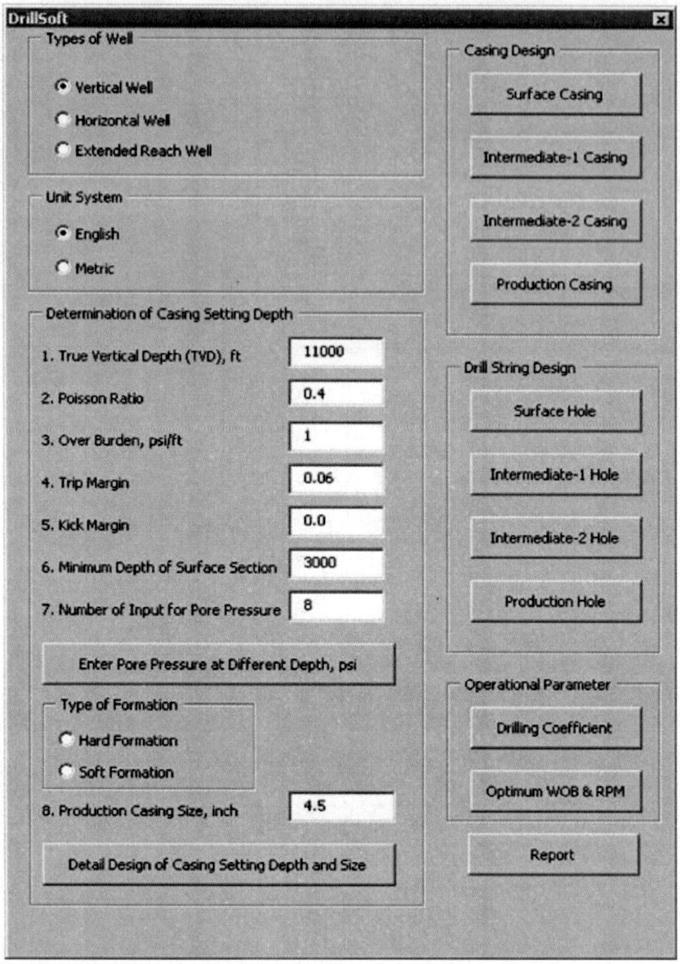

Fig. 3 DrillSoft's main user interface

depth range by using the next break point minus casing covered, and chooses the smaller value between the minimum casing section and the remaining range.

Next, DrillSoft takes the resulting depth value and adds it to the casing length. Then it checks again whether the casing string can support the total depth. If it does not achieve the total depth, the casing selection process continues until the whole depth has been achieved.

After the casing selection, the axial load is checked. A rule base is used to calculate the axial load of each section of casing. If any portion of the casing string fails to satisfy the axial load condition, the casing selection starts again from the beginning. If the axial load with the designed casing is satisfactory, the conceptual design of the casing is complete. This conceptual design will be stored in a data sheet

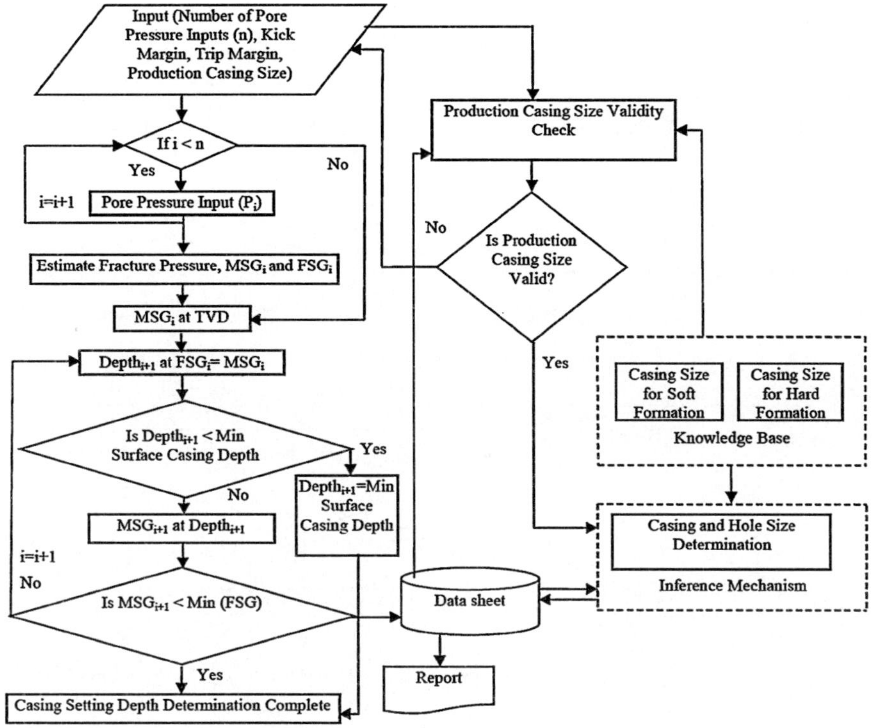

Fig. 4 Flowchart for determining casing setting depth and size

and can be further used by other modules. At the end of casing design, a formal report is generated. This process can be repeated for other sections of the hole.

4.2 Drill-String Design Module

The success of a drilling job is very much dependent on the design of the drill-string. It is a well-known fact that drill-string failure represents one of the major causes of "fishing" operations that likely lead to millions of dollars in loss for the industry. It is therefore crucial to validate the design beforehand by doing FEA or simulation. In addition, a 3D drill-string model will be very helpful in carrying out such analyses.

The DrillSoft program is connected to a CAD system. A knowledge-driven CAD modeling approach has been followed. To eliminate repetitive modeling tasks, a parametric and smart oil well drill-string) module has been prototyped, which enables generation of 3D models with built-in engineering rules, constraints, and controls on various application cases with changing situations

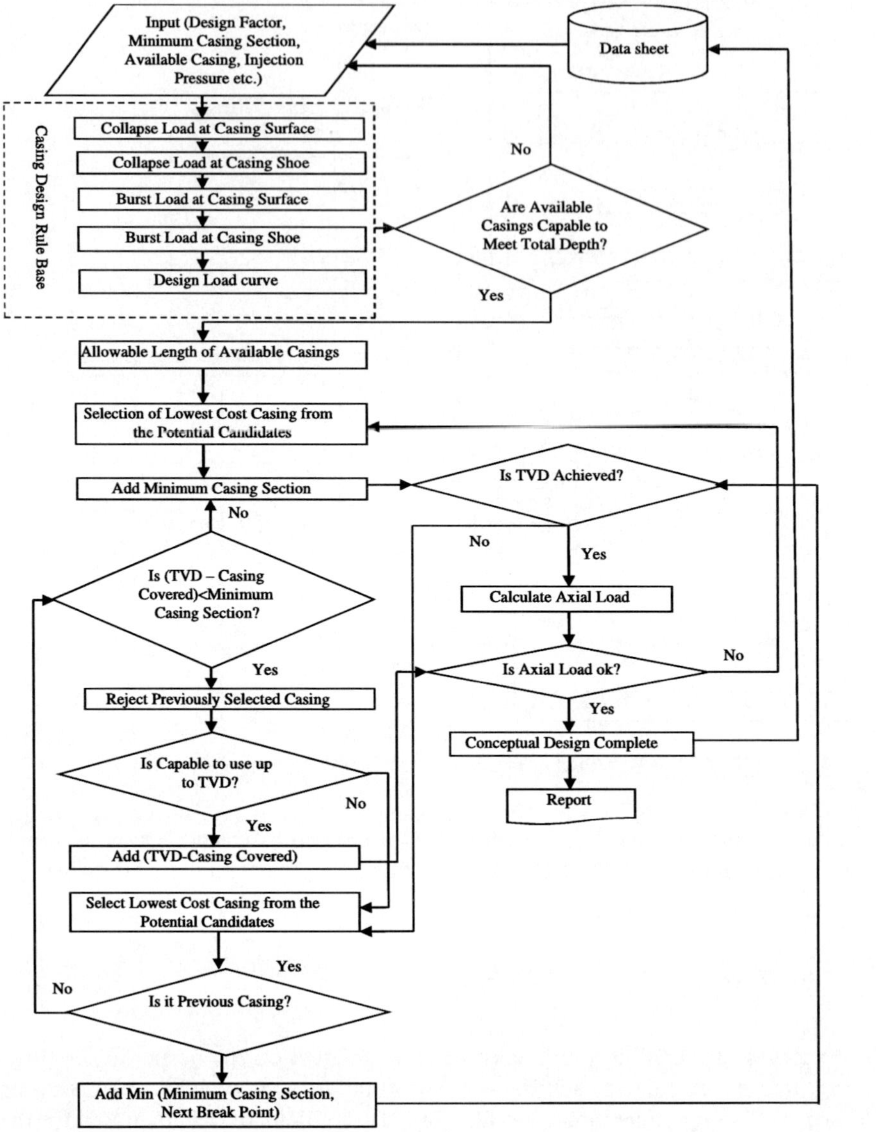

Fig. 5 Flowchart of casing selection

throughout a well-drilling life cycle. A common part data sheet has been integrated into the system so that standard parts can be reused from a well-defined library. Figure 6 shows the steps of the drill-string design process, from conceptual design to 3D model realization.

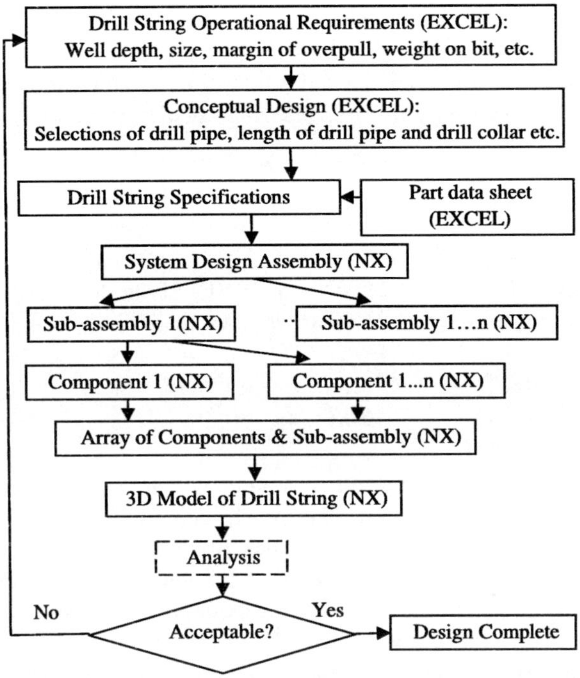

Fig. 6 Drill-string design module

Drill-string design begins after the operational requirements have been defined based on the casing design output and customer input. The operational requirements include type of the well, depth, mud-specific gravity, maximum WOB, margin of over pull, safety factors for collapse, type of drill pipe available in the inventory, drill collar, and heavyweight drill pipe size.

The drill-string design module of the DrillSoft system generates the conceptual design based on a set of built-in engineering rules embedded in the module that follow the recommended practice for drill-stem design standards [3]. One such rule is that the "drill pipe should always be under effective tensile stress; neutral point of buckling should be in the drill collar." In the conceptual design stage, calculating the length of the drill collar requires WOB data. The system then selects the cheapest (in most of the cases also the weakest) drill pipe type from the available inventory and checks it against the load criteria. If the type is not safe to run the whole length of the drill-string, the allowed maximum length of that drill pipe type is worked out. The drill pipe selection cycle continues until the whole length of the drill-string is achieved. Hence, the algorithm selects the cheapest drill-string assembly based on the lowest grade and the unit weight of the pipes in the inventory.

Once the conceptual design of the drill-string is complete, the next step is to determine the drill-string component specifications and configurations. The configuration design determines the number and types of components and their orientation and position in the drill-string assembly. As the drill-string is a vertical column, all components of the string possess the same origin for the x and y coordinates, i.e., (0, 0). Only the vertical coordinate changes when a new part is added to the assembly. Rules have been created to determine the origin or position of a new component. The following rule determines the z-coordinate of the drill pipe:

$$\textit{Origin}\ (z - \textit{coordinate})\ \textit{of Drill pipe} = \textit{HWDP\_o} + \textit{Drill\_collar\_o} + \textit{Bit\_sub\_o} + \textit{Drill\_bit\_o}$$

here,

HWDP_o,	Number of HWDP*Length of HWDP + Origin *(z-coordinate)*
Drill_collar_o,	*Number of Drill_collar * Length of Drill_collar + Origin (z-coordinate),*
Bit_sub_o,	*Origin (z-coordinate),*
Drill_bit_o,	*Origin (z-coordinate).*

The positions of other components are similarly determined. A part data sheet prototype has been developed, containing the geometric and non-geometric specifications of each component. For example, a drill pipe has length, outer diameter (OD) , inner diameter (ID) and tool-joint diameter, upset diameter, and so on. Figure 7 shows a drill pipe sub-assembly with its components.

Figure 8 shows part of the drill pipe data sheet prototype. In this data sheet, each drill pipe is defined by six unique factors: size, class, nominal weight, grade, type of upset, and connection. The values of these six factors must be provided, but the specification generation method will retrieve the rest of the specifications from the data sheet that requires generation of the 3D model.

The system automatically retrieves the necessary data according to the library specifications of each component, and generates the 3D CAD model as well as a

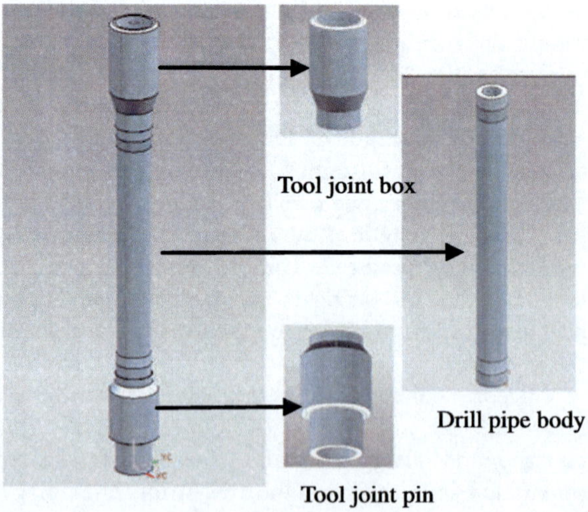

Fig. 7 Drill pipe sub-assembly and components

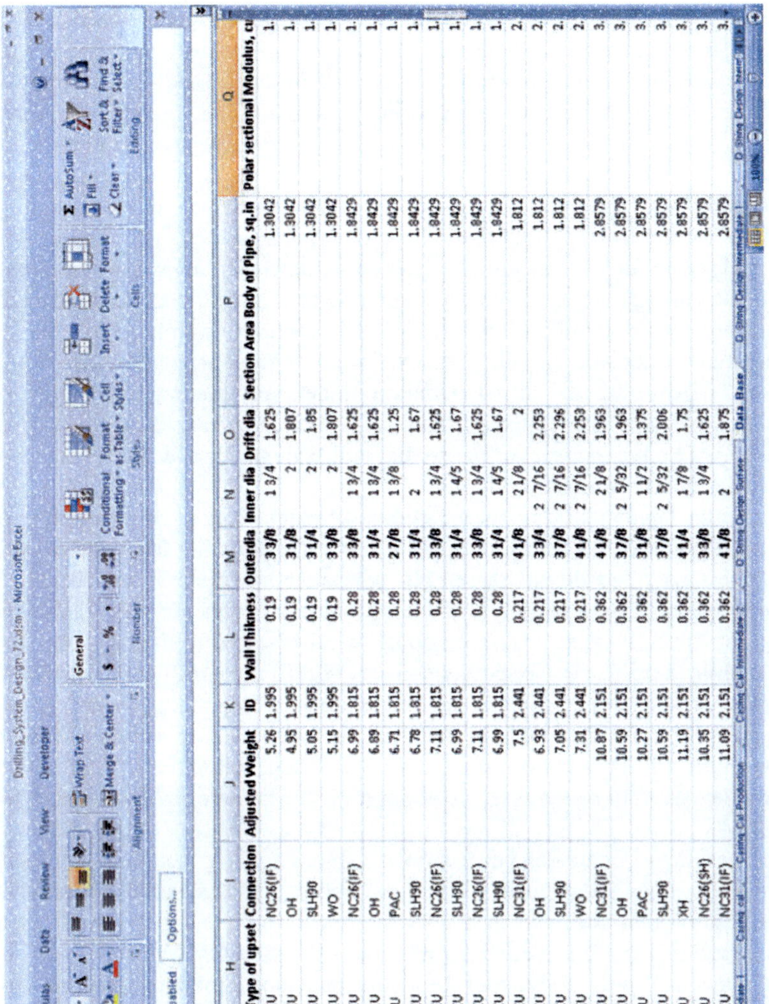

	Class	Size:	Nominal Weight	Grade	Type of upset	Connection	Adjusted Weight	ID	Wall Thikness	Outerdia	Inner dia	Drift dia	Section Area Body of Pipe, sq.in	Polar sectional Modulus, cu
102	1 New	23/8	4.85	E	EU	NC26(IF)	5.26	1.995	0.19	33/8	1 3/4	1.625	1.3042	1.
103	2 New	23/8	4.85	E	EU	OH	4.95	1.995	0.19	31/8	2	1.807	1.3042	1.
104	3 New	23/8	4.85	E	EU	SLH90	5.05	1.995	0.19	31/4	2	1.85	1.3042	1.
105	4 New	23/8	4.85	E	EU	WO	5.15	1.995	0.19	33/8	2	1.807	1.3042	1.
106	5 New	23/8	6.65	E	EU	NC26(IF)	6.99	1.815	0.28	33/8	1 3/4	1.625	1.8429	1.
107	6 New	23/8	6.65	E	EU	OH	6.89	1.815	0.28	31/4	1 3/4	1.625	1.8429	1.
108	7 New	23/8	6.65	E	IU	PAC	6.71	1.815	0.28	27/8	1 3/8	1.25	1.8429	1.
109	8 New	23/8	6.65	E	EU	SLH90	6.78	1.815	0.28	31/4	2	1.67	1.8429	1.
110	9 New	23/8	6.65	X	EU	NC26(IF)	7.11	1.815	0.28	33/8	1 3/4	1.625	1.8429	1.
111	10 New	23/8	6.65	X	EU	SLH90	6.99	1.815	0.28	31/4	1 4/5	1.67	1.8429	1.
112	11 New	23/8	6.65	G	EU	NC26(IF)	7.11	1.815	0.28	33/8	1 3/4	1.625	1.8429	1.
113	12 New	23/8	6.65	G	EU	SLH90	6.99	1.815	0.28	31/4	1 4/5	1.67	1.8429	1.
114	13 New	27/8	6.85	E	EU	NC31(IF)	7.5	2.441	0.217	41/8	21/8	2	1.812	2.
115	14 New	27/8	6.85	E	EU	OH	6.93	2.441	0.217	33/4	2 7/16	2.253	1.812	2.
116	15 New	27/8	6.85	E	EU	SLH90	7.05	2.441	0.217	37/8	2 7/16	2.296	1.812	2.
117	16 New	27/8	6.85	E	EU	WO	7.31	2.441	0.217	41/8	2 7/16	2.253	1.812	2.
118	17 New	27/8	10.4	E	EU	NC31(IF)	10.87	2.151	0.362	41/8	21/8	1.963	2.8579	3.
119	18 New	27/8	10.4	E	EU	OH	10.59	2.151	0.362	37/8	2 5/32	1.963	2.8579	3.
120	19 New	27/8	10.4	E	IU	PAC	10.27	2.151	0.362	31/8	11/2	1.375	2.8579	3.
121	20 New	27/8	10.4	E	EU	SLH90	10.59	2.151	0.362	37/8	2 5/32	2.006	2.8579	3.
122	21 New	27/8	10.4	E	IU	XH	11.19	2.151	0.362	41/4	1 7/8	1.75	2.8579	3.
123	22 New	27/8	10.4	E	IU	NC26(SH)	10.35	2.151	0.362	33/8	1 3/4	1.625	2.8579	3.
124	23 New	27/8	10.4	X	EU	NC31(IF)	11.09	2.151	0.362	41/8	2	1.875	2.8579	3.

Fig. 8 Partial view of drill pipe data sheet

parametric data file known as a CAD expression file. Figure 9 shows the steps involved in the parametric approach for designing the drill pipe tool-joint box. First, the design parameters are identified and initiated by an engineer in Excel format, as shown in Fig. 9a. Engineering constraints are modeled as embedded formulas across different cells and are checked interactively or semiautomatically by using Excel functions, such as a “goal seeking” algorithm. Next, after verifying the accuracy and constraints involved, the parameter names and the result values are exported into a text format in the form of expressions, as shown in Fig. 9b; this is an acceptable format that allows CAD software NX to import into it directly as data input corresponding to built-in expressions. Third, the CAD solid model is constructed by programming NX Open API functions, which are available as a development extension for the CAD software. The source codes are shown in Fig. 9c. Fourth, once the initial template models are generated, the expressions associated with the predefined parameters in the CAD models are then imported into the CAD environment for the given well drill-string and the casing. A CAD part model is shown in Fig. 9d.

For better appreciation of the procedure, the next part of this section explains how NX Open API functions are used with C programming language to generate 3D models. A “top–down” assembly approach has been followed. First, the structure of the whole assembly of the drill-string is created; the generic configuration of the drill-string assembly contains all possible drill-string components. For example, a drill-string assembly consists of drill pipe, drill collar, heavyweight drill pipe, bit sub, cross-over sub, drill bit, and so on. Depending on the operational requirements, some components may not be required. For example, sometimes heavyweight drill pipe is not used in the drill-string; in that case, the module that creates the assembly structure will suppress the heavyweight drill pipe in the assembly.

The next level of model generation is sub-assembly generation. The program first finds out which member of the assembly contains sub-assembly from a configuration definition, and then fires the rule to initiate sub-assembly creation. Refer again to Fig. 7, in which a drill pipe sub-assembly was shown to contain three parts: drill-pipe body, tool-joint pin, and tool-joint box. Such a structure generation algorithm continues iteratively until the configuration of the whole assembly is completed. It is worth mentioning that, in the “top–down” approach, though the conceptual assembly tree structure is created first, in each of the structure members, no physical component geometry entities are created until the program reaches the next stage, in which the component geometry entities are created.

Components have been created using feature modeling methodology. A component is a collection of several features, such as a drill pipe composed of cylinder, cone, hole, chamfer, and so on. Each component therefore requires its own parametric feature-based program. Hence, individual parametric programs have been written for each component to generate a generic 3D model. To reflect the topological variations, the algorithm selects which generation functions to run. For

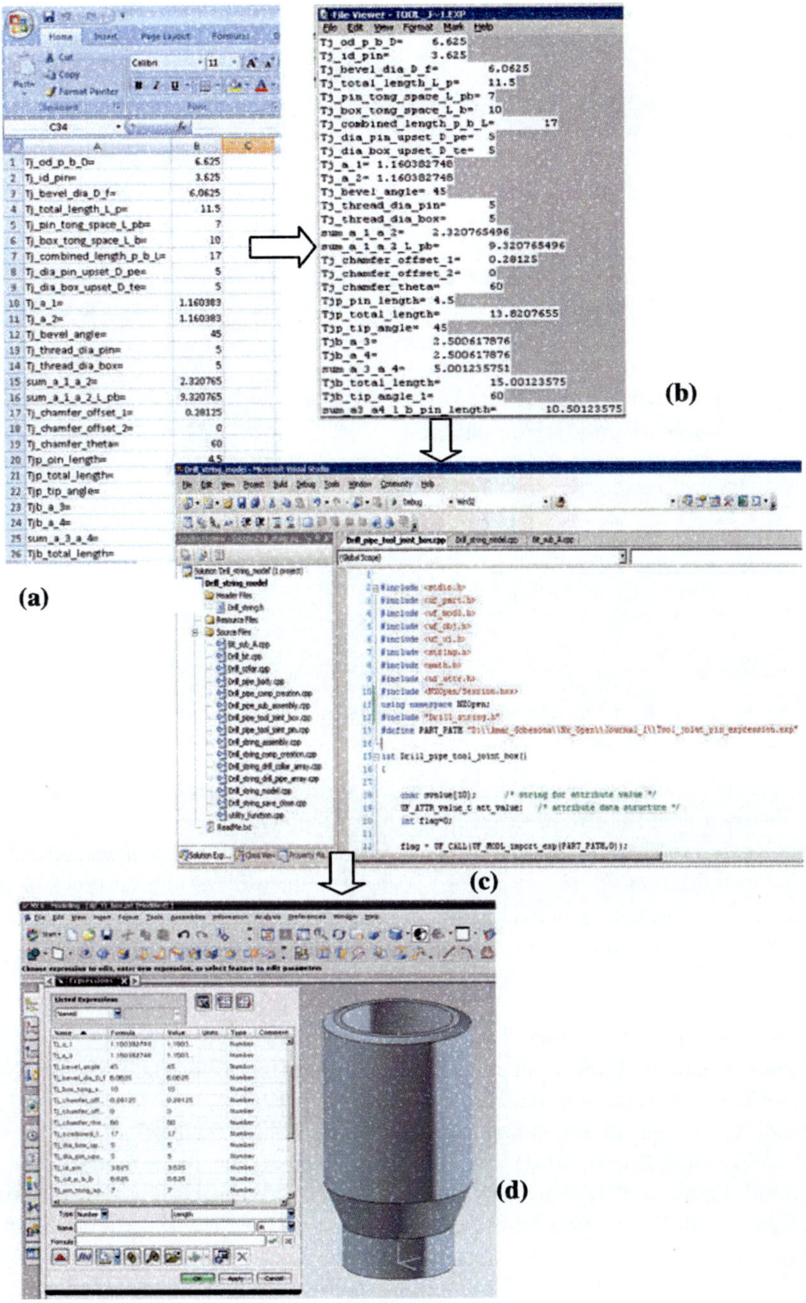

Fig. 9 Steps involved in 3D modeling. **a** Data file. **b** Imported expressions. **c** NX open programming. **d** CAD model realization and expression file

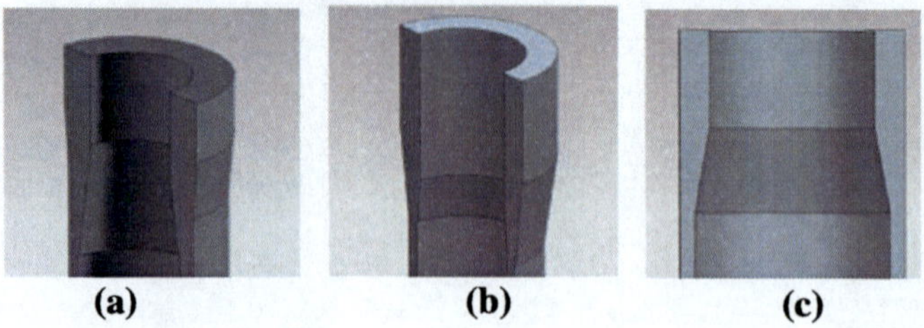

Fig. 10 Drill pipe features. **a** Internal-external upset. **b** External upset. **c** Internal upset

```
for(i=0; i<dptotal; i++) {
 {  if (i == 0) continue;
    origin[2]=pos[2]+i*dst_btn_dp;
    strcpy(iname,cname);
    sprintf(iname+strlen(cname),"_%d",i);
    flag=UF_ASSEM_add_part_to_assembly(*parent, pname, refset, cname,
          origin, matrix, layer, &newinst, &status);
    if(flag!=0) return(flag);
 }
 return(0);    }
```

Fig. 11 Pseudo code to generate drill-string component and sub-assembly array

example, as shown in Fig. 10, a drill pipe body may have three different topologies: internal-external upset, external upset, and internal upset.

An *array* method has been implemented to repeat the component and subassembly instantiation. As shown in Fig. 10, a large quantity of similar types of drill pipe may exist in the drill-string; this *array* method helps to reproduce the drill pipe and other components throughout the assembly.

The pseudocode shown in Fig. 11 is created to execute the *array* method for drill pipe. Here, the *dptotal* is the number of drill pipes in the array, and "*dst_btn_dp*" is the distance between two neighboring drill pipes. These two variables depend on both conceptual design and user input. Similar programs have been written for other components that require repetition.

After the arrays of components and sub-assemblies are carried out, the 3D model of the whole drill-string assembly is realized. This can be used to perform FEA and simulation. If the result of the analysis is unsatisfactory, the program redesigns the drill-string by following the same steps mentioned earlier; the whole design loop is then integrated. Due to the time constraints of this research, the drill-string module and analysis module have not been integrated, although more work is to be done in the future.

4.3 Operational Parameter Module

DrillSoft selects the best combination of WOB and RPM by calculating the cost per foot for a given situation to predict bit performance. The program can be useful for selecting the bit that yields the lowest cost per foot by repeating a set of WOBs and drill-string rotation for several combinations of bits. The operational parameter module contains two parts. First, drilling coefficients are determined by using offset data; second, these coefficients are used to determine the optimum WOB and RPM. This module is constructed based on Bourgoyne and Young's multiple regression approach. According to Bourgoyne et al. [7], eight types of primary drilling variables are required for the regression analysis: depth (D), penetration rate (R), weight per inch of bit diameter (w/d), rotary speed (N), fractional tooth wear (h_f), Reynolds number parameter (N_{RE}), mud density (ECD), and pore pressure gradient (g_p). These eight operational parameters are used to determine the drilling coefficients. The drilling coefficients are then used to predict the behavior of the field and are combined with other input to determine the optimum WOB and RPM.

4.4 Report Generation

The developed program is capable of generating formal reports of design outcome. These reports are very useful for future references and also helpful for further analysis. For example, the operational parameter module generates a cost-per-foot table. This table can be used as a tool for selecting the best bit for a particular interval. Every module has a separate report-generation option; whenever a design is complete the system can generate the corresponding report.

5 Demonstration of the System and Procedure with a Case Study

The casing design module calculates casing setting depth by using formation pore pressure and formation fracture pressure, and determines the size of hole and casing of each section of the well. It then selects the optimum combination of casing string from the available inventory. The drill-string design module will be discussed in more detail later in this chapter. The operational parameter module works out drilling coefficients, the optimum WOB, and the drilling rotary speed (RPM). Drilling coefficients are determined according to the regression analysis procedure of Bourgoyne and Young et al. [7]; at least 30 offset drilling data sets are required. Optimum WOB and RPM are determined for the minimum cost. This program also generates six different tables of economic performance as a function of WOB and RPM as an operational guide, and produces formal reports for design details.

5.1 Casing Setting Depth and Size Determination

The following inputs [38] are provided:

True vertical depth	TVD = 11,000 ft;
Poisson's ratio	$\upsilon = 0.4$;
Over burden	$\sigma_v = 1$psi/ft;
Trip margin (SG)	0.06;
Kick margin (SG)	0;
Minimum depth of surface section	3,000 ft;
Number of pore pressure input	8;
Type of formation	Hard;
Production casing size	4.5 in.

The procedure first estimates the fracture pressure and determines the pore pressure-trip margin (MSG) and fracture pressure-trip margin (FSG) based on the input provided. Table 1 shows the estimated values.

After processing the input, the software tool calculates the mud density at the true vertical depth (TVD). According to the theory, this is the density of mud required to drill the well to the final depth. From here, the design stage is referred to as production section design for the well. The next step is to determine the depth at which the fracture gradient is equal to the mud-specific gravity; in other words, the depth should reach the point at which the vertical line drawn from the mud density curve touches the fracture gradient curve (Fig. 12). Once the depth of the next section is known, the program determines the mud density required to drill up to this depth. In this way, the process continues until the mud density becomes smaller than the minimum fracture gradient or the depth becomes smaller than the minimum casing setting depth. After the determination of casing setting depth, the information is stored in a data sheet and can be shared with other modules.

Mud density at the TVD is $MSG_1 = 2.195$. The depth at which $MSG_1 = FSG_1$ can be found by linear interpolation. In this program it is assumed that the connecting line between the two neighboring points is linear. The program now determines the depth at which $FSG_1 = 2.195$. The two neighboring points of FSG_1

Table 1 Fracture pressure, FSG and MSG estimation

Input number	Depth (ft)	Pore pressure (psi)	Pore pressure (SG)	Mud density (MSG)	Fracture pressure	FSG
1	3,000	1,320	1.01	1.076	1.878	1.878
2	5,000	2,450	1.131	1.191	1.916	1.916
3	8,300	4,067	1.131	1.191	1.916	1.916
4	8,500	4,504	1.223	1.283	1.947	1.947
5	9,000	5,984	1.535	1.595	2.051	2.051
6	9,500	6,810	1.655	1.715	2.091	2.091
7	10,000	7,800	1.801	1.860	2.139	2.139
8	11,000	10,171	2.135	2.195	2.251	2.251

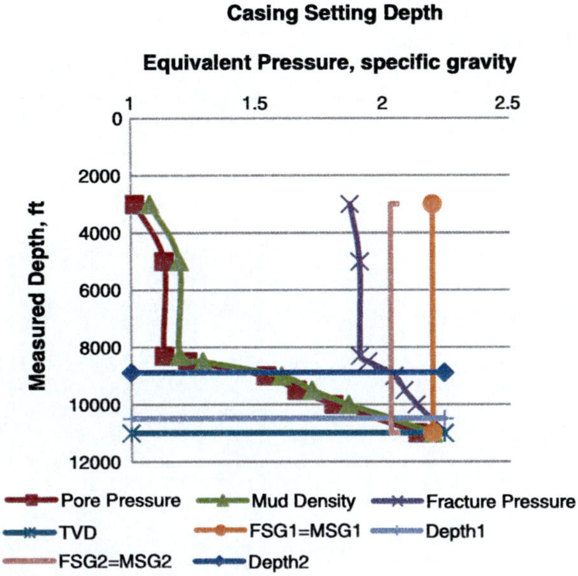

Fig. 12 Casing setting depth

are $FSG_2 = 2.250$ and $FSG_3 = 2.139$ and the corresponding depths are $Depth_2 = 11{,}000$ ft and $Depth_3 = 10{,}000$ ft, respectively.

$$\mathrm{Depth}_1 = \frac{(\mathrm{FSG}_1 - \mathrm{FSG}_2)(\mathrm{Depth}_3 - \mathrm{Depth}_2)}{\mathrm{FSG}_3 - \mathrm{FSG}_2} + \mathrm{Depth}_2 \tag{2}$$

$$\mathrm{Depth}_1 = 10,496\,\mathrm{ft}$$

The next casing setting depth should thus be at 10,496 ft. The mud density above 10,496 ft must now be determined. Again, the program uses linear interpolation; the following equation should be used:

$$\mathrm{MSG}_1 = \frac{(\mathrm{MSG}_3 - \mathrm{MSG}_2)(\mathrm{Depth}_1 - \mathrm{Depth}_2)}{\mathrm{Depth}_3 - \mathrm{Depth}_2} + \mathrm{MSG}_2 \tag{3}$$

Here, $Depth_1 = 10{,}496$ ft, $Depth_2 = 11{,}000$ ft, $Depth_3 = 10{,}000$ ft, $MSG_2 = 2.195$, and $MSG_3 = 1.861$. Thus, $MSG_1 = 2.026$. These processes continue until the mud density becomes smaller than the minimum value of the fracture-specific gravity or the depth becomes smaller than the minimum surface casing depth.

Figure 12 provides the graphical method of casing setting depth determination. Figure 13 shows the partial view of the well schematic based on the output provided by the system. The casing and hole sizes are provided in Table 2.

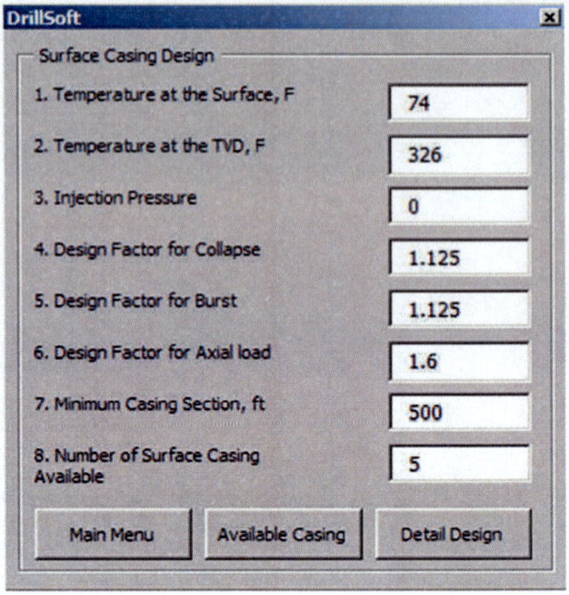

Fig. 13 Surface casing design user interface

5.2 Case Study for Casing Selection

Surface casing is designed as per the procedure specified by Byrom [8]. The necessary inputs are Depth = 3,000 ft, Mud density = 1.11, and Casing size = 13 3/8. Figure 13 shows the surface casing design user interface filled with input. It should be noted that the options for depth and mud density are not provided in the user interface. This is owing to the integration of different parts of the software, which helps the program to automatically retrieve necessary information from the system database. In this particular case, the program automatically retrieves the depth and mud density values from the casing setting depth data sheet. The user needs to provide the specifications of the available casing. The system takes in such casing specifications as input data in a text file format, from which it retrieves the required information. A sample file with the available surface casing specifications is tabulated in Table 3.

Table 2 Casing setting depth and size

	Surface casing	Intermediate 1 casing	Intermediate 2 casing	Production casing
Depth (ft.)	3,000	8,882	10,496	11,000
Mud-specific gravity	1.521	1.521	2.026	2.195
Casing size (in.)	13.375	9.625	7	4.5
Bit size (in.)	17.5	12.25	8.5	6.15

Table 3 Specification and priority sequence of available casing

Casing number	OD(inch)	ID(inch)	Weight (Kg/ft)	Grade	Connection
1	16	12.615	54.5	K-55	ST&C
2	16	12.515	61	K-55	ST&C
3	16	12.415	68	K-55	ST&C
4	16	12.415	68	N-80	ST&C
5	16	12.347	72	N-80	ST&C

Table 4 Casing selection break points and potential candidates

No.	Break points (ft.)	Potential candidates
1	0	1,2,3,4,5
2	2,092	2,3,4,5
3	2,852	3,4,5
4	3,000	3,4,5

After receiving the input, the system calculates the collapse and burst rating at the surface and at the casing shoe. In this particular case, the design collapse pressure at the surface and shoe are 0 and 2,220 psi, respectively, and the design burst pressure at the surface and shoe are 2,180 and 750 psi. The program then determines the break points and the potential candidates that satisfy both the collapse and burst ratings (as shown in Table 4).

In the course of developing this project, the researchers have created a knowledge base containing various configurations of casing sizes for hard and unconsolidated formation. Casing size usually depends on the formation types, the number of casing subsections, and the production casing size. After receiving input from the user, the inference mechanism sorts out the specific sizes for each section of the well. Various combinations of casing sizes are possible. If the specific production casing size is not available, then the system will recommend sizes from the knowledge base.

According to the algorithm, the first break point (0 ft) contains all five available casings as potential candidates. As available casings were listed according to the priority of the user, the system assumes that the first casing is more economical and then the next one, and so on. The system selects the first candidate, casing number 1, with grade K-55 and weight 54.5, and adds the minimum casing section (500 ft). The program then determines if the total depth has been achieved; if it has not, the program again selects casing number 1 from the potential candidates. As it is similar to the previous casing, this time the program will add a minimum value of [Minimum casing section or (Next break point-Casing covered)], i.e., Min [500 (2,092−500 = 1,592)]. The program determines that the minimum casing section (500 ft) is the smaller of these two values. Another 500 ft will then be added to the previous casing. Its length is now 1,000 ft. As the closest break point is at 2,092 ft, this process continues until the casing length has reached 2,000 ft. At this point, the system determines that casing number 1 is still a potential candidate and is similar to the previous casing. The minimum value here is (500, 2,092–2,000 = 92) = 92 ft. Based on these conditions, the system selects casing number 1 and adds a length of 92 ft to the existing length. The total length of casing covered is now 2,092 ft. As the desired depth is 3,000 ft, the system checks the available potential candidates, selects casing number 2, and adds 500 ft to the previous length. The new length is 2,592 ft, still less than the total desired depth. The system again selects casing number 2 and works out the minimum value between (500, 2,852–2,592 = 260 ft) = 260 ft. After adding this value, the length becomes 2,852 ft. Another 148 ft of casing is required to complete the casing string for the surface section. However, this remaining section is less than the minimum casing section length and previously used casing (Number 1 and 2) are not allowed to be used up to the full depth. Another casing type should be selected, but if the system selects a new casing type it will not satisfy the minimum casing section length. To solve this problem, a re-evaluation of the design is required. The system re-evaluates the design and concludes that, instead of using casing number 2 from 2,093 to 2,852 ft, casing 3 should be used to the total depth. In this case, all the design criteria will be satisfied. The preliminary design based on collapse and burst is now complete.

The next step is to check whether the casing string design will satisfy the axial load criteria. From the case study presented, it is found that the casing string meets sufficient safety standards with casings 1 and 3, in their relevant sections. This concludes the surface casing conceptual design. Figures 14, 15, and 16 represent the combined casing selection based on collapse, burst, and axial load. Table 5 shows the sample casing selection output as compared with a published result [38].

Fig. 14 Casing selection based on collapse load

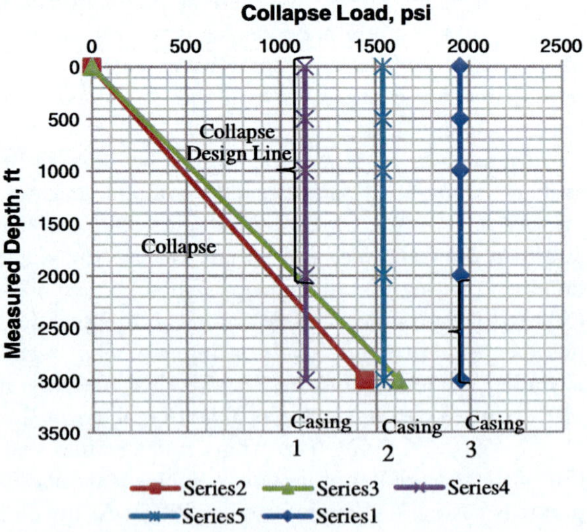

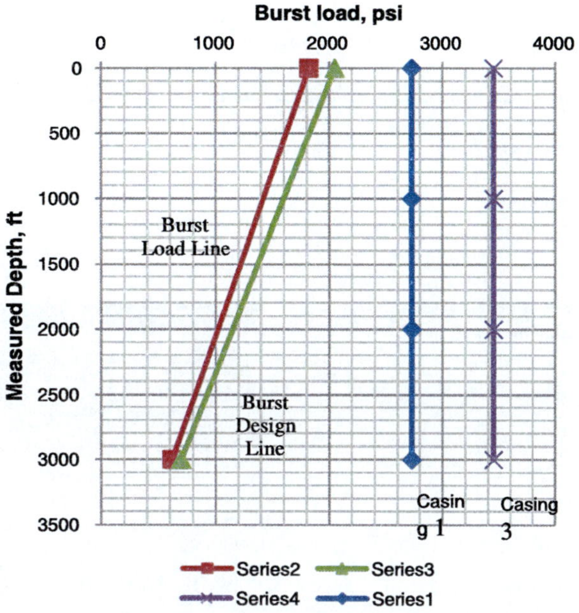

Fig. 15 Casing selection based on burst load

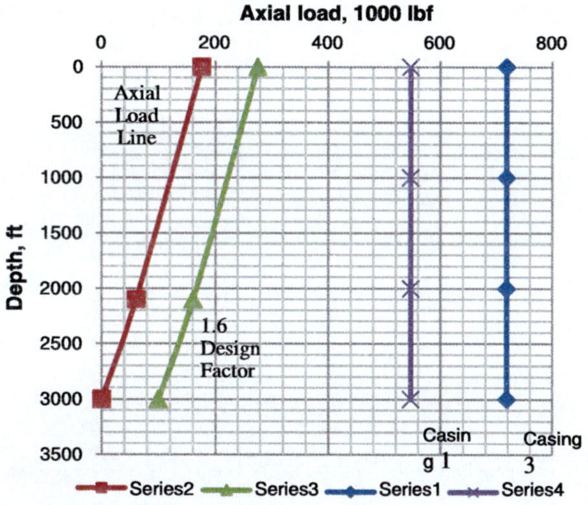

Fig. 16 Casing selection based on axial load

5.3 Drill-String Design Demonstration

Embedded knowledge assists the system in developing the conceptual design of the drill string based on previous modules' output and users' input. This conceptual design is then used to generate the drill-string specifications and configurations, and later this information is converted into expression files. The

Table 5 Comparison of casing setting depth

Section	DrillSoft		Published result [38]	
	Depth (ft)	Mud SG	Depth (ft)	Mud SG
Surface	3,000	1.521	3,000	1.567
Intermediate 1	8,882	1.521	8,850	1.567
Intermediate 2	10,496	2.026	10,500	2.031
Production	11,000	2.194	11,000	2.170

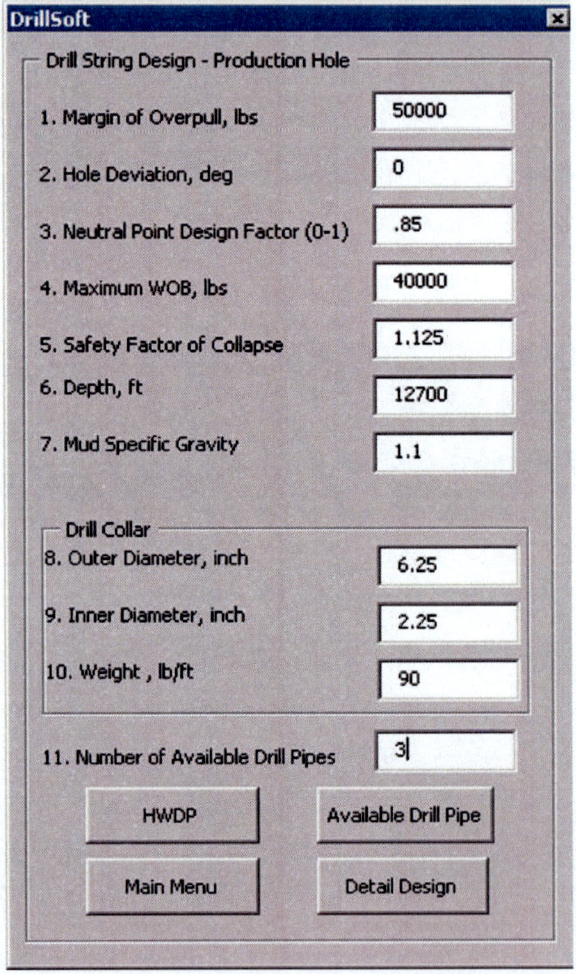

Fig. 17 Drill-string design user interface

conceptual design and the CAD model are bridged through the expression files, as shown above in Fig. 9. Any changes in the conceptual design will thus change the expression files, and will be reflected in the 3D model.

Table 6 Conceptual design parameters for a drill-string

Drill-string components	Length (ft)	Number of array
Drill collar: 6 ¼″OD × 2 ¼″ID	630	21
Drill pipe type 1: 4 ½″ × 16.6 lb, grade E75, class2	6,750	225
Drill pipe type 2: 4 1/2″ × 16.6 lb, grade X95, Premium class	5,320	178

To prove the concept of the developed module, a drill-string has been designed as a test case. The case is taken from the API standard handbook. Figure 17 shows the drill-string module user interface filled with input. Table 6 shows two of the drill pipe types available in the inventory. User-friendly interfaces have been created so as to integrate all the modules. The interfaces guide the user to develop the well plan with less effort and in an organized sequence. More UIs are to be introduced in the following sections.

Based on operational input, the rule-based system designed a drill-string that uses two different types of drill pipes. As mentioned above, the program first considers the most economical drill pipe among the available three. It first chooses grade E75 and determines the safe length of 6,750 ft. After accounting for the length of 21 drill collars, the drill-string has reached a length of 7,380 ft, which is less than the required depth of 12,700 ft. The program then considers the second type of pipe, grade X95, and decides to use this type for the remaining 5,320 ft. When considering the collapse load, the program generates messages for the user; in this case it is worked out to be 10,267 ft. The drill-string should not be run dry below this depth, as doing so may cause damage in the string. Based on this conceptual design, the system generates the necessary configuration and specification files and converts these files into expression files.

In this work, drill-string design is based on the input of casing design as introduced previously and the authors have developed a parametric well structure model, which works out the casing setting depth and casing sizes and generates different 3D casing sections. Figure 18 shows a partial drill-string assembly model.

5.4 Operational Parameter Optimization

The program provides two options for operation optimization depending on the input in order to determine the optimum WOB and RPM. Option 1 assumes that the abrasive constant, bearing constant, and drillability are unknown. Option 2 assumes that they are known. In Option 1, abrasive constant, bearing constant, and drillability should be calculated first from the offset bit data; these values will then be used in the rest of the calculations. The list of required input data for this option is shown in Fig. 19. In Option 2, abrasive constant, bearing constant, and drillability are given; the other required inputs are shown in Fig. 20.

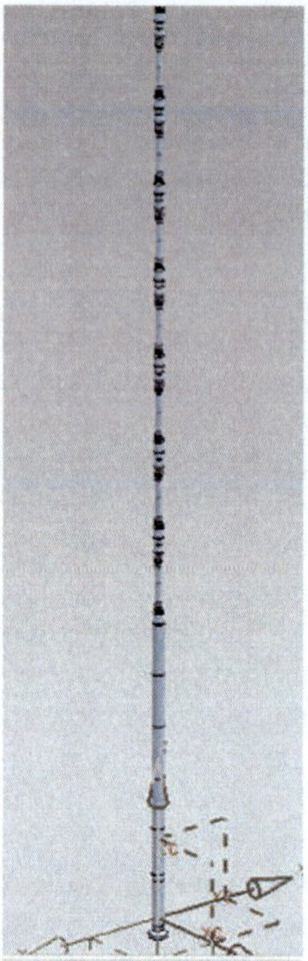

Fig. 18 Partial drill-string assembly [41]

The program also generates six tables, which include cost per foot, bit life, footage drilled, final tooth wear, final bearing wear, and penetration rate. A cost-per-foot table can be used to quickly identify (1) the best combination of bit weight and rotary speed; (2) the best rotary speed for a given bit weight; and (3) the best bit weight for a given rotary speed [7].

6 Comparison of the Generated Results with the Published Sources

Validation of the various modules in the DrillSoft system was based on previously published data. For example, the casing setting depths and mud-specific gravities were compared with the published data from Rabia's work [38]. The casing selections generated were checked against Byrom's [8]. The values of drilling coefficients calculated by DrillSoft were benchmarked against the values from the work of Bourgoyne et al. [7]. From these comparisons, it can be safely concluded that the results produced by the DrillSoft program are satisfactory.

The program has the following advantages: (1) it automatically selects the casing that meets all the loading criteria; (2) different modules are integrated with

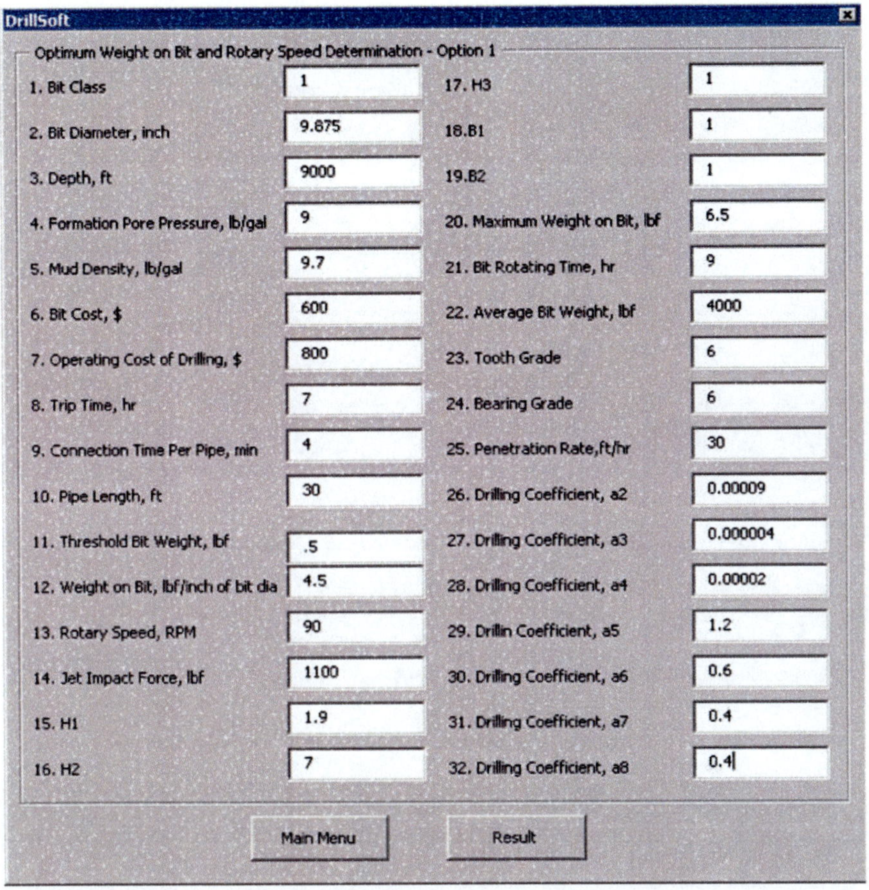

Fig. 19 Operational parameter optimization: option 1

each other and the flow of information from one module to another module is very smooth; (3) design parameters obtained from the program are used to generate a 3D CAD model, i.e., drill-string and casing; and (4) the parametric CAD model (drill-string) is connected with DrillSoft through automatically created expression files.

DrillSoft has some limitations; however: (1) the determination of casing setting depth and mud weight based on formation pore pressure and formation fracture pressure does not always guarantee well bore stability; and (2) biaxial effects of loading were not considered during the selection of casing. More research is still required.

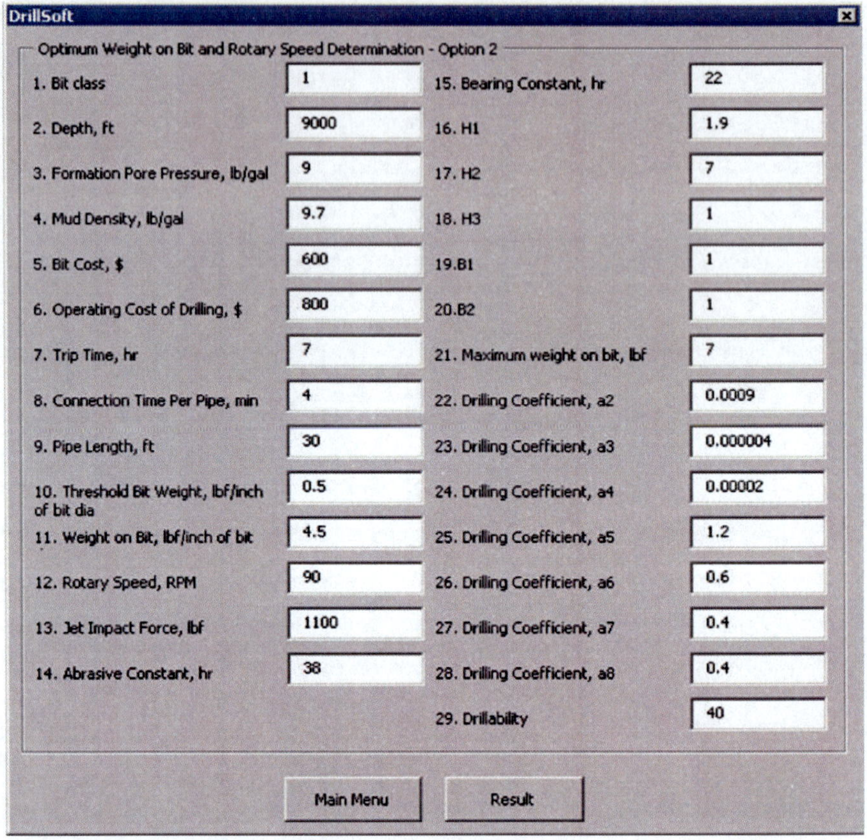

Fig. 20 Operational parameter optimization: option 2

7 Summary

This chapter described a feature-based well-drilling system design approach through a detailed case study. A prototype system that integrates three important well-drilling planning stages (casing design, drill-string design, and operational optimization) has been demonstrated. The software design concepts and the generative algorithms are presented. The results are promising. The prototyped software tool can help the drilling engineer to interactively design and model the well casing and the drill-string. The design modules are parametric and feature-based; hence, they are capable of handling different configurations and topologies of drilling components according to varying application conditions. At this moment, the reported software tool can only handle vertical oil wells. More research should be carried out to develop a generic model that can be equally applicable to horizontal, extended reach, and multilateral wells. The approach proposed here could also be applied to integrate other well-planning modules, such as hydraulic program, bit program, and time and cost estimation, into a comprehensive system. It is expected that, with full implementation in the future, the prototype system will be a very useful support tool for engineering decision making in drilling companies.

Acknowledgments The authors would like to thank the Natural Sciences and Engineering Research Council of Canada (NSERC) for its Discovery grant support (No. 355454-09) and the University of Alberta for the faculty startup grant.

References

1. Akpan HO (2005) Efficient computational method for casing string design. Paper SPE 98790-MS presented at the annual SPE international technical conference and exhibition. Abuja, Nigeria. doi:10.2118/98790-MS
2. Al-Yami AS, Schubert J, Medina-Cetina Z, Yu OY (2010) Drilling expert system for the optimal design and execution of successful cementing practices. Paper SPE 135183-MS presented at the IADC/SPE Asia Pacific Drilling Technology conference and exhibition. Ho Chi Minh City, Vietnam. doi:10.2118/135183-MS
3. American Petroleum Institute (1998) Recommended practice for drill stem design and operating limits 7G, 16th edn
4. Anderson RA, Ingram DS, Zanier AM (1973) Determining fracture pressure gradients from well logs. J Pet Technol 25:1259–1268
5. Austin AZ (1993) Benefits of 3D visualization to reservoir simulation. SPE western regional meeting, Anchorage, USA
6. Bjornsson E, Hucik B, Szutak G, Brown LA, Evans H, Curry D, Perry P (2004) Drilling optimization using bit selection expert system and ROP prediction algorithm improves drilling performance and enhances operational decision making by reducing performance uncertainties. Paper SPE 90752-MS presented at the SPE Annual Technical conference and exhibition, Houston, Texas. doi:10.2118/90752-MS
7. Bourgoyne AT Jr, Young FS Jr (1974) A multiple regression approach to optimal drilling and abnormal pressure detection. SPE J 14:371–384. doi:10.2118/4238-PA
8. Byrom TG (2007) Casing and liners for drilling and completion. Gulf Publishing Company, Houston
9. Chiu TJ, Caudel FLW (1993) Development of an expert system to assist with complex fluid design. Paper SPE 24416-PA presented at the SPE Computer Application, Houston, USA. doi:10.2118/24416-PA
10. Chuna JC (2002) Drill-string and casing design for horizontal and extended reach wells, Part 1. Paper SPE 79001-MS presented at the SPE international thermal operations and heavy oil symposium and international horizontal well technology conference, Calgary, Canada. doi:10.2118/79001-MS
11. Chu CH, Song MC, Luo CS (2006) Computer-aided parametric design for 3D tire mold production. Comput Ind 57:11–25
12. Devereux S (1998) Practical well planning and drilling manual. Penwell Publishing Company, Tulsa
13. Dupriest FE, Koederitz W (2005) Maximizing drill rates with real-time surveillance of mechanical specific energy. Paper IADC/SPE 92914-MS presented at the SPE/IADC drilling conference, Dallas, USA. doi:10.2118/92194-MS

14. Eaton BA (1969) Fracture gradient prediction and its application in oilfield operations. J Pet Technol 21:1353–1360. doi:10.2118/2163-PA (SPE 2163-PA)
15. Eren T, Ozbayoglu ME (2010) Real time optimization of drilling parameters during drilling operations. Paper SPE 129126-MS presented at the SPE Oil and Gas India conference and exhibition, Mumbai, India. doi:10.2118/129126-MS
16. Fear MJ, Meany NC, Evans JM (1994) An expert system for drill bit selection. Paper SPE 27470-MS presented at the IADC/SPE drilling conference, Dallas, USA. doi:10.2118/27470-MS
17. Halal AS, Warling DJ, Wagner RR (1996) Minimum cost casing design. Paper SPE 36448-MS presented at the SPE Annual Technical conference and exhibition, Denver, USA. doi:10.2118/36448-MS
18. Hayes-Roth F (1987) Part 1: Expert systems applied to the petroleum industry upstream portion. Paper SPE 22411 presented at the 12th World Petroleum Congress, Houston, USA
19. Heinz LR (1993) CHES: Casing hydraulic expert system. SPE Comput Appl 4:26–31. doi:10.2118/24420-PA (SPE 24420-PA)
20. Jellison MJ, Klementich EF (1990) An expert system for casing string design. Paper SPE 20328-MS presented at the 5th SPE petroleum computer conference, Denver, USA. doi:10.2118/20328-MS
21. Kasravi K (1994) Understanding knowledge-based CAD/CAM. Comput Aided Eng 13:72–78
22. Koo DY, Han SH (1998) An object-oriented configuration design method for paper feeding mechanisms. Int J Exp Sys App 14:283–289
23. Kulakofsky D, Henry SR, Porter D (1993) PC-based dement job simulator improves primary job design. Paper SPE 26110-MS presented at the SPE western regional meeting, Anchorage, USA. doi:10.2118/26110-MS
24. Lee RS, Hsu JYH, Su SL (1999) Development of a parametric computer-aided die design system for cold forging. J Mater Process Tech 91:80–89
25. Lin BT, Chan CK, Wang JC (2008) A knowledge-based parametric design system for drawing dies. Int J Adv Manuf Tech 36:671–680
26. Ma YS, Tor SB, Britton GA (2003) The development of standard component library for plastic injection mould design using an object oriented approach. Int J Adv Manuf Tech 22:611–618
27. Ma YS, Britton GA, Tor SB (2007) Associative assembly design features: concept, implementation and application. Int J Adv Manuf Tech 32:434–444
28. Mabile CM, Hamelin JPA (1989) An expert system helps in formation recognition. Paper SPE 19132-MS presented at the petroleum computer conference, San Antonio, USA. doi:10.2118/19132-MS
29. Maidla EE, Ohara S (1991) Field verification of drilling models and computerized selection of drill bit, WOB, and drillstring rotation. SPE Drill Eng 6:189–195. doi:10.2118/19130-PA (SPE 19130-PA)
30. Marsh GL, Smith JR (1984) Exploratory Well Design for 5,000- to 7,500-Ft Water Depths, U.S. East Coast. Offshore Technology Conference, Houston, Texas. ISBN 978-1-61399-077-3
31. Martinez E (1992) Directional drilling expert system. Paper SPE 23664-MS presented at the SPE Latin America petroleum engineering conference, Caracas, Venezuela. doi:10.2118/23664-MS
32. Mattiello D, Sansone A (1992) CASCADE: a knowledge-based drilling engineering software tool. Paper SPE 24273-MS presented at the SPE European petroleum compute conference, Stanvanger, Norway. doi:10.2118/24273-MS
33. Mattiello D, Piantanida M, Schenato A, Tomada L (1993) Casing shoe depths accurately and quickly selected with computer assistance. Oil Gas J 91:86–93
34. Menand S, Sellami H, Tijani M, Stab O (2006) Advancements in 3D drillstring mechanics: from the bit to the topdrive. IADC/SPE drilling conference, Miami, USA
35. Mendes JRP, Morooka CK, Guilherme IR (2003) Case based reasoning in offshore well design. J Pet Sci Eng 40:47–60. doi:10.1016/S0920-4105(03)00083-4

36. Morooka CK, Guilherme IR, Mendes JRP (2001) Development of intelligent systems for well drilling and petroleum production. J Pet Sci Eng 32:191–199. doi:10.1016/S0920-4105(01)00161-9
37. Myung S, Han S (2001) Knowledge-based parametric design of mechanical products based on configuration design method. Expert Syst Appl 21:99–107
38. Rabia H (1985) Oil well drilling engineering principles and practice. Graham and Trotman, London
39. Rabia H (1988) Discussion of minimum cost casing design for vertical and directional wells. J Pet Technol 40:504–506
40. Rahman SMR, Ma YS (2011) Knowledge driven generic drill string modeling. Proceedings of CANCAM, 23rd Canadian congress of applied mechanics, Vancouver, BC, Canada
41. Rahman SMR, Ma YS (2011) Smart modeling of drilling-well in an integrated approach. MSc Thesis, University of Alberta, https://era.library.ualberta.ca/public/view/item/uuid:e8fdf9d3-0144-4a6d-bb30-644d3bf6cf27
42. Rashidi B, Harland G, Nygaard R (2008) Real-time drill bit wear prediction by combining rock energy and drilling strength concepts. Paper SPE 117109-MS presented at the Abu Dhabi international petroleum exhibition and conference, Abu Dhabi, UAE. doi:10.2118/117109-MS
43. Roh MI, Lee KY (2006) An initial hull structural modeling system for computer-aided process planning in shipbuilding. Adv Eng Softw 37:457–476
44. Roque JL, Maidla EE, Wagner RR (1994) Casing cost optimization for complex loading situations. SPE Comput Appl 12:24–29. doi:10.2118/28224-PA
45. Shokouhi SV, Skalle P, Aamodt A, Sørmo A (2009) Integration of real-time data and past experiences for reducing operational problems. Paper SPE 13969-MS presented at the international petroleum technology conference, Doha, Qatar. doi:10.2523/13969-MS
46. Siddique Z, Yanjiang Z (2002) Automatic generation of product family member CAD models supported by a platform using a template approach. ASME DETC conference, Montreal, Canada
47. Wojtanowicz AK, Maidla EE (1987) Minimum-cost casing design for vertical and directional well. J Pet Technol 39:1269–1282. doi:10.2118/14499-PA
48. Zha XF, Du HJ, Qiu JH (2001) Knowledge-based approach and system for assembly-oriented design, Part II: the system implementation. Eng Appl Artif Intell 14:239–254

Fundamental Concepts of Generic Features

S.-H. Tang, Gang Chen and Y.-S. Ma

1 Introduction

To achieve information integration among CAx applications, a shared common product model is crucial. Such a multi-view product engineering model should support different disciplinary views for various applications. Here, the term *view* refers to the context-dependent and self-contained interpretation data set (subset) of the entire product model (EPM) related to one particular engineering domain or aspect of the product.

In this chapter, a four-layer information integration infrastructure is presented based on Tang's work [28] for building the shared product model. Tang's product feature model is built on the Standard for the Exchange of Product model data (STEP) framework [23], because STEP is the international standard and has been widely accepted by both vendors and users. However, using only STEP-based product specification cannot ensure feature model integration, because STEP does not define interrelationships and constraints between applications. In this product model, the STEP framework is extended with a new concept, the unified feature model [4], under which a generic feature representation schema is given. Next, design and manufacturing feature models are described based on the new concept. The different definitions of *slot* features in both applications are analyzed as

S.-H. Tang
Guangdong University of Technology, Guangzhou, Guangdong, People's Republic of China
e-mail: fwei@scut.edu.cn

G. Chen
Tianjing University of Science and Technology, Tianjin, People's Republic of China
e-mail: chengang@tust.edu.cn

Y.-S. Ma (✉)
University of Alberta, Edmonton, Alberta, Canada
e-mail: yongsheng.ma@ualberta.ca

Y.-S. Ma (ed.), *Semantic Modeling and Interoperability in Product and Process Engineering*, Springer Series in Advanced Manufacturing,
DOI: 10.1007/978-1-4471-5073-2_4,

examples. For feature-based modeling processes, the concept of *operation* is introduced, followed by its representation schema. Finally, the cellular geometric model, shared by different applications, is described.

As reviewed in "A Review of Data Representation of Product and Process Models", most of the current software tools for product development are already feature based, but are limited to individual applications. In the authors' previous work [28], the proposed information integration infrastructure supports multi-stage applications throughout the product lifecycle, owing to the adoption of a unified feature modeling scheme [4]. This infrastructure is centered with a core model representation of a basic feature type, *generic feature*, which is defined based on the associative feature constraint management method.

The definition of *generic feature* was first introduced in Chen et al.'s work [5], which used unified modeling language (UML) and in which it was dubbed a "unified feature." The authors felt it necessary to reconsider the name convention for the related feature terminology. As shown in Fig. 1, the authors decided to give a clear definition of *generic feature*, which is supposed to be the smallest grain element in the so-called feature-based informatics domain. In fact, the only change that has been made from the original publication is that the name "unified feature" has been changed to *generic feature*. The reason for this renaming is to maintain the consistency of conceptual understanding about the proposed feature-based theoretical informatics model. *Generic feature* is defined as the most basic feature entity template, or a common *class* as defined in an object-oriented software engineering approach; that is, the ultimate bottom-level engineering characterization data structure. *Generic feature* is expected to reflect the reusability and abstracting capability of the engineering semantic patterns for different engineering applications. The term *unified feature* is reserved to refer to the systematic

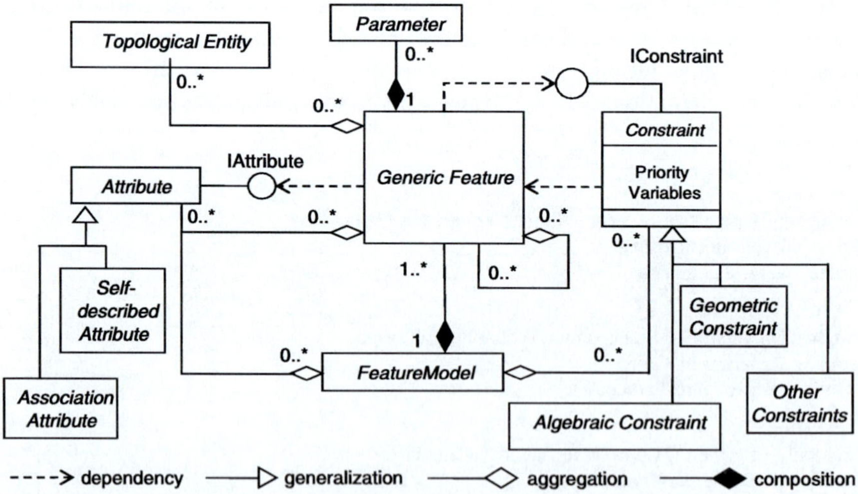

Fig. 1 Generic feature model enhanced from Chen et al. [5]

framework that has been developed for implementing application systems based on the *generic feature* definition.

As shown in Fig. 1, *generic feature* consists of four main fields: Attributes, parameters, constraints, and topological entity pointers. *Attributes* here refer to the properties of the feature that do not specify a feature's shape, dimension, orientation, or position, but rather such attributes as material and surface finish (which are self-describing attributes), and non-geometric entities such as functions, rules, and machining operations (which are association attributes). On the other hand, *parameters* are the key to describing the shape, dimensions, orientation, and position of entities. *Topological entities* are those that can be shown to the user on the screen, such as a point, line, cylinder, or cube.

The major fields and methods defined in the generic feature class are described in Table 1 below with reference to Ma et al.'s recent work [13]. Again, note that there has been a name change, such that the common generic class definition, abstracted from different features, has been renamed from the original *unified feature* to *generic feature*.

2 Generic Feature Model

Theoretically, the unified feature model allows different applications to define specific features in a unified approach. Application features are modeled as the child class of the *generic feature*. In other words, unified feature modeling allows for the coexistence of specific views for different applications. However, although from an application point of view it is essential that each feature type has a well-defined meaning, or semantics, as a base class, a *generic feature* definition that enables common mechanisms, such as data storage, searching, validation, updating, and information sharing, must be modeled and developed. For more details about the class definition and properties of generic features, see Ma et al. [12].

The generic feature class includes the structured description of all common properties and methods of application feature types. Such properties include feature shape representation with parameters, constraint types, reference mechanisms, and validity methods. For example, all types of constraints are used for capturing design intent in the context of product design models. The generic feature representation schema in EXPRESS-G for database design and implementation can also be found in the literature [12]. The generic feature model provides a template for application-specific feature definition.

2.1 Feature Shape Representation

Representing the shape of a feature means defining feature geometry, topology, and their associated entities, such as Attributes, parameters, and feature

Table 1 Major fields and methods of the generic feature class [13]

Class section	Element types	Member element lists	Description
Fields	Attributes	Association attributes	Identities of the associated objects, such as functions and behaviors in a conceptual design, machines and cutters in a process plan, other features, etc.
		Self-describing attributes	Material, surface finish, belonging application, etc.
	Parameters		Variables used as input to geometry creation methods
	Constraints	Geometric constraints	Dependency relations among the feature's geometrical and topological entities
		Algebraic constraints	Engineering equation relations among the feature's self-defined attributes and parameters, mainly applied for physics and mechanism principle formulas
		Rule-based constraints	Identities of rules that the feature or its self-describing attributes, parameters, or numerical constraints participated in
	Geometric references		Topological entities
Methods	Geometry construction	createGeometry()	Generate the feature geometry
	Interface to geometric model	getCell()	Retrieve the feature's member cell entity properties by a pointer or a name
		setCell()	Assign a topological entity as the feature's identity
		insertGeometry()	Notify the geometric model to insert the feature geometry
		deleteGeometry()	Notify the geometric model to delete the feature geometry
	Interface to expert system	getFact(), setFact()	Retrieve or create the corresponding facts
		getRule(), setRule()	Retrieve or assign the corresponding rules
		checkRule()	Check whether the related rules are satisfied or not
	Interface to relation manager	addToJTMS()	Add a node to the graph of JTMSJustification-based truth maintenance systems (JTMS) managed by the system to track the feature, and its constraints, attributes, and parameters for validation-checking purposes
		validityChecking()	Call the relation manager for feature validation
	Interface to database	saveFeature(), retrieveFeature()	Store a feature in or retrieve a feature from the database

manipulation (e.g., creation, modification, and deletion) functions. Feature parameters support user interfaces to create and modify features in modeling operations. To explicitly maintain the shape of a feature in a part model, shape representation in the research discussed here is based on the cellular topology of

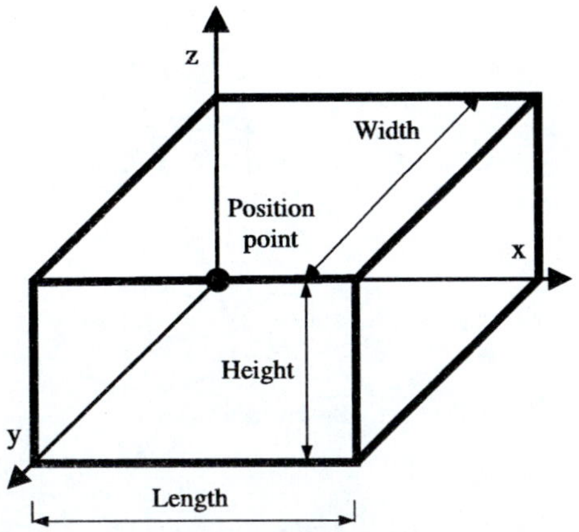

Fig. 2 Block feature

ACIS, which is on the top of the common B-rep model. For example, a *block* feature may have four parameters: *length, width, height*, and *position point* (see Fig. 2). Creation of a *block* feature is associated with the function *api_solid_-block()*, which creates a primitive solid block with two positions. With these four parameters and feature creation schema, the shape of the *block* feature can be determined. Note that the length parameter is along the *x*-axis; the width parameter is along the *y*-axis; and height is along the *z*-axis. Other primitive features with parameters, such as *cone, cylinder, wedge*, and *sphere*, are shown in Fig. 3.

Table 2 lists other types of features with their basic parameters. Note that in the authors' opinion, datum entities (which include datum plane, datum axis, and datum point) are also regarded as a kind of feature.

2.2 Validity Condition (Constraint) Definition

Validity conditions, namely constraints, must be explicitly defined in the unified feature model to specify relationships among features and geometric or topological entities, and provide invariant characteristics in the model.

Constraints may have various types. Some classifications, such as Dohmen's [7] and Bettig and Shah's [2], are reviewed below. In this work, we follow the classification by Dohmen (which is also used in most current CAD systems). Constraints can be classified as geometric constraints, dimension constraints, algebraic constraints, or semantic constraints, a constraint schema defined by Ma et al. [12].

Geometric constraints specify the geometric relations between feature elements; they can be classified into two categories, dimensional and semantic. Dimensional constraints specify distances between two feature member entities. Semantic constraints specify the topological properties of feature elements. For a vertex, edge, or face, a semantic constraint specifies the extent to which the

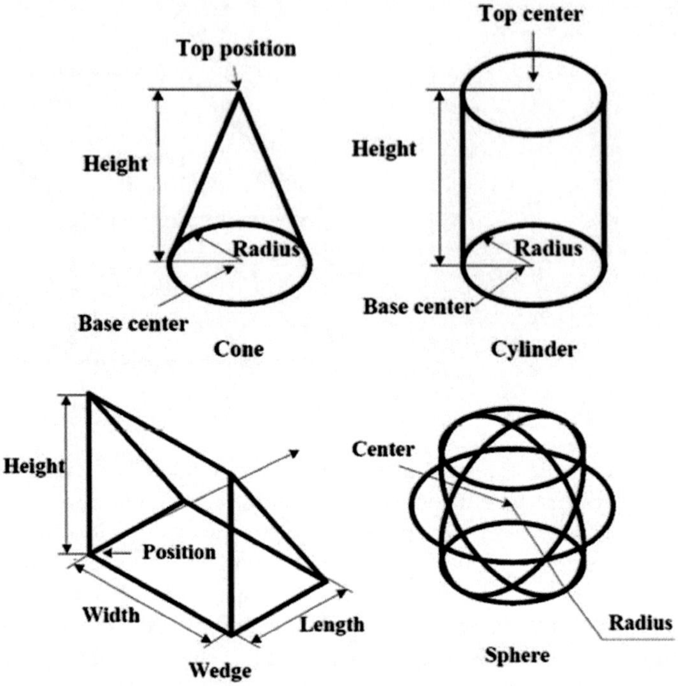

Fig. 3 Other primitive features with parameters

element must lie on the product boundary. For a volume, a semantic constraint specifies the extent to which the volume is allowed to be intersected by other feature volumes instantiated later [7]. For example, a *through_hole* feature has semantic constraints in that the cylindrical side face must at least be partly on the material boundary, and the top and bottom faces of the cylinder must not be on the material boundary. If in the later design stage, a boss, which is a solid with material, is placed just over the hole, the semantic constraint on the top face is violated, and the *through_hole* feature is no longer valid. It becomes a new *blind_hole* feature. If both the top and bottom of the cylinder space are blocked with other material features, the feature then is transformed into a *hollow_space* with no accessibility to the open space.

In the generic feature definition, constraints are modeled as Attributes attached to topological entities or sub-features with the associative validation methods defined in the feature definition. Although different types of constraints have different attributes, some attributes are common:

- *Constraint_ID* is the identifier of a constraint instance.
- *Constraint_name* specifies the name of a constraint instance.
- *Owner_ID* uniquely identifies which feature a constraint belongs to.
- *Constraint_expression* represents the relationship between the constrained elements and referenced elements.

Table 2 Other features with parameters

Feature type	Primary feature parameters	
Slot	1. A 2D profile (e.g., U profile)	
	2. A path	
	3. Slot end type (if it is not through slot)	
Hole	Simple hole →	1. Radius
		2. Depth
	Countable hole →	1. Countable_hole radius
		2. Countable_hole depth
		3. Hole radius
		4. Hole depth
Pocket	1. The pocket profile (rectangular or circular)	
	2. Depth	
	3. Corner radius	
	4. Floor radius	
Extrusion	1. A 2D sketch profile	
	2. An extrusion path (or direction with distance)	
Revolution	1. A 2D sketch profile	
	2. An axis	
	3. Revolution angle	
Sweep	1. A 2D sketch profile	
	2. A sweep guide (path)	
Chamfer	1. An edge (chain of edges) or connected faces	
	2. Two distances (or a distance with angle)	
Fillet	1. An edge (chain of edges) or connected faces	
	2. Radius	
Array	Rectangular →	1. Arrayed objects
		2. Column offset with number of instances
		3. Row offset with number of instances
	Polar →	1. Arrayed objects
		2. Axis of polar
		3. Number of instances (with fulfilled angle)
Offset	1. Offset objects	
	2. Offset distance with direction	
Mirror	1. Mirror objects	
	2. Mirror plane	
Datum	Datum plane →	1. A point
		2. A plane normal (vector)
	Datum axis →	1. A point
		2. A direction (vector)
	Datum point →	A point

- *Constrained_entity_ID* is used to specify a list of pointers of the constrained entities.
- *Referenced_entity_ID list* can be used to uniquely identify referenced entities. In modern CAD systems, the *reference_entity*, which is the existing geometry (a face, edge, or vertex), is regarded as a kind of datum for positioning (or orienting) a new feature.

- *Constraint_strength* has an enumeration data type, which may include several levels, such as *required, strong, medium*, or *weak*. It represents the extent to which the constraint needs to be imposed when constraints conflict with one another.
- *Constraint_sense* is used to specify the direction between constrained entities and referenced entities.
- *Constraint solving functions* are responsible for solving constraints according to constraint types.
- *Other manipulation functions* may include attribute access functions, behavior control functions, and so on.

The definition of *constraint_strength* is for handling over-constrained situations. Such constraint attributes were used in another external constraint solver, SkyBlue, to solve the over-constraint problem [21, 22]. The use of *constraint_strength* is supported by this solver. SkyBlue constraints each have an associated priority, or strength, indicating how important it is to satisfy the constraint. A constraint of lower priority is said to be weaker than a constraint of higher priority, which is called stronger. The highest strength is "*required*;" the lowest is "*weak*." An arbitrary number of strength levels may be defined. If the SkyBlue constraint graph has conflicting constraints, SkyBlue will always determine a solution such that no unsatisfied constraint can be satisfied by making a weaker constraint unsatisfied. An example is shown in Fig. 4. Note that in the figure, a box represents a constraint, and a circle represents a variable or parameter. In the first graph, the strong constraint C2 has just been added. The second graph shows a possible chain of constraints without conflicts; in order to satisfy C2, weak constraint C1 is left unsatisfied. Another solution would be to leave C4 unsatisfied instead of C1. Since

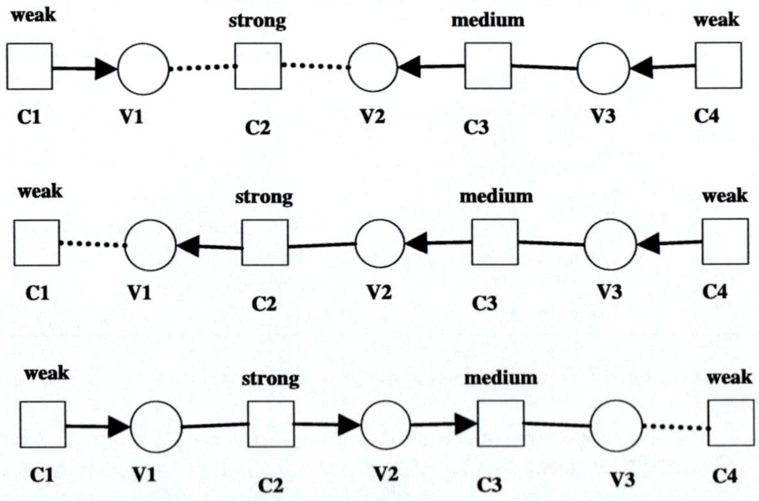

Fig. 4 *Constraint_strength* for constraint solving

C1 and C4 have equal strength, SkyBlue will arbitrarily choose one of these solutions. The third graph shows the solution that results when the strength of the constraint has been set to medium. Here, C4 is left unsatisfied. For details of constraint solving in SkyBlue, please refer to Sannella [21, 22].

The *constraint_sense* attribute can be assigned with two string options, *directed* and *undirected.* A constraint is *directed* if any of the members of the constrained entities is constrained with respect to a sequence of evaluation, where those referenced entities must exist and be evaluated first. A constraint is undirected if there is no required sequence of evaluation among referenced entities, and the constraint is mutually applicable among member-constrained entities. Stated differently, in the undirected constraint, there is no difference between constrained entities and referenced entities [12].

2.3 Other Generic Feature Properties

Other properties defined in the generic feature schema can be defined as follows:

- *General feature Attributes.* General feature attributes such as *feature_name* and *feature_id* are defined to serve as the index for searching during feature modeling operations.
- *Feature type.* Feature type is essentially determined by the instance feature class name derived from a generic feature class, e.g., *block* feature or *slot* feature.
- *Depended_feature_id list.* To maintain feature relationships, feature dependency relations should be kept during the modeling procedure [26]. The feature dependency relation is described by Bidarra et al. [3]: "feature f1 directly depends on feature f2 whenever f1 is attached, positioned or, in some other way, constrained relative to f2." The feature dependency graph illustrates the feature dependency relations with a simplified constraint graph. In the graph, each of the edges of the graph is directed. The direction of each edge in the feature dependency graph runs from one feature to another feature that depends on it. For example, the part in Fig. 5 can be expressed as both a constraint graph and feature dependency graph as shown in Fig. 6. In these graphs, the slot and two holes are *depended_feature* of the base block feature. *Depended_feature_id* records the feature dependency relation. It plays an important role in maintaining the feature dependency graph, as well as in maintaining feature relationships during feature modeling operations. Modern CAD systems also retain feature dependency relations. For example, in Siemens NX 7.0, users can query all such information.
- *Feature label* is an *entity_list* in the feature definition, used to record feature elements. Feature labels are attached as attributes to feature member entities, e.g., faces, edges, and vertices.
- *Domain specification.* As the proposed feature model must support collaborative feature-based modeling among multiple applications [30], *domain* is used to

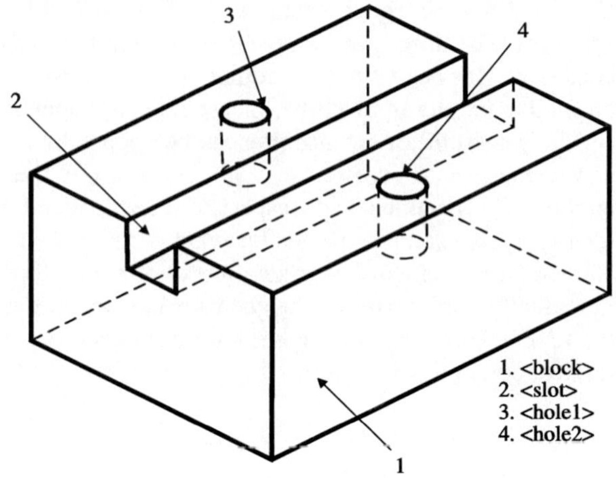

Fig. 5 A simple example part

designate which application a feature belongs to. *Domain* has the enumeration data type; values can be "design," "manufacturing," "assembly," and others. By specifying different domains, multiple application features that refer to the same product geometry can coexist in the feature-oriented database.

- *Nature.* The nature feature is the enumeration data type, which is either *additive* or *subtractive*. *Additive* means that the feature is to be instantiated by adding material. *Subtractive* means that the feature must be obtained by subtracting material.

2.4 Member Functions

- *Attributes access functions* need to be defined to manage a feature's attributes. Most of these functions are common to all types of features, e.g., *backup()*, *findOwner()*, *findConstraint()*, *getParameter()*, *setParameter()*, and so on. Other specific attribute methods for individual application features will be addressed at the application level.
- *Modeling operation functions.* These functions are used to control the behavior of a feature during a modeling operation, e.g., creating, editing, deleting, splitting owners, merging owners, or translation.
- *Feature validation functions.* Whenever a feature operation is activated via the user interface, the product model needs to be modified and updated. This process requires feature evaluation, which ensures the consistency of the geometrical model at low levels. In the work discussed here, the run-time product model is generated via an integrated solid modeler and managed based on the database records. All feature evaluation functions triggered by such operations call the solid modeler's APIs to access and determine the geometrical procedures; they are implemented separately in the feature classes. In this way, the details of

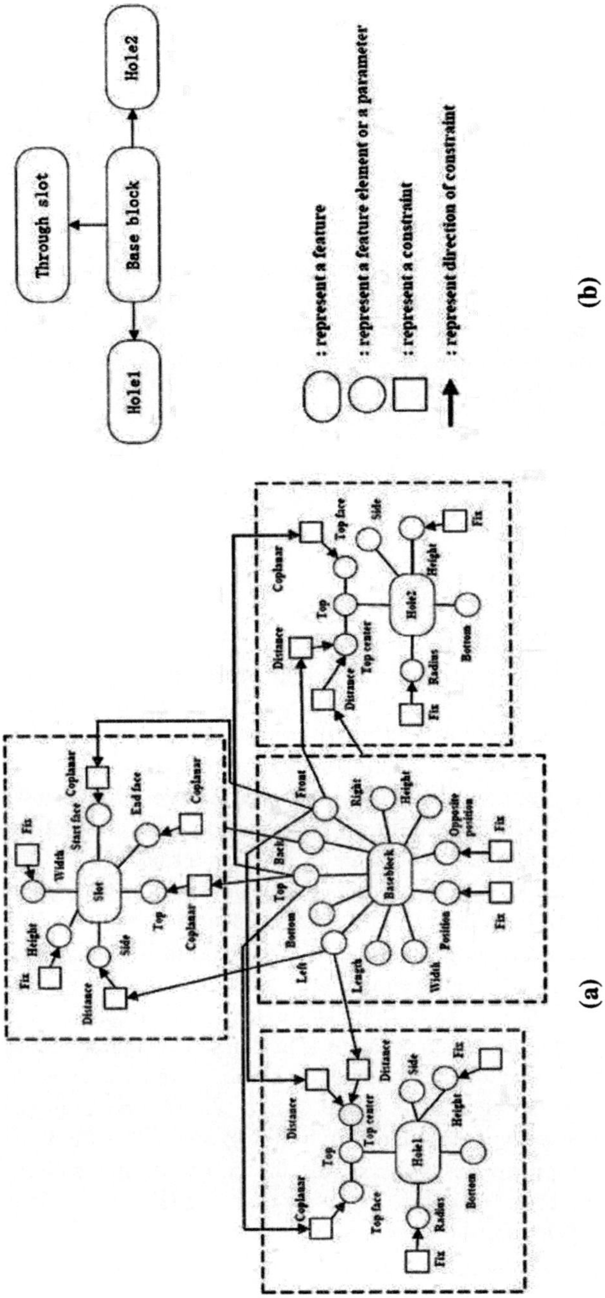

Fig. 6 Constraint graph and feature dependency graph of sample part (**a**) Constraint graph, (**b**) Feature dependency graph

geometrical operations are maintained by the solid modeler; hence, the development effort is significantly reduced. Theoretically, feature process functions can be classified into two kinds, those dealing with the geometry and those managing constraints. With the incorporation of a solid modeler, feature process functions rely on the solid modeler for manipulating and validating feature geometry. Constraint solving functions need to call on specific algorithms defined in the individual constraint sub-classes to solve different associative relations according to their types.
- *Save and restore function.* For repository purposes, feature saving and restoring functions, which are the interactions between the run-time feature model and the database, must be defined in the unified feature model classes, because these functions have to organize information for different application views according to users' requirements.

3 Advanced Feature-Based Engineering Modeling: A Prospect of Advanced Design and Manufacturing Methodology

It is expected by the authors that with the generic feature implementation and the related database schemas [14, 16], the implementation methods of different feature-based engineering applications can be unified and developed within a systematic framework. A holistic feature-based engineering informatics modeling scheme that is based on the generic feature concept presents a complete product and process information repository, and supports high-level feature information integration across different engineering application software tools; this is dubbed *unified feature modeling*. This approach is to be introduced in "Unified Feature Paradigm". The authors suggest a unified information infrastructure model that has been published by Tang et al. [29] with reference to Zha and Du's work [31]. To briefly introduce the concept, a partial schema-level EPM representation is defined as shown in Fig. 7. A design feature model and a manufacturing feature model are represented as sub-models in the application layer; they need to be integrated with application-specific functional modules. The commonly shared feature information model below the application modules contains all components defined with a unified feature modeling scheme supporting the entire product model (EPM) with *generic features* as the basic semantic building units. The EPM describes information across applications, and contains the domain classification ontology and metadata. This layer contains assembly-part models, product geometry and topology, the related attributes, and so on. This chapter is dedicated to presenting the fundamentals of the *generic feature* concept.

All EXPRESS-G representations in this work follow a convention defined by the ISO standard [9], as shown in Fig. 8. Note that in this work, those entities shown in EXPRESS-G diagrams with page reference "#, #" have been defined in the standard.

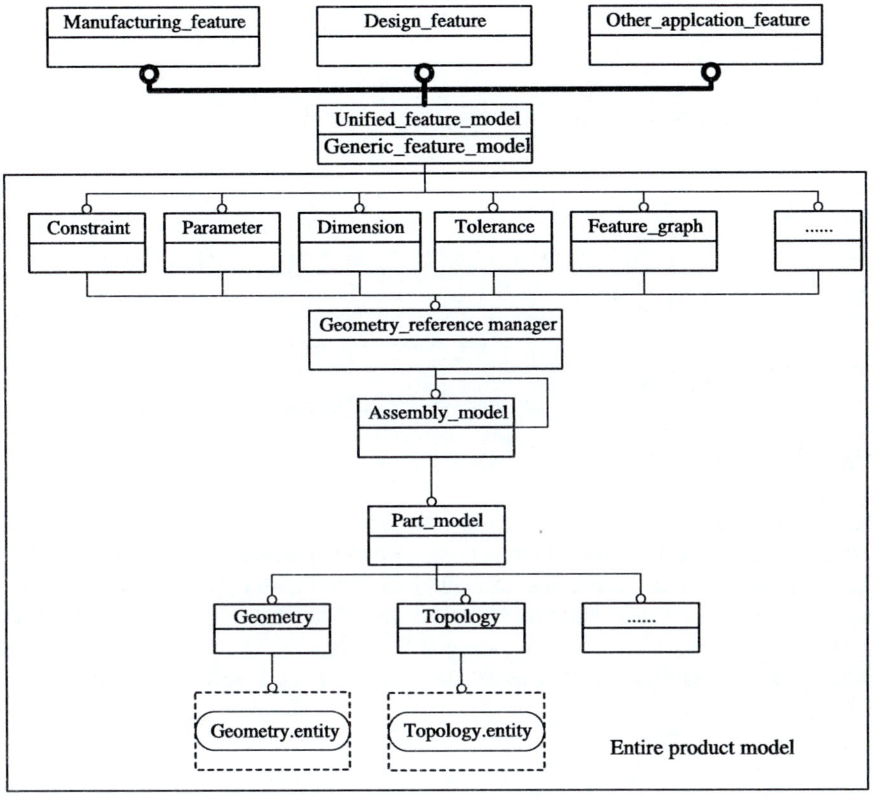

Fig. 7 Partial schema-level EPM (enhanced from Tang [28])

In this chapter, we focus only on the feature-based design and manufacturing models of a product, which includes geometry, constraints, parameters, and dimensions. Other related information, such as product-related documents and categories, are not discussed here. In the following section, design and manufacturing feature models are described.

4 Application-Specific Feature Models

Much research effort has been directed toward feature classification. Shah and Rogers have classified features according to three basic forms: form features, precision features, and material features [25]. They considered that features can represent other logical information sets, such as assembly relations and functional entities. Rossignac clarified the distinction between intentional features and their geometric embodiment, and between volume features and surface features [20].

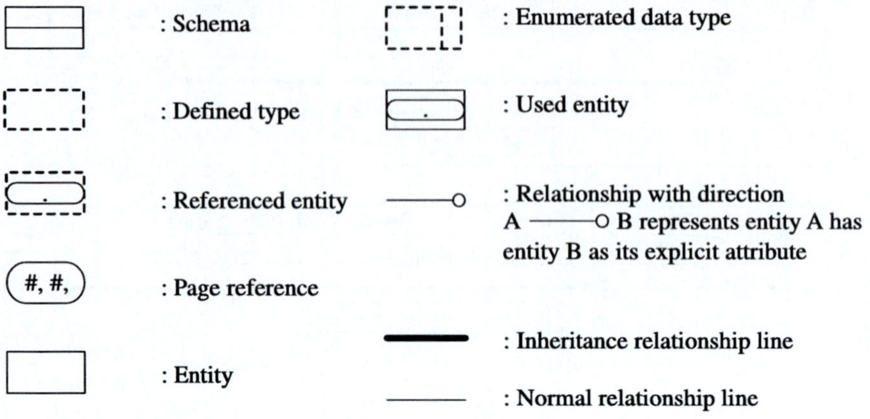

Fig. 8 Symbol convention of EXPRESS-G [9]

Juri further classified the form features into primary and secondary, and also into external and internal. Primary features correspond to cylindrical and conical shafts, whereas the secondary are holes, threads, fillets, and so on [10]. In this work, we consider only design and manufacturing features. The design feature classification used here is similar to those in commercial CAD systems (such as Pro-E, NX, and others). The classification of manufacturing features is based on the AP 224 of STEP.

4.1 Design Feature Representation

4.1.1 Design Feature Representation Schema

In this section, design features are used as an example subgroup to illustrate how application-specific feature models can be defined. A design feature model can be expressed as shown in Fig. 9.

The primitive feature type is separated into two subtypes, additive and subtractive features. Additive features include all instances of features formed by adding material, such as *cylinder, taper, sphere, boss, block, torus*, and so on. Subtractive feature types represent all features such as *hole*, *pocket*, and *slot* that are formed by subtracting material. The transition feature type includes *chamfer*, *edge_round*, and *fillet*, which are always associated with other primitive features. The compound feature type is a union of several primitive features. Datum, which is used as the reference for feature-based modeling, is also regarded as a kind of design feature in the authors' opinion. Datum has three subtypes, namely datum plane, datum axis, and datum point. Additional feature types such as extrusion, revolution, sweep, and others are also accommodated in this schema. Each specific

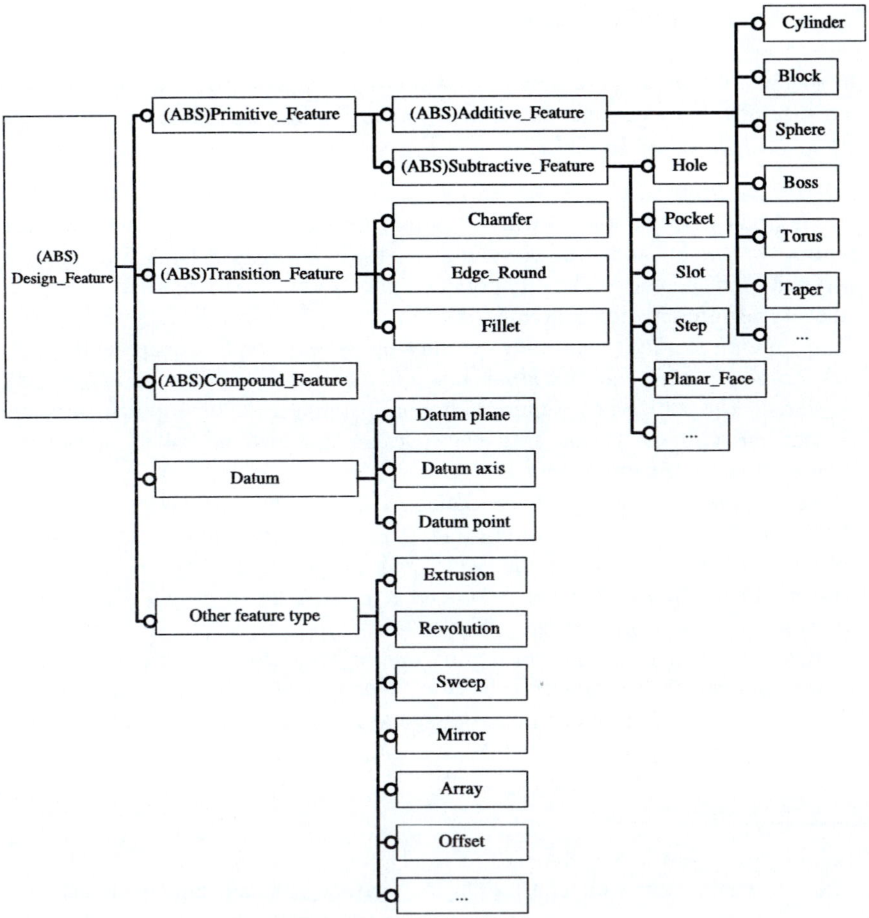

Fig. 9 Design feature representation schema

design feature type has predefined explicit geometry, topology, parameterization, and constraints specifications. Note that not all the feature types are included in this schema, because the number of feature types is infinite [24]. But by using the generic feature model, feature definitions are extensible.

4.1.2 Example of a Design Feature Definition: Slot

Based on the generic feature definition, design features such as *slot* can be defined in EXPRESS-G according to STEP AP 224 [9], as shown by Ma et al. [12]. The *Slot* feature class inherits all the common properties and methods from the generic feature class.

Generic Shape Representation of Slot Feature

The shape of the *slot* feature is expressed as swept depression volume with a cross-section profile and a continuous curve of travel. The member elements associated with a *slot* feature are listed below:

- *Course_of_travel*. This member entity specifies a 3D space curve (e.g., line and circle), that when combined with a "profile," creates the shape of the *slot*. The *course_of_travel* can be represented as a *path* in EXPRESS-G. A *path* shall be defined as the geometrical entity pointer, which serves as the input parameter for the *slot* feature creation function.
- *End_conditions*. *End_conditions* specifies the type of implicit shape at the ends of the *slot*, which can be *blind_slot_end_type* or *open_slot_end_type* [28]. Different *slot* end types require different parameters. These parameters are associated with the *create_slot_end* function that will be called in the *slot* feature creation function.
- *Sweep_shape*. The *sweep_shape* defines the sectional 2D sweeping profile. When combined with the course of travel, the sweep operation creates the shape of a *slot*. For the *slot* feature, the *sweep_shape* is represented as an *open_profile* that could be *square_u_profile*, *rounded_u_profile*, *linear_profile*, *vee_profile*, *partial_circular_profile*, or *tee_profile*. Each type of 2D profile has its own initializing parameters. For example, *square_U_profile* has two parameters, *length* and *height*, which will be used in *create_profile()* functions to create the 2D profile. This 2D profile will be defined as an entity pointer and will serve as an input parameter for the *slot* feature creation function.

Constraints

In the *slot* feature definition, constraints are regarded as an attribute list attached to the *slot* feature, and are therefore defined as an attribute list. Different types of constraints (e.g., *distance* and *angle* constraints) are defined first. All the constraints are treated as common attributes in the feature's attribute list and, to maintain feature validity, are accessible for the validity check.

Other Feature Properties

- feature name: slot,
- depended_feature_id: entity_list,
- domain: "design,"
- nature: "negative."

Given the *slot* feature definition described above, instantiating a design feature *slot* shall be carried out in two steps: defining the shape of the *slot* and positioning the *slot* feature. Using the *through_slot* feature with *square_U_profile* and a straight line path shown in Fig. 10 as an example, the details are shown as follows:

(a) Specify the type of the *slot* feature. This is to define the sweep shape (such as *square_U_profile*, *T_profile,* or *round_U_profile*) of the *slot* feature by specifying the required parameters. In the instantiation function of *slot* feature, a *square_U_profile* entity is then parametrically created.
(b) Specify the *slot* end type. A *slot* feature may have a number of end types. The shape of the *slot* end will be created and combined with the main body of the *slot* feature to complete the shape of the *slot.* A *slot* end type with parameters will be recorded as the entity pointer. In the example cited here, for *open_-slot_end_type*, no action will be taken for the creation of the *slot* end shape.
(c) Define a course of travel for the *slot* feature: it can be a line, a circle, or a 3D space curve. By default, a straight line perpendicular to the sweeping profile will be taken as the course of travel. This course of travel will be stored as an entity pointer. Here, a course of travel with direction (0, −1, 0) is created by specifying the start and end face of the *slot.*
(d) Create the body of the *slot* feature by sweeping the profile along the path. This kind of operation will result in a solid as the main body of the *slot* feature. In this example, the given *square_U_profile* is swept along the direction (0, −1, 0) for a distance equal to the distance between start face and end face of the *slot*. The result is shown in Fig. 10.
(e) Position the *slot* feature. To position the slot feature on a planar surface, dimension constraints, which are used to define *constrained_entity* (the geometry of feature to be created) and *referenced_entity* (existing geometry on the model), are used. In this case, for the definition of *through_slot* feature, two coplanar constraints (C_1 and C_2) are defined to determine the start and end of the *slot* feature. There is thus no need to set such a constraint along the *y*-axis. In addition, a distance constraint, D, is used to dimension the distance between the *slot_left* face (constrained_entity) and *block_left* face (reference_entity or datum).

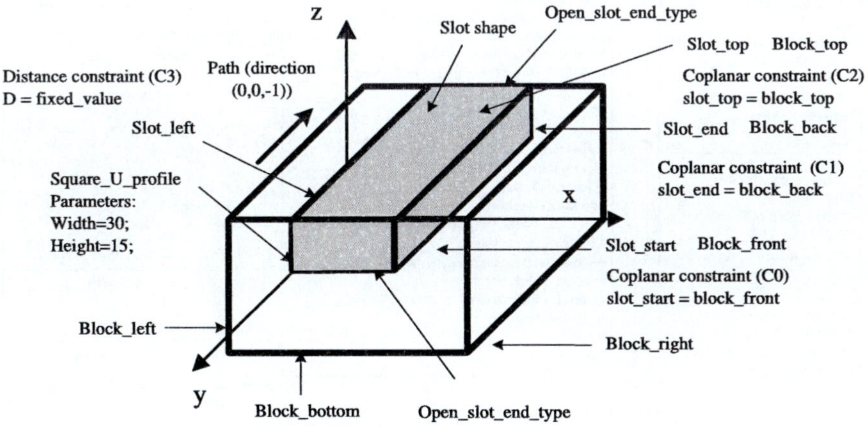

Fig. 10 An open-ended *slot* design feature with the *square_U_profile*

(f) Generate the 3D cell to the shape of the feature on the basis of cellular topology, and insert the shape into the part by carrying out a non-regular Boolean operation. Details for cellular topology will be described later in this chapter. Here, the *slot* shape will be *Boolean union* with the *base_block* and will result in the final part shown in Fig. 10.

The slot feature instance, in the example, can therefore be expressed as shown in Fig. 11. Note that upon cellular decomposition, there are two cells (3D cells) in the cellular model of the part shown in Fig. 10. The shaded cell represents a cell of slot shape, while the remainder represents the cell of the base_block.

4.2 Manufacturing Feature Representation

4.2.1 Manufacturing Feature Representation Schema

A manufacturing feature represents a geometric shape that is associated with a manufacturing process to produce the associated part faces as designed. STEP AP 224 [9] has categorized manufacturing features into three groups: machining features, replicate features, and transition features.

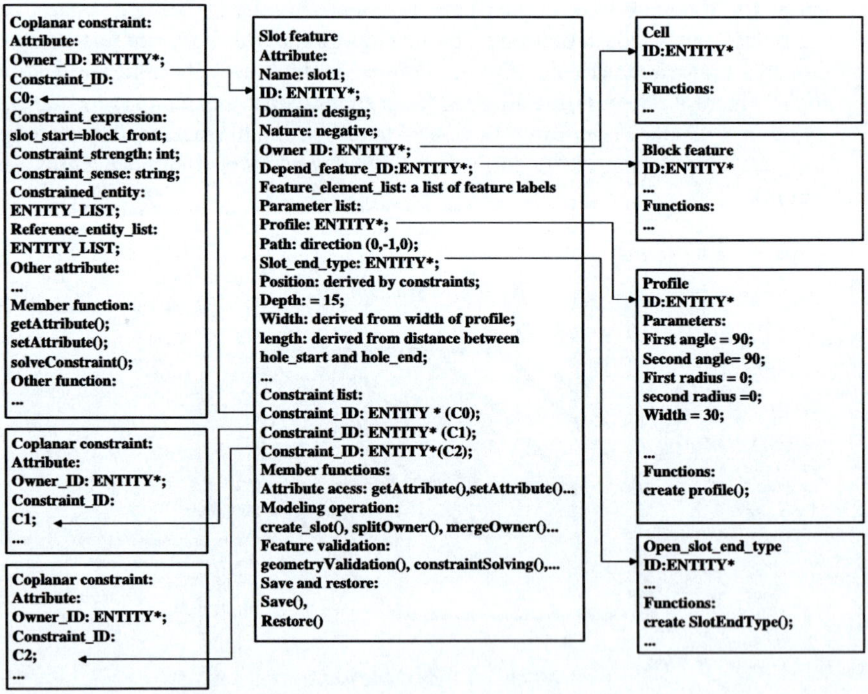

Fig. 11 *Slot*, a sub-class of design geometry feature

A machining feature is a subtype of manufacturing feature that is formed by removing solid materials from the initial stock in order to obtain the target part geometry. According to STEP, machining features can have the following subtypes: *knurl, multi_axis_feature, thread, marking, spherical_cap, outer_round, revolved_feature*, and *compound_feature*. For details of the definition of the above-mentioned feature types, please refer to the standard [9].

Each machining feature requires direction and position to place it on a part. Therefore, a *placement* data structure is defined. *Placement* specifies the position and orientation of a machining feature relative to the base shape of a part. The data associated with a machining feature also includes the attribute *usage_name*. The *usage_name* specifies a user-defined name that contains additional information about the use of a feature. The *usage_name* is optional; it does not need to be specified for every machining feature.

A *compound_feature* unites one or more machining feature objects to create a more complex feature definition. The placement of a *compound_feature* is relative to the part, another *compound_feature*, or a *replicate_feature* which uses a *compound_feature* as the base feature. Features that are elements of the *compound_feature* have their placement defined relative to the *compound_feature* placement.

A *multi_axis_feature* usually identifies milling features for a part, such as *boss, general_removal_volume, hole, rounded_end, planar_face, pocket, profile_feature, protrusion, rib_top, slot*, and *step*.

In the authors' view, manufacturing features, unlike design features, depend on the process plan, although manufacturing features can have predefined geometry. This means the geometry of a manufacturing feature can be determined only after a manufacturing operation has been determined by the process planner. In order to generate a manufacturing feature model from the part model, a predefined generic feature template library can be used for feature recognition. This procedure can be implemented automatically or interactively, or as a combination of the two. After feature recognition) and selection of appropriate machining operations, the shape of the manufacturing feature can be determined. Each manufacturing feature has an associated *machining_operation*, defined and stored as the attributes of the relevant geometrical entities. Candidate machining operations are those combinations of the machine tool and the cutting tool whose capabilities (shape, size, tolerance, surface finish) and accessibility satisfy the required manufacturing specifications. Therefore, the machining operation schema has four major components: machining method, machine tool, cutting tool, and machining operation.

The shape of a machining feature contains two volumes: an accessing volume and a removal volume [19]. In a traditional machining operation, material is removed by a moving cutting tool, which is attached to the machine tool. The moving cutting tool together with its chuck, which is driven by the machine tool, will sweep a volume in space. The cutting portion of this swept volume is known as the removal volume, i.e., the effective removal volume. The remainder of the swept volume is referred to as the accessing volume.

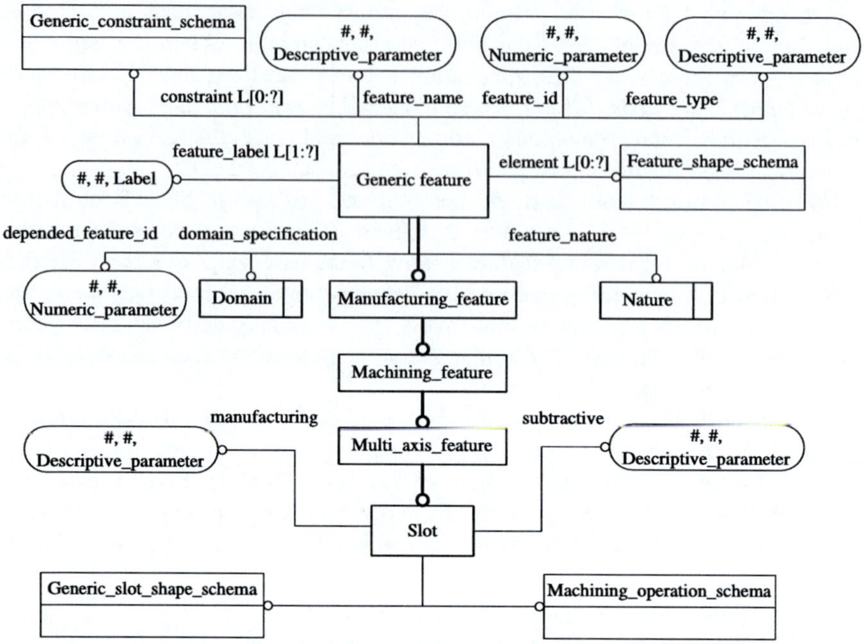

Fig. 12 Machining feature *slot* definition in EXPRESS-G (adopted from Ma et al. [12])

4.2.2 Example of a Machining Feature Definition: Slot

A machining feature *slot* can be represented in schema format as shown in Fig. 12.

Generic Shape of the Machining Feature Slot

The machining feature *slot* can be used to achieve many kinds of design features. When a design feature is achieved by applying a slot manufacturing feature, they become associated; note, however, that they are neither identical nor overlapping. There are two key differences between design features and manufacturing features. The first is that in the manufacturing domain, accessing volumes and raw work piece volume should also be considered for manufacturability analysis. These volumes are evaluated when a machining operation is decided upon. The other difference is the associated relationship between the removal volume and the design feature. The design feature represents the design requirements, while the removal volume represents machining steps. The removal volumes of a machining feature are the chunks of material must be machined away with each machining steps in order to achieve the ideal design features. Unfortunately, machining features and design features are not corresponded in a one-to-one manner; rather, they are associated by the critical faces which are defined by the both types of features.

Design features specify the resulting requirement about critical faces while machining features define the steps, or how those critical faces are produced.

Validity Condition Definition

- The surface type of the *slo*t feature must match the surface type of at least one machining operation in a given manufacturing environment.
- Tolerance and surface finish specifications of the *slot* feature surface must match the tolerance/surface finish capability of at least one set of machining operations in a given manufacturing environment. It should be noted that a feature machining process is usually an ordered set of machining operations, whose total effects are equal to or better than the tolerance/surface finish specifications of the finished surface.
- The effective removal volume of the *slot* machining feature cannot intersect with the final design part volume.
- The accessing volume of the slot shape must not intersect with the blank or work piece, fixtures, or other machine tool elements at any time.
- When machining the *slot* with an end-milling tool, the minimum corner radius of the *slot* must not be smaller than the radius of the selected milling tool.

Other Feature Properties

- Feature name: slot
- Depended_feature_id: entity_list
- Domain: manufacturing
- Nature: negative.

Instantiating a *slot* feature in the manufacturing domain often requires automatic or interactive feature recognition with the input of users. Automatic feature recognition is very complicated and is an entire research area unto itself; here, due to space limitations, it will not be discussed. The corresponding features in different domains are associated with the final product model. Given the design feature *slot* shown in Fig. 10, the corresponding machining *slot* feature as defined in Fig. 12 can be automatically identified. Note that the example contains two features represented as 3D cells, namely, the *slot* and the *base_block* (workpiece). The detailed properties of a manufacturing feature *slot* instance are shown in Fig. 13.

5 Operation for Multi-Application Interoperability

Owing to the large sizes of CAD files, data transmission among various CAx applications over the Internet is quite time-consuming and unreliable, causing intolerable wait times when updating large CAD models across networks. To reduce the network load, an appropriate way to represent CAD data is needed. Incremental transfer is one

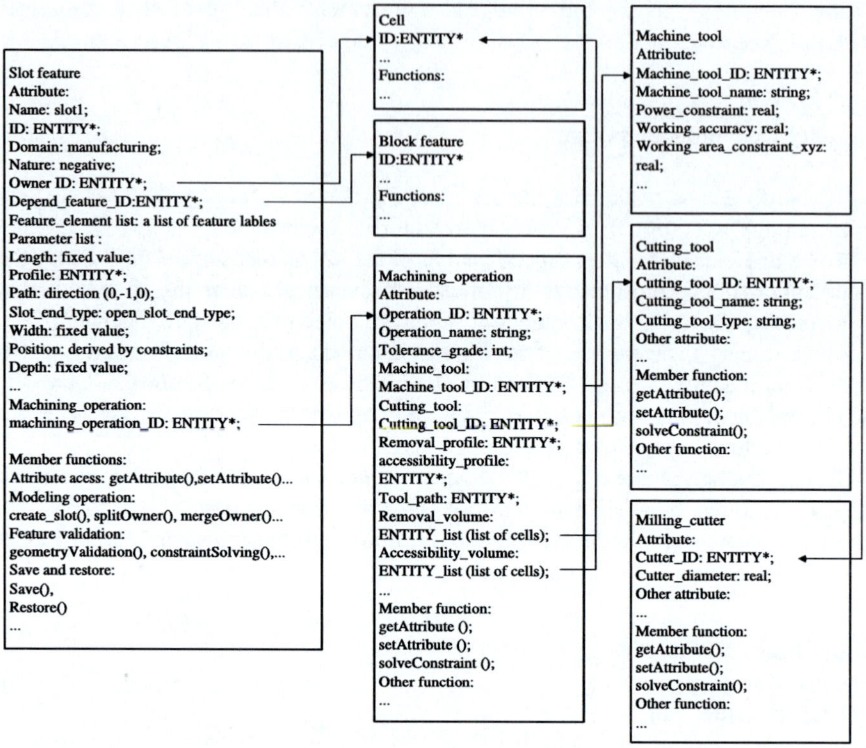

Fig. 13 Sub-class definition of a machining *slot* feature

way to do this. Only the modifications, instead of the whole CAD model, are transferred incrementally during the design process. In this method, an *operation* is used to incrementally transfer model modifications to reduce the communication load.

Operation is defined as a set of related commands that are responsible for functional manipulation of entities. It is directly used to support the interface of the CAx system. As categorized by Chen et al. [6], operations have two types: geometry- and non-geometry-related operations. An operation can be represented using a schema such as the one shown in Fig. 14. The geometry-related operations can be further classified into feature-related and low-level operations according to the entities that they manipulate. Low-level operations create or modify low-level entities, such as points, lines, and faces. Feature-related operations (feature operations) include instantiating a feature or modifying a feature. Non-geometry-related operations can be divided into "auxiliary" and "additional" operations. "Auxiliary" operations mainly facilitate geometric modeling but do not affect the geometry, such as layer management and view manipulation. Other non-geometric operations can be classified into an "additional" group, such as those related to file management. Supporting operations are a basic requirement for generic feature definition; it is important for the manipulation of the feature model, especially for distributed

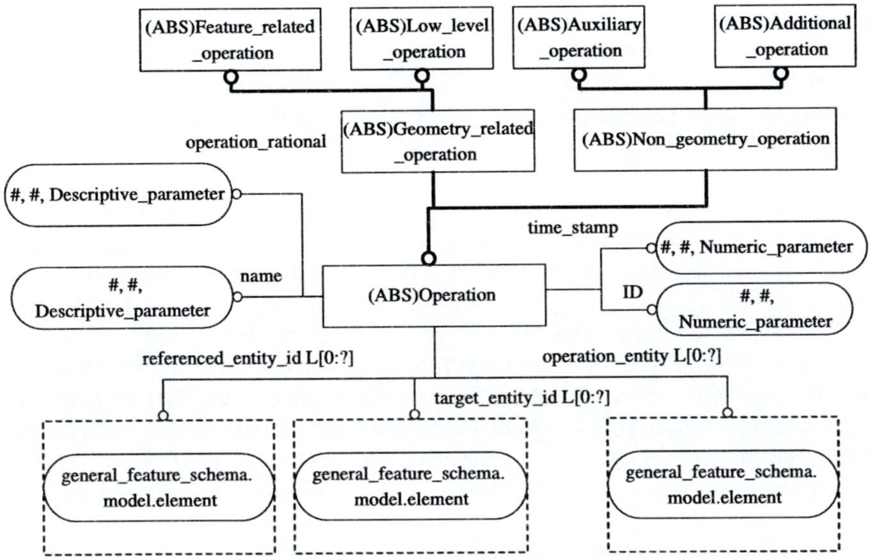

Fig. 14 Operation representation schema

collaboration over the web [30], because the communication data loads between distributed clients and database servers can be maximally reduced by using operation command-based messages.

An operation entity has a *name* and an *ID*. An attribute named *time_stamp* is used to record the time sequence during a collaboration session. An operation records the entities to be created or modified in an *operation_entity_list*. In the *referenced_entity_list*, entities that are related to a particular operation are recorded. For example, when an operation that reconstrains a feature with reference to an element of another feature is sent, the old and new *constrained_entities* and *referenced_entities* are recorded in the *referenced_entity_list* such that the application, which receives this operation, can easily match the corresponding entities in its application. Such matched entities in the receiving application are recorded in the *target_entity_list*, which is used for model updates according to the operation. An *operation_rational* specifies what kind of action the operation will do to the operation entity, e.g., for a feature-related operation, *operation_rational* specifies the actions such as *add*, *delete*, or *modify*.

6 ACIS Cellular Geometrical Representation Schema: Multi-Application Geometry Interoperability Model

A unified feature model allows different applications to define features in different ways, but they are associated with the same master product model. As reviewed in "A Review of Data Representation of Product and Process Models", explicitly maintaining feature shapes in the product model has many advantages. In this research, the feature-oriented product structure generation is modeled in a neutral format, which is designed to be an extension of ACIS [1]. A cellular topology-based geometrical representation schema is adopted as the basic multi-application oriented geometry model.

The cellular model represents a part as a connected set of volumetric quasi-disjoint cells [1]. By cellular decomposition of space, cells are never volumetrically overlapped. As each cell lies either entirely inside or outside a shaped cell, a feature shape can be represented explicitly as one cell or a set of related cells in the part.

The special characteristics of cellular topology require a special Boolean operation. This is also directly supported by the geometric modeling kernel ACIS. ACIS allows the use of non-regularized Boolean operations. This is generally implemented by specifying a special argument to a Boolean function. In ACIS, thc API function *outcome api_boolean* has an optional argument *Bool_type* to specify the type of the Boolean operation, which can be *Union, Intersection, Subtraction, Nonreg_union, Nonreg_intersection*, or *Nonreg_subtraction.*

Specifying a non-regularized Boolean type essentially adds three new conditions to the Boolean operations:

1. When *single_sided* faces become *double_sided*, *both_inside* faces, they remain in the resulting body.
2. Any face-face coincident region remains in the resulting body.
3. No edge or vertex merging is performed at the end of the Boolean.

Owing to the first condition, the union operation always keeps all face regions from the two bodies (though it may split them into separate faces). Owing to the second condition, the intersection of two blocks that share a coincident face always leaves the face instead of deleting it. Owing to the third condition, subtracting a sheet from a non-coincident sheet leaves the imprint of the subtracted sheet on the other sheet [1].

The cellular model-based geometrical representation schemas adopted in the authors' research are further discussed in "Unified Feature Paradigm". Since the data structure of feature entities is neutral, in order to support different feature definitions used by different applications, application-specific feature schemas need to be mapped onto a set of common schemas. As a default implementation solution, the common feature schemas are interfaced with the ACIS cellular model, as currently, only ACIS supports cellular geometrical models. ACIS provides an intermediate data format, which is used for information integration from the application angle. Theoretically, any other solid modeler supporting a neutral format can be adopted as long as it supports multiple application views in a consistent manner. This has also been addressed by other researchers such as Owen [15] and Rappoport [17, 18]. In addition, ACIS also provides lower-level geometric modeling functions, which can greatly reduce development efforts. At this stage, we believe that it is feasible to evaluate features via a single solid modeler with a neutral format.

Collaboration with other feature-based systems requires translators, which are used to translate both feature-level and geometrical data between proprietary application formats and the neutral intermediate format. The adoption of ACIS does not affect our investigation of feature-level association and sharing among different applications based on a common database structure [14, 16].

Regarding pure geometry data, similar research has been done. Kim and Han [11] describe an interface (OpenDIS) between the geometric modeling kernel and the database management system (DBMS) for the implementation of a CAD system that uses the STEP database as the native storage [11, 27]. A prototype system was developed using OpenCascade geometric modeling kernel and ObjectStore. Bidarra's team also uses OpenCascade and its geometric modeling kernel for information integration purposes [3].

A commercial system, OneSpace by Cocreate [8], allows multiple-system input, and uses SolidDesigner as its modeling engine to support collaborative product design, but the system is not feature-based. Feature-level information-sharing has not been reported thus far in the literature.

7 Summary

In this chapter, a conceptual information infrastructure has been proposed to integrate product information in EPM and support multi-view applications with its underlying sub-models. To this end, a *generic feature* representation schema has been presented, which includes feature shape representation and constraint representation. The generic feature model provides a template for different application feature definitions. In order to maintain feature relationships, a *depended_feature_id_list*, used to maintain feature dependency relations, has been defined; this is to be further addressed in "Unified Feature Paradigm". In addition, feature labels have been defined to record feature elements and support history-independent model re-evaluation. On the basis of the generic feature model and STEP AP 224, design feature and manufacturing feature models have been described. Examples (design *slot* feature and manufacturing *slot* feature) were given to illustrate how a specific feature type can be defined. In order to effectively communicate between distributed clients and geometry management servers with centralized databases during a collaboration session, an *operation* schema has been developed. Finally, a detailed geometrical representation schema was investigated based on cellular topology.

Acknowledgments The authors would like to acknowledge that the following research grants have supported the presented work: The Natural Science Foundation of Guangdong Province, China (Grant No. 10451009001004484), Canada Natural Sciences and Engineering Research Council of Canada (NSERC) discovery grant (No. 355454-09), and the University of Alberta GRF (G121140079) grant.

References

1. 3D ACIS Modeling (2012). http://www.spatial.com/products/3d-acis-modeling. Accessed 25 Oct 2012
2. Bettig B, Shah JJ (2001) Derivation of a standard set of geometric constraints for parametric modeling and data exchange. Comput Aided Des 33:17–33
3. Bidarra R, de Kraker KJ, Bronsvoort WF (1998) Representation and management of feature information in a cellular model. Comput Aided Des 30:301–313
4. Chen G, Ma YS, Thimm G, Tang SH (2004) Unified feature modeling scheme for the integration of CAD and CAx. Comput Aided Des Appl 1:595–602
5. Chen G, Ma YS, Thimm G, Tang SH (2006) Associations in a unified feature modeling scheme. ASME Trans J Comput Inform Sci Eng 6:114–126
6. Chen JY, Ma YS, Wang CL, Au CK (2005) Collaborative design environment with multiple CAD systems. Comput Aided Des Appl 2:367–376
7. Dohmen M (1998) Constraint-based feature validation. PhD thesis, Delft University of Technology, Netherlands
8. Emmel J (2000) OneSpace integrating collaboration technology and enterprise PDM. Tech Whitepaper of CoCreate Software GambH
9. Industrial Automation Systems and Integration (2000) Product data representation and exchange, Part 224: Mechanical product definition for process planning using machining features. ISO Document ISO TC 184/SC4/WG3 N854
10. Juri AH, Saia A, De Pennington A (1990) Reasoning about machining operations using feature-based models. Int J Prod Res 28:153–171
11. Kim J, Han S (2003) Encapsulation of geometric functions for ship structural CAD using a STEP database as native storage. Comput Aided Des 35:1161–1170
12. Ma YS, Tang SH, Chen G (2007) A fine-grain and feature-oriented product database for collaborative engineering. In: Li WD, Ong SK, Nee AYC, McMahon CA (eds) Collaborative product design and manufacturing methodologies and applications. Springer, England
13. Ma YS, Chen G, Thimm G (2009) Fine grain feature associations in collaborative design and manufacturing—a new modeling approach. In: Wang LH, Nee AYC (eds) Collaborative design and planning for digital manufacturing. Springer, London
14. Mittra SS (1991) Principles of relational database systems. Prentice Hall, Englewood Cliffs
15. Owen J (1993) STEP: an introduction. Information Geometers, Winchester
16. Ramakrishnan R, Gehrke J (2000) Database management systems. McGraw-Hill, Boston
17. Rappoport A (2003) An architecture for universal CAD data exchange Proceedings of Solid Modeling' 03, ACM Press, New York
18. Rappoport A, Steven S, Michal E (2005) One-dimensional selections for feature-based data exchange. Proceedings of Solid Modeling'05. Massachusetts Institute of Technology
19. Regli WC, Gupta SK, Nau DS (1994) Extracting alternative machining features: an algorithmic approach. Tech Rep 94–95, The University of Maryland, Institute for Systems Research, College Park, MD, USA
20. Rossignac JR (1990) Issues on feature-based editing and interrogation of solid models. Comput Graph 14:149–172
21. Sannella M (1993a) The SkyBlue constraint solver. Tech Rep 92-07-02, Department of Computer Science and Engineering, University of Washington
22. Sannella M (1993b) The SkyBlue constraint solver and its applications. First Workshop on Principles and Practice of Constraint Programming
23. SCRA (2006) STEP Application Handbook ISO 10303 Version 3. http://www.uspro.org/documents/STEP_application_hdbk_63006_BF.pdf. Accessed 25 Oct 2012
24. Shah JJ (1991) Assessment of features technology. Comput Aided Des 23:331–343
25. Shah JJ, Rogers MT (1988) Feature based modelling shell: design and implementation. In: Proceedings of ASME computers in engineering conference. San Francisco, USA

26. Sheu LC, Lin JT (1993) Representation scheme for defining and operating form features. Comput Aided Des 25:333–347
27. Shin Y, Han SH, Bae DH (2000) Integration of homogeneous CAD databases using STEP and internet. Decis Support Syst 28:365–379
28. Tang SH (2006) The investigation for a feature-oriented product database. PhD Thesis, Nanyang Technological University, Singapore
29. Tang SH, Ma YS, Chen G (2004) A feature-oriented database framework for web-based CAx applications. Comput Aided Des Appl 1:117–125
30. Tang SH, Ma YS, Chen G (2004b) A web-based collaborative feature modeling system framework. In: Proceedings of the 34th International MATADOR Conference
31. Zha XF, Du H (2002) A PDES/STEP-based model and system for concurrent integrated design and assembly planning. Comput Aided Des 34:1087–1110

Unified Feature Paradigm

Zhengrong Cheng, S.-H. Tang, Gang Chen and Y.-S. Ma

1 An Overview of Feature Modeling Methods and Trends

Product development consists of several lifecycle stages, including conceptual design, detailed design, process planning, machining, assembly, and so on. Concurrent engineering, a method based on the parallelization of tasks, takes into account of the whole lifecycle of the product, in order to reduce the time needed to develop a new product at the early design stage and quickly stabilize production to better reflect changes in market demand during the product lifecycle. Collaborative engineering, an extension of concurrent engineering, is a technological approach that supports distributed, multi-disciplinary, and multi-organizational teams during the product development and manufacturing processes [33]. The effective adoption of concurrent engineering and collaborative engineering requires a high level of integration both among engineering domains and between company departments, such as business and engineering [15]. Within the engineering domains, more research is needed regarding the efficacy of data sharing and the associated control processes related to the product model changes according to complex engineering knowledge. Regarding the business management aspect, mass customization effort (which is usually in the form of dynamic company responses to customer

Z. Cheng · Y.-S. Ma (✉)
University of Alberta, Edmonton, AB T6G 2G8, Canada
e-mail: yongsheng.ma@ualberta.ca

Z. Cheng
e-mail: zcheng3@ualberta.ca

S.-H. Tang
Guangdong University of Technology, Guangzhou, People's Republic of China
e-mail: fwei@scut.edu.cn

G. Chen
Tianjing University of Science and Technology, Tianjin, People's Republic of China
e-mail: chengang@tust.edu.cn

Y.-S. Ma (ed.), *Semantic Modeling and Interoperability in Product and Process Engineering*, Springer Series in Advanced Manufacturing,
DOI: 10.1007/978-1-4471-5073-2_5,

requirements) would identify what specific functions of the product need to be delivered and/or prioritized during the tailored conceptual design stage [30, 31].

In concurrent and collaborative engineering, multiple computer-aided tools are utilized to support diverse activities throughout the product lifecycle, aiming to achieve a reduction in product development time as well as better quality results. Feature technology has played a vital role in the practice. This chapter begins with a discussion of the limitations of current feature-based modeling systems, including the loss of engineering intent, "hard-coded" feature semantics, unsatisfactory compatibility, inconsistency, and limited scalability for information transfer and sharing among different CAx systems [15]. Such limitations cause the chaos during engineering change propagation within and across different product development stages. The authors proposes a unified feature system which is expected to provide the required scalability and flexibility. The validity of the unified feature modeling scheme in satisfying the demands of modern concurrent and collaborative engineering is further confirmed in the subsequent sections.

2 Informatics Challenges for Modern Concurrent and Collaborative Engineering

It is widely acknowledged that feature modeling, unlike traditional geometric modeling, triumphs in its capability to associate shape information with functional and engineering information in a product model. This advantage leads to the wide use of features for modeling products [2, 17, 47]. However, new system integration needs have emerged that make the traditional feature technology no longer satisfactory.

In modern concurrent and collaborative engineering environments, it is common that engineers are geographically distributed while working on the same engineering task [9]. Thus, the lack of high interoperability, resulting from the fact that engineers are using different modeling systems with different ways of representing products [13], hinders the speed of product development and engineering processes [6, 27, 52]. Moreover, even within a single modeling system, the poorly defined feature semantics are still a problem; for example, current feature-based modeling systems typically define the geometric features only and leave the non-geometric features largely undefined, which limits the capability of capturing design intent in the model. Further, the semantics are not maintained consistently and coherently, which causes the previous design intent to be unusable and have to be overruled [2, 22, 34]. In addition, the associations between feature attributes and constraints among different stages of product development, such as conceptual design, detailed design, assembly design, and process planning, are neither well modeled nor maintained, be they geometric or non-geometric, which makes the design change propagation tremendously difficult across product development; this, in turn, impairs the integrity and consistency of the product model [12]. For instance, CAE models are usually derived and simplified from the CAD model; the disadvantage of this approach is that constructing CAE models has to be repeated every time changes are made in the design model [49]. In addition, inter-stages data transfer and sharing are not well supported too. This is illustrated by the fact that until now, modeling systems have been either primarily part-oriented or primarily assembly-oriented [41]. Clearly, the current CAx technology fails to consider full product lifecycle, especially for, the conceptual design stage, due to the lack of well-defined features that are self-contained and flexible enough to facilitate communication among different stages.

3 Basic Requirements for a Unified Platform of Engineering Information Systems

The proposed unified feature modeling system is based on the generic feature concept; its definition, and implementation scheme have been presented in "Fundamental Concepts of Generic Features". As a foundation class, generic feature is supposed to represent the common properties and methods of different features throughout the product lifecycle and encapsulate geometric and non-geometric relations generically [33, 34].

As Sec. 2 discussion indicates, considering the requirements of modern concurrent and collaborative engineering, feature-based systems should be reinvented. Ideally, the new paradigm would fulfill the integration of CAx systems with a generic feature semantic definition scheme. Figure 1 shows the proposed unified engineering information system infrastructure concept. This infrastructure model is expected to better convey engineering intent, facilitate data transfer and sharing, and to support both the well-modeled geometric and non-geometric associations across different stages of the product lifecycle. This approach will allow the process of engineering change propagation to be systematically managed.

As shown in Fig. 1, the unified feature modeling architecture uses a four-layer structure. The top layer is the application model layer, which consists of different functional application models, such as conceptual design, detailed design, CAE, process planning, etc.

The second layer is the information layer, which can be further decomposed into four sub-layers. The first sub-layer is the macro EXPRESS-based semantic specification layer. This layer keeps a semantic consistency management module and provides system level data structure definitions; it is like a set of header files in C/C++ programming. Second, the application feature sub-layer offers services specifically required for each application with tailored extension of feature modeling support. All application-specific feature definitions with their necessary member functions (which are usually specialization classes of the lower-level generic features) are implemented [35]. Third, a unified feature sub-layer systematically deals with the geometry, topology, dimensions, tolerances, constraints,

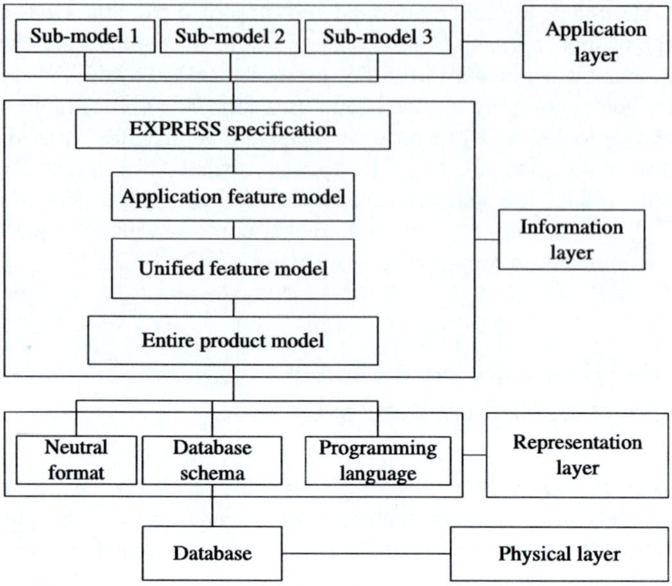

Fig. 1 A four-layer unified feature information modeling infrastructure [51]

and parameters as its attributes via generic features as discussed in "Fundamental Concepts of Generic Features". The unified feature model materializes generic features whose properties and methods interface with the fourth sub-layer, entire product model (EPM). By using generic feature as the basic and common type of feature semantic modeling, high-level semantic attributes and constraints [25] are associated with geometric and topologic entities managed by a geometry modeling kernel via the predefined geometry referencing mechanism. This mechanism can be implemented by kernel level services for connecting higher-level feature specifications with EPM entities to solve persistent problems, such as creating, registering, naming, indexing, inquiring, editing, consistency checking, etc. The unified feature model also contains the domain classification ontology and meta-data. It manages product-, assembly-, and part-related information and inter-application feature models, respectively.

The third layer of Fig. 1 is the data structure representation layer. In this level, all the feature-related data structures are mapped into database-oriented schemas, where the physical database repository system [54] is formulated and interfaced.

The bottom layer is the database management, which is the platform infrastructure technology to be adopted. The database in the physical layer provides repository for many kinds of information, such as geometrical data, feature information, and others [54]. Feature information is stored as data records across database tables. Database structures make information sharing natural, flexible, and with levels of granularity. Currently, the authors can envision the use of massive collaborative database solutions supported by cloud computing or other well-known commercial packages such as Oracle [51]; however, database support is not the focus of this research work.

3.1 Introduction of Unified Feature Modeling Scheme

As a continuation of associative feature modeling theory [31, 34], feature unification was proposed to provide a generic definition to represent common characteristics and methods for various application features. A unified feature modeling scheme, which aims at a systematic informatics engineering framework, was originally proposed by the authors [7, 8, 50]. By using a generic feature template, their unified feature modeling scheme was intended to provide a framework for different engineering application feature models. Although engineering applications define features in different ways, their features have the common types of data entities, such as geometry, topology, dimensions, tolerances, constraints, and parameters. Furthermore, different application features refer to the same master product geometry. These commonalities provided the framework for the research team to propose the unified feature model [8].

Ideally, high-level feature objects are organized by different sub-models in the application feature layer. Application feature sub-models are used to define specific views for different applications. In Tang's work [50], the EPM, application feature model, and unified feature model were described in EXPRESS language. For future implementation, such EXPRESS-defined information models need to be mapped onto database schema (for data storage), expressed in a programming language (*workform* format), and represented in a neutral data format for data communication.

In Tang's scheme [50], features were modeled as an intermediate information layer for the propagation of modifications and to maintain information associations and consistency. To identify and extract the common characteristics of different application features and achieve the required reusability in the feature modeling system, the generic feature concept was used as the basic type of information grain elements, which was expected to be a highly abstracted class in object-oriented programming terminology. As can be seen in Fig. 2 [9], the unified feature model was suggested as the parent class of the multiple application feature models. Concept design and detail design are two example child feature models. From concept feature model (CDFM) to detail feature model (DDFM), a series of refinements is necessary which can be managed by a refine functional module. For feature transformation from detail design DDFM to CAE analysis (ASFM), a simplification process is involved; therefore, the simplify module is necessary to be implemented.

All these feature models use the geometry model (GeoM) for entity creation, editing, data references, and constraint evaluation [25]. The intricate relationships among the features of the whole unified feature model are managed by the relation model. It was the authors' vision that such a hierarchical feature type abstracting

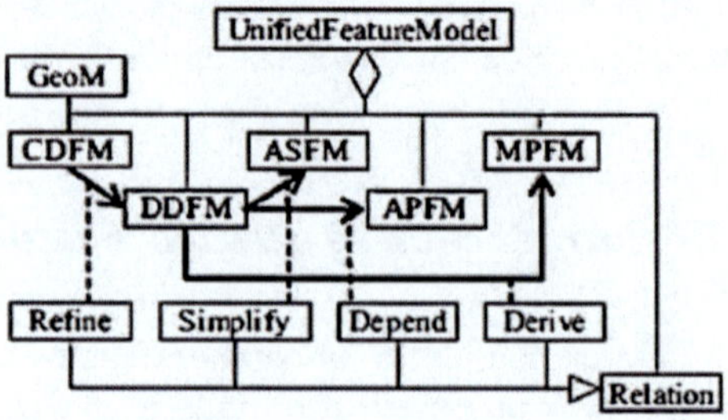

Fig. 2 Unified feature model [9], *GeoM* geometry model, *CDFM* conceptual design feature model, *DDFM* detail design feature model, *ASFM* analysis feature model, *APFM* assembly planning feature model, and *MPFM* machining planning feature

scheme could support a polymorphic definition of features under the umbrella of object-oriented philosophy such that systematic processing methods—including, but not limited to, repository services, validating checking routines, and change propagation—could be shared and automated [33, 34]. The graphical block diagram shown in Fig. 2 follows the UML data representation convention [14]. The member feature of each application feature model can be built from the generic feature class.

3.2 *Expected Capability of Unified Feature System*

Ideally, the unified feature system that is defined in a flexible and consistent manner should have the following components to support modern concurrent and collaborative engineering:

- Integration of different stages of the product lifecycle among multiple CAx systems with high data interoperability [8, 15];
- Geometric and non-geometric association modeling via generic features [10];
- Capture of engineering intent [9];
- Smooth modification propagation with highly maintained information consistency and integrity [11, 33]; and
- Scalability of representing diverse product characteristic engineering patterns.

4 Unified Feature Paradigm

A unified feature system framework consists of an embedded expert system, a unified cellular model (UCM), and an association and change propagation mechanism. Feature, as an intermediate information layer, must interface upward to the knowledge base and downward to the GeoM. In the object-oriented approach, feature is modeled as classes [13], consisting of four types of member variables (geometric entities, Attributes, constraints, and parameters), which embody the "what" part of the feature, and a set of member functions (such as

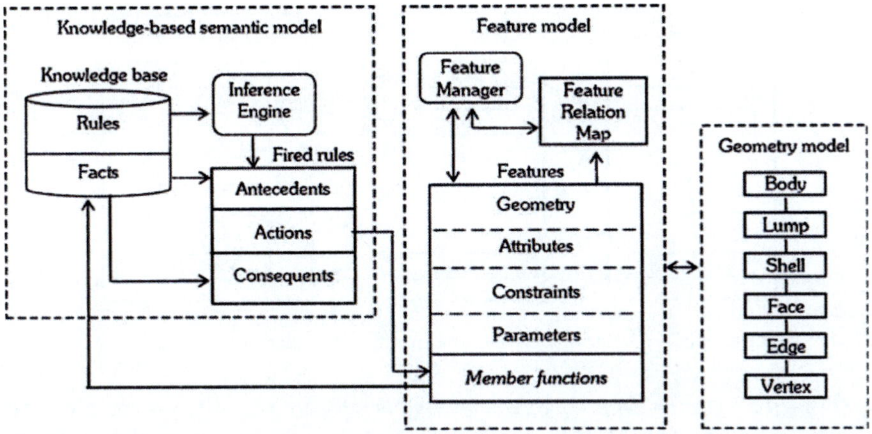

Fig. 3 Structure of a three-part product information model [10]

geometry-creation, query, validity-check, etc.), which represent "how" the feature works, namely the processes a feature is capable of carrying out [10]. This process is illustrated in Fig. 3.

The next section introduces the UCM, knowledge reasoning process, and association and changes propagation mechanism in the unified feature system.

4.1 Unified Cellular Modeling Process

The development of this unified cellular modeling method is intended to achieve the capability of cross-referencing the cellular models among different application features which are defined separately in their application schemes. This mechanism is necessary to achieve useful information sharing across different applications which commonly refer to the same master product model in both semantic and geometric levels. Regarding the geometry aspect, cross-referencing means the core GeoM must be able to deal with multiple representations of associative features with mutual associations automatically. The proposed unified cellular modeling method is based on non-manifold boundary representation; relations between the feature and its corresponding topological entities can be modeled in a complete and comprehensive way (see Fig. 4).

According to Chen et al. [12], the UCM must support the following mechanisms:

- *Boolean attribute mechanism.* Two different Attributes are defined in a cellular model: cellular nature is either additive or negative, depending on whether it adds or subtracts material to a form feature.

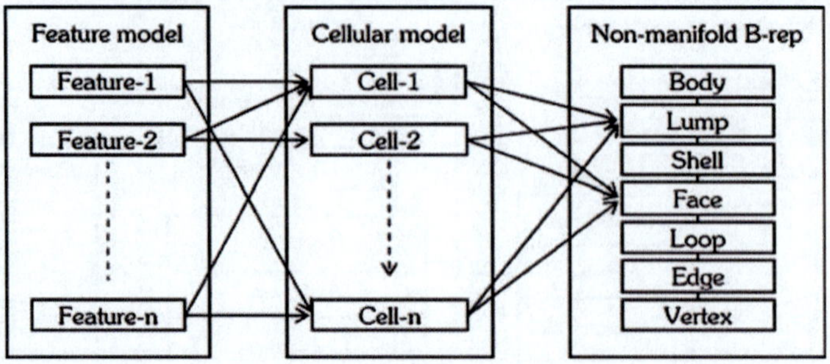

Fig. 4 Feature model, cellular model, and non-manifold boundary representation [12]

- *Owner recognition and management*. Owner keeps track of the shape elements that the cell belongs to. The owner list, as well as the sequence of cells in the owner list, should be maintained according to each application scenario.
- *Decomposition mechanism*. No cells should ever be volumetrically overlapped. When this is the case, a different cell should be created to represent the overlapping part of the shape, as well as the parts lying in either of the shapes.
- *Topology construction mechanism*. New topologies of faces, edges, and vertices are generated before classifying the topological entities as in, out, or on the boundary when cell operations are carried out.

The cell is a data type defined in the cellular model, which represents every element in the set of all the shape entities of the geometrical model throughout the product development process [4]. The UCM is capable of integrating different aspects or abstraction levels of a product in different stages of product development; for example, abstracted lines, faces, curves, or surfaces in the conceptual design, two-manifold solid model in the detailed design, or solid representation with surface manipulation support in the process planning stages. The unified feature model is the parent class of each specific application feature, the relations among which can be seen in Fig. 5.

A few remarks can serve as highlights for the proposed scheme. An application feature model (AFM) contains application features (AF) and non-geometric entities (NGE). An application cellular model (ACM) is created at runtime and consists of multiple application cells. A set of unified cells (UC) makes up the unified cellular model , which is mapped to the components of an application model, and might be shared among different applications. Interested readers can refer to the original sources [3, 12] for more details. The working mechanism of the UCM provides a good basis for accommodating different geometry representation requirements for different application modules.

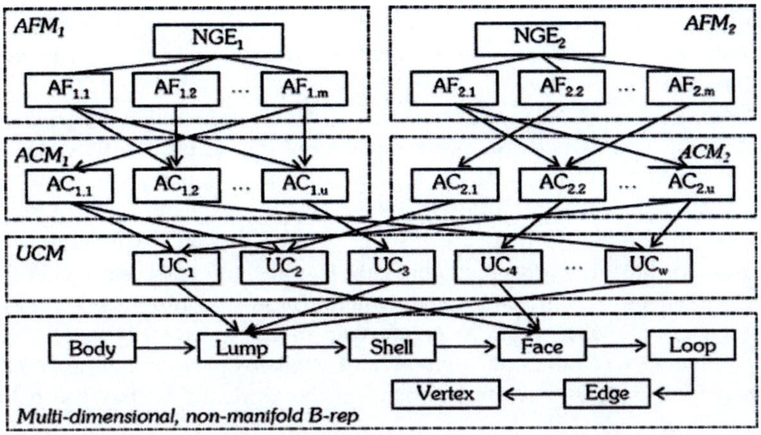

Fig. 5 Hierarchical structure of the unified cellular model [12]

4.2 Knowledge-Based Reasoning

Engineering knowledge is represented in the form of "if… then…" rules in the knowledge base [43]. The alignment between the knowledge base and the feature models, knowledge-based constraint-driven product modeling, knowledge updating, and extraction throughout the unified feature modeling are the unalienable parts in the unified feature system framework.

As has been discussed above, a feature class consists of four types of member variables (geometric entities, attributes, constraints, and parameters) and a set of member functions (such as geometry-creation, query, validity-check, etc.), from which attributes and constraints, together with parameters, comprise a feature's interface with the knowledge bases. Rules, facts, and actions make up the basic information that comprises knowledge stored in the knowledge base. Rules represent the human element of engineers' abstract reasoning patterns; facts are represented by data on which rules can operate by reasoning; and actions are defined operations to be carried out for reasoning which derives from the knowledge states.

Geometric and topological descriptions of a product are, by contrast, contained in the GeoM [10]. Feature operations [13], when they create, modify, or remove facts, involve the functionality of the knowledge base and indirectly activate rules [36]. The knowledge evolution process can thus be described as the sequenced sets of the consequent facts associated with "fired" rules. They are transformed into the corresponding attributes and/or constraints of feature instances, where relations between the knowledge base and the feature model must be indexed and maintained consistently during the feature updating process. The same concept applies to the intra- and inter-feature relations, as well as the relations between the GeoM and the feature model. Consequently, such processes and procedures should be well defined to support the unified feature system with the well-defined functional operations, such as feature validity checking within the knowledge-based reasoning process.

A three-level relational mechanism has been proposed [10] among the knowledge, feature, and GeoMs, as represented in Fig. 6. Briefly, the three major classes in the knowledge model are defined: rule, action, and fact. A list of predefined rule objects and existing fact objects are maintained as entities of a knowledge model [43], which are responsible for forward-chaining reasoning, including, but not limited to, matching rule patterns and selecting a rule to fire. Further, dependency relations are maintained in a justification-based truth maintenance system (JTMS), which helps to avoid infinite processing loops in large-scale problems by evaluating general and domain-specific validity. Detailed implementation information can be found in the works of Forbus and de Kleer [19], and Chen et al. [12].

4.3 Association and Change Propagation

Associations are essential in the unified feature system when evaluating and validating the entities and variables to which constraints refer to, which further validate the corresponding features and support change propagation. That is, whenever changes are made to the shared geometric or topological entities,

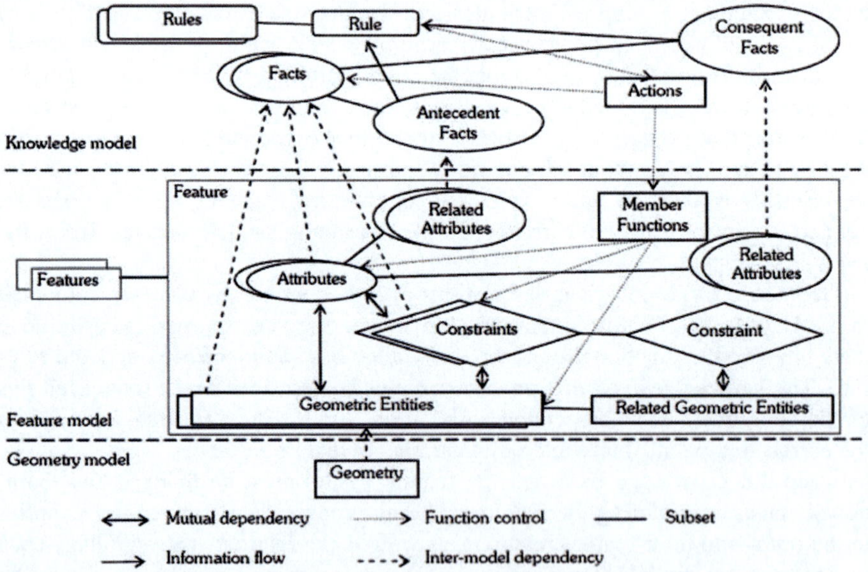

Fig. 6 Partial relations between the knowledge base and the feature model [10]

notification and validation will be carried out on all the dependent features automatically through assigning and associating a feature's geometric reference with the related geometric or topological pointers [11].

Maintaining the feature validity is the process of monitoring each modeling operation to ensure that all features conform to the validity criteria specified in their respective classes [11]. Two types of association have been introduced: sharing and dependency. Sharing association is for common geometry-based feature association. Dependency associations describe the geometric and non-geometric relationships, and can be further divided into three sub-types: (a) dependencies between the knowledge model [43] and the feature model; (b) single-stage intra- or inter-feature dependencies; and (c) inter-feature dependencies across multiple stages [11]. A JTMS is necessary for implementing the dependency associations with a generic data structure and algorithms, which allows it to handle geometric and non-geometric constraints uniformly [12, 19, 35]. In terms of implementing the sharing associations, two methods must be developed with the help of the UCM discussed previously, in order to maintain geometric consistency among lifecycle stages. One of these methods is to generate a new application feature, and the other is to modify an application feature uniformly. An algorithm has been developed to support change propagation in the unified feature system, which consists of several processes: checking the intra-stage associations; checking the inter-stage associations; and retaining or retracting all changes temporarily made.

A detailed feature change management procedure is described in Ma et al. [33]. With such an association scenario and change propagation algorithms built within the unified feature system, validity and consistency of product models could be achieved.

5 Constraint Modeling in the Unified Feature System

Constraint modeling is essential in the unified feature system [25]. This section introduces the broader and more flexible definition of constraint in the unified feature paradigm and its representations. Well-defined constraint, be it geometric constraint, algebraic constraint, or rule-based constraint, can be created and enforced when an application feature is instantiated, or dynamically created through interactive user interfaces, such as during the design process, to achieve the function of validating the unified feature model [25].

5.1 Constraints Definition and Representation

The scope of constraint is much wider than simply geometric in the unified feature system. According to [23], constraints cover the following items:

- Geometric relationships (concentricity, perpendicularity, parallelism, etc.) and metric dimensional constraints (length, angle, etc.);
- Equation constraints (equal length, etc.) and/or technical variables such as torque;
- Semantic constraints that determine the conditions in which a shape is valid or not;
- Topological relations among entities in a model.

In a sense, those constraints mainly deal with geometric elements. However, in the unified feature system, constraints are defined more broadly. Anything that restricts the possibility of a given existence can be seen as a constraint. For example, in an assembly feature, which is usually an inter-part associative feature with a coherent definition for assembly purposes and/or overall functionalities, different levels of constraints are involved, including:

- Engineering constraints, or the fundamental law of physics behind the mechanical system;
- System mechanism constraints, which define the constraints within the sub-assembly systems that make the system workable in terms of fundamental physics principles. Such sub-assembly systems are the means to an end instead of the end itself;
- Mechanical assembly constraints, which define the part and/or part relations within the assembly and subassembly;
- Component constraints, which primarily address the dimensional constraints of a specific component;
- Geometry modeling constraints, which vary with different modeling methods (such as modeling from scratch, like line and arc, or modeling parametrically with a feature-based technique);
- Manufacturing constraints, including geometric and dimensional tolerances. Such constraints are fundamental for manufacturing engineers.

Such non-geometric constraints reflect essential design intent information in conjunction with geometric constraints. After constraints are identified, the next step is representing them. Fortunately, a variety of approaches are available for constraint representation; these include equations/formulas, symbolic relations/logics, optimization processes, tables, curves, dependency graphs, and references.

In the generic feature definition, constraints are modeled as a special procedural object type with associated attributes of other diverse objects such as geometric and NGE or features [34]. Furthermore, specialized functions should be included in the objects to handle those procedural methods, including defining constraints, constraints solving, and other constraints behavior control functions. For example, because a class is defined in the object-oriented approach, *data_type* could be seen as one kind of objects or a *class*, i.e., integer, floating point, or char.

5.2 Constraints in Unified Feature System

Constraints can be solved with a numerical constraints solver, a geometric constraints solver, or a rule-based expert system solver with the help of appropriately defined functions of features to call on the services of these solvers. Similarly, constraints strength, one kind of constraints behavior, should be set with the corresponding function.

There are a variety of approaches to constraint solving, such as the graph-based approach, the propagation approach, and evolutionary methods, among which graph-constructive solving techniques have become dominant [1]. In this approach, a graph should be developed first to represent the constraint problem at hand, and then the constraint should be addressed by a searching technique.

In terms of searching strategies, exhaustive search and heuristics are applicable according to the nature of the constraints problem at hand. Search direction and search order also matter a great deal when choosing a search technique to solve a specific problem. On the one hand, search order could either be data-driven, which is called forward chaining, or goal-driven, which is called backward chaining, depending on the properties of the problem, i.e., the "shape" of the state space, the complexity of the rule, and the nature and availability of the problem data. On the other hand, searching methods could be further categorized as *depth-first*, *breadth-first*, or *depth-first iterative* for deepening the search. If the size of the search space is overwhelming large, it would be more practical to apply heuristics, a strategy for selectively searching a problem space. A heuristic algorithm consists of two parts: the heuristic measure and an algorithm that uses it to search the state space, which could be either of those mentioned above, for example, A-Star search. A good introduction to these classic searching techniques can be found in the work of Luger [29].

Two steps are involved in the graph-based geometric constraints solving techniques. In the first step, the geometric problem is translated into a graph with vertices representing the set of geometric elements and edges constraints. In the second step, the constraint problem is handled by decomposing the graph into a set of sub-graphs with each representing a standard problem that is solvable by a specific solver [21, 26]. To illustrate, a 2D profile that has been fully constrained is shown in Fig. 7.

In this example, A, B, C, and D are four end points, whereas a, b, c, and d indicate four edges connecting those points. As indicated by the graph, the explicitly specified constraints include the length between point C and point D, the perpendicularity between lines c and d, the tangential constraint between arc d and line a, the radius of d, of which the end points are A and D, and the angularity between lines a and b. This is a solvable and well-constrained problem as the solution is unique and all the possible positions, orientations, and dimensions of the primitives can be determined based on the given constraints. For example, the angularity between lines c and b, which is 45°, can easily be derived. Additionally, the length of edge a, between points A and B, can be derived as 2. If more constraints are added to this case, it will become an over-constraint problem. In such a case, those redundant constraints should be determined and eliminated; otherwise, the problem is unsolvable. Sometimes, even if a problem is fully constrained, it might still be unsolvable due to the conflict between the inner complexity of the problem and the limited technical ability of the solver. However, such an unsolvable problem does not mean that there is no solution.

Technically, an under-constrained problem has an infinite number of solutions, in which case, more constraints are needed to make the problem fully constrained and solvable. Take the example of the constrained problem above: if the length between points C and D were left unspecified, then the problem would not be fully constrained and would thus have no unique solution. As shown in this case, even if only one constraint is missing, the possible solutions are numerous. One such possible variation is shown in Fig. 7: the length of the undefined edge c might be any value. The tree decomposition analysis technique, based on graph theory, is reportedly capable of analyzing under-constrained problems and determining the needed constraints that, once added, could make the problem fully constrained and solvable [20].

Since features are essentially sets of associative relationships among engineering entities and hence not standalone, they must be dynamically evaluated due to their interaction with one another. This characteristic thus creates the need to maintain feature relationships. It is also desirable to distinguish between features

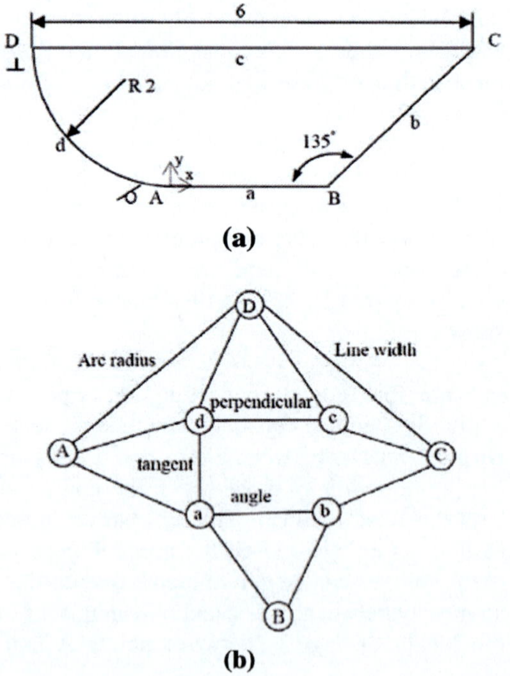

Fig. 7 Constraint graph of a 2D example, **a** Fully constrained 2D profile. **b** The corresponding constraint graph. *Legend* Constraints. ⊥ Perpendicular. Tangent. Coordinate system. *Points* A, B, C, D. *Edges* a, b, c, d. *Default* Line segments between connected points

that determine other features and those that are being determined; dependency has hierarchical relationships: feature A might depend on feature B, which is further determined by feature C, and so on. Graph theory has been found to be capable of describing such dynamic dependency relationships [3]. Dynamics here means that entities are prone to be modified such that feature relationships will evolve correspondingly. Thus, inter-feature dependencies create another kind of constraint, which require maintenance to provide consistency.

To understand how modular integration works in the unified feature system, it is best to have some knowledge about the interfacing functions that have been defined in the unified feature modeling approach among the geometric module, knowledge engineering module, relation manager, and database as shown in the form of intra-stage associations in Fig. 8 [33].

Interfacing functions can be briefly specified as the following [30]:

1. Creating and editing feature geometry;
2. Supporting knowledge embedment [43];
3. Supporting data associations and validity maintenance.

Further detailed information about those functions can be found in the work of Ma et al. [33]. Here the authors want to stress those elements pertaining to constraints modeling. When an application feature is created for supporting data associations and validity maintenance, a resultant node is generated and inserted into a relation manager by a certain mechanism, which is responsible for managing the dependency relationships (or dependency constraints). At the same time, the

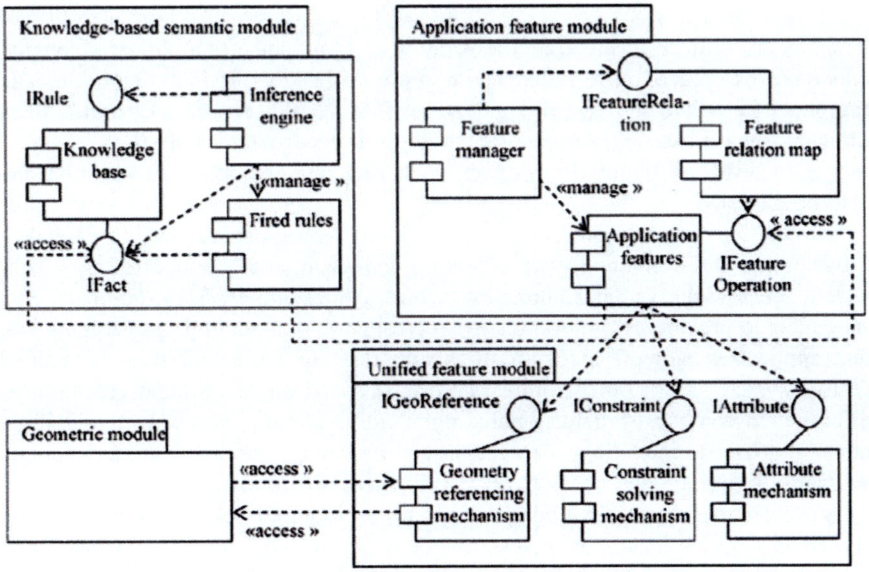

Fig. 8 Intra-stage associations in the unified feature modeling scheme [33]

constraints, which are in charge of the feature's presence and the values of feature parameters or self-describing attributes, are also put into the relation manager and, more importantly, are associated with the corresponding feature node [30]. Thus, it could be said that one of the functions of a relation manager is working as an intermediate agent between feature modification and constraint validation. The agent is called when a change happens to any feature and change propagation requires the constraints to validate the change.

It is necessary to determine whether or not a constraint is satisfied by evaluating and validating the entities and variables to which the constraint refers. This mechanism is called constraint-based association and is made possible by using a JTMS and a numerical constraint solver or a rule-based expert system [24]. Constraint-based association is just one of the associations that evaluates the validity of an information model among its composing entities and includes sharing and dependency. In fact, satisfying all constraints is one of the significant criteria for evaluating information validity, be it of a feature, a model, or the consistency between conceptual and detail design [12]. Moreover, the embedded constraints in the feature model are in themselves of a highly abstract data type that allows inheritance by all instances and subclasses of the parental feature. In addition, constraints, together with attributes and parameters specified in the conceptual design feature model, are transformed into the detail design features correspondingly as Attributes, parameters, and constraints [12].

6 Implementation Methods for the Unified Feature Model

Features are utilized to illustrate the geometric significance of a part or assembly, and have exerted a huge influence in product design, product definition, and reasoning for a diverse range of applications [46]. Moreover, much effort has been directed toward extending feature technology to encapsulate more than just geometric information about the product, such that non-geometric information can also be modeled.

The unified feature modeling system is multi-view-oriented with an extended, flexible, and self-contained generic feature definition, and is supported by a consistent and associative data repository system without unnecessary duplicate data. Its goal is to provide a common feature representation, modeling, implementation, and application support platform. It is important to remember that the unified feature system is not built from scratch; it is based on many other researchers' efforts in a variety of fields, including solid modeling, parametric modeling, constraint-based modeling, feature-based modeling, multi-view technology, semantic feature modeling, artificial intelligence, just to name a few.

Since engineering knowledge first has to be acquired before it can be represented, the unified feature system should be well developed so that it is capable of capturing engineering knowledge in multiple domains across the product lifecycle with both predefined and user-extendable functionalities [43]. However, the existing modeling techniques of various engineering domains are not universal. While it is understandable that different domains choose the technique that is best suited to a particular problem, it is not ideal from a unified feature system development standpoint, as integration becomes almost unachievable in the absence of a common modeling technique among diverse domains. In order to address this difficulty, the unified feature system uses the object-oriented approach as the key unifying mechanism which will be pursued throughout the process of modeling.

This section is organized by starting with an introduction to the basic object-oriented engineering modeling approach, followed by a briefing on the feature concept and its variety of applications in the field of engineering informatics. Next, the data representation and servicing functions of a typical feature class are described. Finally, following a review of the emerging demands from industries for more consistent and efficient systems, the ongoing development of a unified feature system is presented. The unified feature system is aimed at supporting a multi-view application framework. As the basic cell of the proposed framework, a "generic feature" class has been introduced in the system which provides a set of common interfacing mechanisms for data storage, sharing, and manipulation that is required for the polymorphism of applications.

6.1 Object-Oriented Approach

Over, the past decades, the object-oriented paradigm has become almost universally familiar [53]. Object-oriented applications have also been reported in different engineering domains, for example, in conceptual process design, automation engineering, computer-aided process planning, computer-aided engineering, etc. [28, 37–39, 55], which creates opportunities to integrate CAx systems from diverse engineering domains using this approach.

Objects and *classes* are fundamental to the object-oriented paradigm. Objects encapsulate pieces of data in a way that is self-contained, with communicating messages that can interact with other objects. Classes are applied to organize objects hierarchically with properties of inheritance and polymorphism. Indeed, the key concepts of object-oriented modeling are abstraction, hierarchy, encapsulation, modularity, concurrency, and persistence [53], which have been widely utilized in the implementation of the unified feature system. For instance, take the example of one of the object-oriented programming languages, C++. The unified feature is equivalent to the C++ generic feature in that the unified feature is just what "object" is in C++, and the features and sub-features with hierarchical structure in the unified feature correspond to "class" in C++. In the C++ programming language, "class" describes both the properties (data) and behaviors (functions) of objects; similarly, features in the unified feature system are capable of modeling both static attributes and dynamic functions. In the following sections, some of these attributes and methods will be introduced.

6.2 Feature Representation

Feature may be represented at various levels [46]. For example, one may just create a feature, such as a hole, to describe the geometric property of a hole, the same as chamfer, cylinder, slot, and so on. Most features of this kind can be expressed parametrically. However, in the unified feature system, the information contained in the feature is richer. Knowledge (in the form of parameters, rules, procedures, and so on) [43] is also included in the generic feature, which allows features to be handled in a unified way for creation, deletion, and manipulation through methods or functions defined in the generic feature [46]. The generic feature, as an encapsulation of an information unit containing data and functions (procedures to manipulate the data), should model the following information:

- Geometry;
- Information on dimensions, tolerances, materials, etc.;
- Design intent, engineering principles, and manufacturing methods, which are modeled as functions with constraints imposed on the geometrical entities and associated with a form of cross-linked relation graph.

In the object-oriented approach, it becomes natural that the unified feature system is capable of the following feature manipulation mechanisms:

- Inserting or deleting an entire feature;
- Changing feature attributes;
- Invoking feature function.

The design of a unified feature modeling system should keep track and update those attributes of the design model for certain qualities, including coupling, cohesion, sufficiency, completeness, and primitiveness [53]. To achieve this goal, the generic feature has to be modeled in a flexible and self-contained way.

6.3 Feature Shape Representation

Feature shape representation defines the geometrical and topological part of the feature. It may be represented with different schemes in different application domains. For instance, in the conceptual design stage it would be appropriate to model the product with critical lines or faces in the form of relative positions, orientations, fit relations, or relative motions, together with other significant dimensions/specifications. However, in the detailed design stage, complete geometry and specifications of the product should be developed with traditional form features, which is richer in terms of information. Thus, methods should be created for features to manipulate the shape of a feature, including, but not limited to, creation, modification, and deletion [31, 32]. Meanwhile, methods should also be available to link such detailed design features to the conceptual design feature.

Take the example seen in Fig. 9. In the conceptual design stage, two connected boxes would be enough to illustrate the design of the spur gear pair. However, in the detailed design stage, it is preferable to represent the product in solid in a 3-dimensional part model. Thus, there exist diverse forms of feature representation for a specific application depending on the different stages of a product's development. Nevertheless, they are just different aspects of the same holistic feature model. These associative form features can be defined from the same generic feature class. Such feature objects can be instantiated with the known information from one real world perspective; their members attributes may or may not be fully initialized; and during the interactions among applications, their member attributes may be further defined or reinterpreted from another perspective by generating derivate objects or applying difference references which changes domain application feature' constituents. In object technology, such behavior is known as *polymorphism*.

6.4 Member Functions

In order to construct the generic feature class, four groups of member functions are needed: attribute access functions, modeling operation functions, feature evaluation and validation functions, and data-saving and -restoring functions to and from the repository [32]. Attribute access functions are defined to manage a feature's attributes, which may be either common to all types of features, e.g., backup(), or feature-specific, e.g., getParameter() or setParameter(). Modeling operation functions are responsible for controlling the behavior of features during the process of modeling the product, including splitting, merging, translation, rotation, and so on.

Conceptual design sketch feature (2D)	Detailed design form feature (3D)

Fig. 9 Example of features in conceptual design and detailed design of a pair of spur gears

Feature evaluation and validation functions are defined to keep the feature consistent when modifications are made to them [34, 35]. Saving and restoring functions are crucial as they define the interactions between the runtime feature models and the data repository, which is capable of representing data uniquely without redundancies. In sum, functions should be defined for features not only to operate within the feature, but also to interact with the geometric model, the expert system, the relations manager, and the database [30].

In the current example, the possible functions for the previous interfacing mechanism class, *spur_gear_pairing*, include but are not limited to the items as shown in Fig. 10. It should be emphasized that each function should be enriched according to engineering needs, and is not as simple as it appears to be in Fig. 10. For example, the function responsible for stress analysis for the gear tooth should take every piece of knowledge needed to fulfill this functionality, including the material properties, pitch, pressure angle, velocity effect, face width, and application factor, among others. All this information is stored in the data repository, from where the function can encapsulate them into features for easy manipulation.

6.5 Multi-View Concept

Feature sets are determined both by product types, such as casting or sheet metal, and by people's viewpoints or engineering tasks, such as design, finite element analysis (FEA), or process planning [46]. Feature space is a set of features that serves engineers' specific purposes when carrying out various engineering tasks. Traditionally, the feature space defines the application-specific features that are related through feature mapping or feature recognition mechanisms [45]. In different application domains, feature definitions become diverse, which creates barriers to maintaining feature consistency across all application domains, because

Fig. 10 Partial list of class functions required for the *spur_gear_pairing* feature

```
-Instantiate();
-Evaluate_gear_ratio();
-Get_gear_ratio();
-Evaluate_pitch();
-Get_pitch();
-Evaluate_teeth_number();
-Get_teeth_number();
-Delete();
-Constraint_solving();
-Save();
-Restore();
-Feature_validation();
-Stress_analysis_of_a_tooth();
-Assembly_analysis();
-Assign_material();
...
```

different data elements in the product model are contributed by different applications.

In the multi-view approach, ideally, each application has its own view of the data, which is unique, and each view is associated with the master (or common) model as a set of features. In the unified feature system, a multi-view technique is applied so that the generic feature will be defined in such a way that different application domains will be able to extract information from the unique generic feature, rather than from different application-specific features through feature recognition or feature mapping. Consequently, modifications made in one specific application can be propagated to other "views" of the same generic feature automatically. For an in-depth discussion of the multi-view technique, see the work of de Kraker, Bronsvoort, and Bidarra [5, 6, 16], who address the problem of editing feature views on a common net-shape model with the help of the cellular approach. Moreover, the unified feature goes further to accommodate not only the form feature but also non-geometric features into the generic feature model. In the *spur_gear_pairing* example, both the application domain of conceptual design and detailed design are manipulating the same data set encapsulated in the same generic feature.

6.6 *Functional Operations and their Processes and Procedures*

Operations in the unified feature system are what make processes and procedures possible. As has been discussed above, features are defined in the object-oriented approach, where operations can be modeled as methods, or functions, in the feature classes. An operation is referred to as a set of Associated commands (or linked functional methods) that are used to realize the functions and manipulations of feature entities within or across multiple engineering application packages. Moreover, operations can be modeled as separate operation classes associated with specific feature functions.

For a functional application, operations exist in various levels of the unified feature system, from the lowest layer of data manipulation in the data structure, to generic feature function, to functional feature functions, to the modular application functions, and to UI functions (see Fig. 11). Lower-level functional operations support the higher-level ones; the application-specific functions should be defined to support the application-specific feature model on the basis of a unified generic feature model.

Take the example of a design feature. Functions like creating, editing, and checking the validity of features such as cylinders, holes, or tapers should be well defined as encapsulated methods for either associated design features or a set of predefined standard features [42]. Further, operations in a CAD module can be classified into two types: geometry-related and non-geometry-related operations,

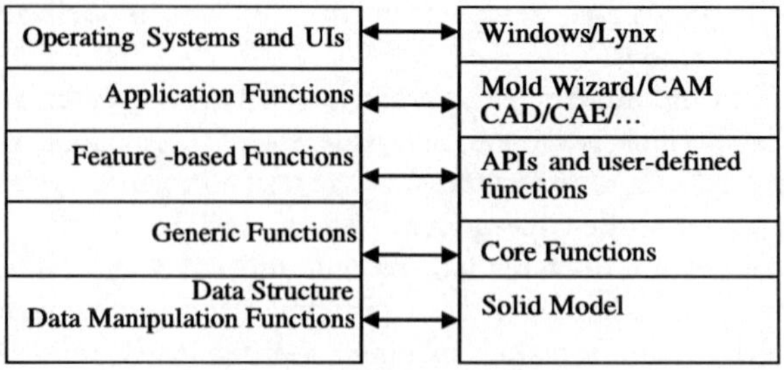

Fig. 11 Layers of functional applications in the architecture

Table 1 Functional application implementation API support resources required

Tools	Functionality	API support resources
CAD	Creating geometrical description of products as well as individual parts with all the necessary dimensions and specifications	A facility to directly evaluate and modify the parametric rules or constraints on the geometry Scalable data/knowledge repository Complementary macro and low-level programmability Programmable functions to generate history-tree-supported geometry Functions to traverse and modify the history tree and the geometry tree Interfaces for geometric entity-level API calls Conflict or interference awareness functions Control of regeneration …
CAE	Pre/post geometry data processing capability dedicated for auto-meshing and manual mesh refinements; finite element analysis functionality	Macro programming control of GUI commands and programmatic interrogation/control of the model properties at the mesh level Math/statistical capability for post-processing enhancements Detailed design optimization Interactive and programmatic capabilities …
CAM	Associative geometry referencing capability with common CAD models; Providing manufacturing specific functions including CNC machine tool path generation; Assisting tools for molding, forming, and fixture design; Interfaces to CAE system for some FEA-based manufacturing analysis such as plastic molding flow analysis	Allowing further development via two-tier support with "user function" type of macro commands and lower-level programming interfaces that complement the basic CAD functionality; Providing access to all databases, process planning rules, feature mapping, and tool path generation; Tying the process definition to the local factory layout and providing some simulation tools …

where geometry-related operations affect the geometric model and non-geometry-related operations can be further classified into "auxiliary" operations and "additional" operations [9].

The implementation is, however, less straightforward than this. For example, different application-specific features are modeled according to different functionality requirements at the user level in the form of specifications, while the realization of such features has to meet the specific application programming interface (API) data structure requirement. Table 1 lists some API resources provided. Thus, it can be seen that the procedures and processes of the unified feature system define how functional operations can work properly among and within different levels of the subsystems and modules. It can be appreciated that processes and procedures are defined within certain modules and can be shared among them from certain common perspectives, or within the scope of some configuration units.

7 Summary

This chapter introduced the unified feature modeling scheme, a systematic approach to address semantics existing in engineering activities from design to manufacturing, which can meet the needs of modern concurrent and collaborative engineering. The framework of the unified feature system, including the modeling scheme, unified cellular modeling process, knowledge-based reasoning, and association and change propagation have been presented. Constraints modeling, defined in a broader sense than simply geometric constraints, has also been illustrated with the help of graph theory embedded in the constraint solver. Multiple-view technique in the unified feature system enables engineers to access data pertaining to specific engineering domains from the generic feature, rather than from specific features that are gained through feature recognition or feature mapping; this is the key to maintaining feature consistency. Different function operations are available in different levels of the system, as well as in different engineering domains, to support the usability of the unified feature system.

The research approach presented here forms the foundation of associative and unified feature-based technology toward information integration for collaborative product development. Compared with the reported works in the literature as reviewed in "A Review of Data Representation of Product and Process Models", this research aims to solve the interoperability of CAD systems at feature level. This approach is unique because interoperability is difficult to realize and common CAx implementation efforts are limited by using a file-based approach that typically represents only geometry. A generic feature concept supporting different computer-aided applications is adopted in this research. It includes associated geometric and NGE and is flexible and scalable because product information is represented in object-oriented feature entities. Such feature objects support polymorphism and make information sharing possible with a finer level of granularity in comparison with the traditional file-based approach. With the generic feature definition, the new concept of "unified feature" scheme is proposed for developing a multi-facet and collaborative engineering informatics implementation method aiming to achieve efficient communication between multiple engineering systems without heavy data load.

It can be appreciated that a database approach is more efficient than a file-based approach [18, 40, 44]. Theoretically, databases can be designed as feature-oriented by using the proposed generic feature definition and the unified feature scheme. To enhance interoperability, some mapping mechanisms from an application to a neutral format product model and from neutral format back to application-specific feature model are required. Feature-level information as well as the related associations among them can be maintained in a hierarchical structure [48]. Feature-oriented databases can be generic for adding, deleting, updating, and querying information. Furthermore, the database approach is expected to reduce duplicated data and potential data inconsistency. Ideally, in a feature-oriented database, product and process information across different associated database tables, which may include geometrical entities, features, security data and others, would be globally managed and shared for multi-users via the network. Based on this feature-oriented database technology, knowledge workers or application software developers can consistently organize the product model for different application views, while the product model repository can be constructed simultaneously with levels of granularity [54]. Such a repository system is expected to enhance the support for concurrent and collaborative engineering. Clearly, much future research is expected to make this idea workable in any real application.

So far, this proposed unified feature scheme is limited to the mechanical product design field, which covers product design and manufacturing. Theoretically, the generic feature model could be expanded to other CAx applications, or other engineering domains such as civil, or electrical/electronic engineering.

References

1. Bettig B, Hoffmann CM (2011) Geometric constraint solving in parametric computer-aided design. ASME Trans J Comput Inf Sci Eng 11:021001–021009
2. Bidarra R, Bronsvoort WF (1996) Towards classification and automatic detection of feature interactions. In: Proceedings of the dedicated conference on mechatronics, 29th ISATA. Florence, Italy
3. Bidarra R, Bronsvoort WF (2000) Semantic feature modeling. Comput Aided Des 32:201–225
4. Bidarra R, de Kraker KJ, Bronsvoort WF (1998) Representation and management of feature information in a cellular model. Comput Aided Des 30:301–313
5. Bidarra R, Madeirab J, Neelsa WJ, Bronsvoort WF (2005) Efficiency of boundary evaluation for a cellular model. Comput Aided Des 37:1266–1284

6. Bronsvoort WF, Noort A (2004) Multiple-view feature modelling for integral product development. Comput Aided Des 36:929–946
7. Chen G, Ma YS, Thimm G, Tang SH (2004) Unified feature based integration of design and process planning. In: Hinduja S (ed) Proceedings of the 34th international MATADOR conference, Springer, London
8. Chen G, Ma YS, Thimm G, Tang SH (2004) Unified feature modeling scheme for the integration of CAD and Cax. Comput Aided Des Appl 1:595–601
9. Chen G, Ma YS, Ming XG, Thimm G, Lee SG, Khoo LP, Tang SH, Lu WF (2005) A unified feature modeling scheme for multi-applications in PLM. In: Sobolewski M, Ghodous P (eds) Proceedings of the 12th ISPE international conference on concurrent engineering (CE2005): Research and applications -next generation concurrent engineering: smart and concurrent integration of product data, services, and control strategies. ISPE, Dallas
10. Chen G, Ma YS, Thimm G, Tang SH (2005) Knowledge-based reasoning in a unified feature modeling scheme. Comput Aided Des Appl 2:173–182
11. Chen G, Ma YS, Thimm G, Tang SH (2006) Associations in a unified feature modeling scheme. ASME Trans J Comput Inf Sci Eng 6:114–126
12. Chen G, Ma YS, Thimm G, Tang SH (2006) Using cellular topology in a unified feature modeling scheme. Comput Aided Des Appl 3:89–98
13. Chen JY, Ma YS, Wang CL, Au CK (2005) Collaborative design environment with multiple CAD systems. Comput Aided Des Appl 2:367–376
14. D'Souza DF, Wills AC (1999) Objects, components, and frameworks with UML—the Catalysis^{SM} approach. Addison Wesley Longman, Boston
15. Dankwort CW, Weidlich R, Guenther B, Blanrock JE (2004) Engineers' Cax education: it's not only CAD. Comput Aided Des 36:1339–1450
16. de Kraker KJ, Dohmen M, Bronsvoort WF (1997) Maintaining multiple views in feature modeling. In: Proceedings of the fourth ACM symposium on solid modeling and applications (SMA 97), ACM, New York
17. de Kraker KJ, Dohmen M, Bronsvoort WF (1997) Multiple-way feature conversion: opening a view. In: Pratt M, Siriram R, Wozny M (eds) Product modeling for computer integrated design and manufacture. Chapman and Hall, London
18. Fortier PJ (1997) Database systems handbook. McGraw-Hill, New York
19. Forbus KD, de Kleer J (1993) Building problem solvers. MIT Press, Cambridge
20. Fudos I (1995) Constraint solving for computer aided design. PhD Thesis, Purdue University
21. Gao XS, Lin Q, Zhang GF (2006) A C-tree decomposition algorithm for 2D and 3D geometric constraint solving. Comput Aided Des 38:1–13
22. Geelink R, Salomons OW, Van Slooten F, Van Houten FJAM, Kals HJJ (1995) Unified feature definition for feature based design and feature based manufacturing. In: Proceedings of the 15th annual international computers in engineering conference and the 9th annual ASME engineering database symposium
23. Hoffmann C, Joan-Arinyo R (2002) Parametric modeling. In: Farin G, Hoschek J, Kim MS (eds) Handbook of CAGD. Elsevier, Amsterdam
24. Hower W, Graf WF (1996) A bibliographical survey of constraint-based approaches to CAD, graphics, layout, visualization, and related topics. Knowl Based Syst 9:449–464
25. Joan-Arinyo R, Soto-Riera A, Vila-Marta S, Vilaplana-Pasto J (2003) Transforming an under-constrained geometric constraint problem into a well-constrained one. ACM Solid Modeling 2003 (SM'03), Seattle, Washington, USA
26. Joan-Arinyo R, Tarrés-Puertas M, Vila-Marta S (2009) Tree decomposition of geometric constraint graphs based on computing graph circuits. In: Proceedings of 2009 SIAM/ACM joint conference on geometric and physical modeling
27. Kim KY, Manley DG, Yang HJ (2006) Ontology-based assembly design and information sharing for collaborative product development. Comput Aided Des 38:1233–1250
28. Law HW, Tam HY, Chan AHS, Hui IK (2001) Object-oriented knowledge-based computer-aided process planning system for bare circuit boards manufacturing. Comput Ind 45:137–153

29. Luger GF (2002) Artificial intelligence structures and strategies for complex problem solving, 4th edn. Addison Wesley, England
30. Ma YS (2009) Towards semantic interoperability of collaborative engineering in oil production industry. Concurrent Eng 17:111–119
31. Ma YS, Britton GA, Tor SB, Jin LY (2007) Associative assembly design features: concept, implementation and application. Int J Adv Manuf Technol 32:434–444
32. Ma YS, Tang SH, Chen G (2007) A fine-grain and feature-oriented product database for collaborative engineering. In: Li WD, Ong SK, Nee AYC, McMahon CA (eds) Collaborative product design and manufacturing methodologies and applications. Springer, England
33. Ma YS, Chen G, Thimm G (2008) Change propagation algorithm in a unified feature modeling scheme. Comput Ind 59:110–118
34. Ma YS, Chen G, Thimm G (2008) Paradigm shift: unified and associative feature-based concurrent and collaborative engineering. J Intell Manuf 19:626–641
35. Ma YS, Chen G, Thimm G (2009) Fine grain feature associations in collaborative design and manufacturing—a new modeling approach. In: Wang LH, Nee AYC (eds) Collaborative design and planning for digital manufacturing. Springer, London
36. Ma YS, Tang SH, Au CK, Chen JY (2009) Collaborative feature-based design via operations with a fine-grain product database. Comput Ind 60:381–391
37. Mackerle J (2004) Object-oriented programming in FEM and BEM: a bibliography (1990–2003). Adv Eng Softw 35:325–336
38. Maffezzoni C, Ferrarini L, Carpanzano E (1999) Object-oriented models for advanced automation engineering. Control Eng Pract 7:957–968
39. Marquardt W (1992) An object-oriented representation of structured process models. Comput Chem Eng 16(Suppl 1):S329–S336
40. Mittra SS (1991) Principles of relational database systems. Prentice Hall, Englewood Cliffs
41. Noort A, Hoek GFM, Bronsvoort WF (2002) Integrating part and assembly modeling. Comput Aided Des 34:899–912
42. Ovtcharova J, Jasnoch U (1994) Featured-based design and consistency management in CAD applications: a unified approach. Adv Eng Softw 20:65–73
43. Penoyer JA, Burnett G, Fawcett DJ, Liou SY (2000) Knowledge based product life cycle systems: principles of integration of KBE and C3P. Comput Aided Des 32:311–320
44. Ramakrishnan R, Gehrke J (2000) Database management systems. McGraw-Hill, Boston
45. Shah JJ (1988) Feature transformations between application-specific feature spaces. Comput Aided Eng J 5:247–255
46. Shah JJ (1991) Assessment of features technology. Comput Aided Des 23:331–343
47. Shah JJ, Mantyla M (1995) Parametric and feature-based CAD/CAM concepts, techniques and applications. Wiley-Interscience, New York
48. Shirinivas SG, Vetrivel S, Elango NM (2010) Applications of graph theory in computer science: an overview. Int J Eng Sci Tech 2:4610–4621
49. Smit MS, Bronsvoort WF (2009) Integration of design and analysis models. Comput Aided Des Appl 6:795–808
50. Tang SH (2007) The investigation for a feature-oriented product database. PhD thesis, School of Mechanical and Aerospace Engineering, Nanyang Technological University, Singapore
51. Tang SH, Ma YS, Chen G (2004) A feature-oriented database framework for seb-based CAx applications. Comput Aided Des Appl 1:117–125
52. Tornincasa S, Monaca FD (2010) The future and the evolution of CAD. In: 14th international research/expert conference: trends in the development of machinery and associated technology, Mediterranean Cruise
53. Turk Z (1993) Object-oriented modeling and integrated CAD. Autom Constr 1:323–337
54. Uddin MM, Ma YS (2011) Towards a feature-based and fine-grain product repository for heterogeneous computer-aided systems. Enabling manufacturing competitiveness and economic sustainability (4th CARV2011 Proceedings). Springer, London
55. Wang Q, Zhu JY et al (1995) An intelligent design environment for conceptual process design. Eng Appl Artif Intell 8:115–127

Features and Interoperability of Computer Aided Engineering Systems

Yanan Xie, Jikai Liu, Hongyi Liu and Y.-S. Ma

1 CAx Systems, Customization, and Application Development

1.1 Introduction to CAx Systems

CAx is a broad term that means the use of computer technology to aid in the design, analysis, and manufacture of products. CAx usually includes computer-aided design (CAD), computer-aided engineering (CAE), computer-aided manufacture (CAM), computer-aided process planning (CAPP), and product data management (PDM) [78].

In the design process, increasingly more tasks have been supported by CAx tools in the last 30 years. Starting with drafting and surfacing, classical mechanical design was gradually replaced by 3D wire frame, solid modeling, and parametric and feature-based design. Today the entire product creation process, including production preparation, is run with CAx. According to the various application stages, CAx systems were developed with different computer solutions, such as computer-aided styling (CAS) [98], computer-aided esthetic design (CAAD) [76], computer-aided conceptual design (CACD) [97], and so on. All these technologies are categorized as different aspects of CAD/CAM and CAE, two of the more important CAx technologies, were developed almost independently. The latter is mainly used in a limited sense for simulation and finite element analysis (FEA). Although started independently as separate packages, both technologies require geometry data input from CAD.

Y. Xie · J. Liu · H. Liu · Y.-S. Ma (✉)
Department of Mechanical Engineering, University of Alberta,
Edmonton, AB T6G 2G8, Canada
e-mail: yongsheng.ma@ualberta.ca

Y.-S. Ma (ed.), *Semantic Modeling and Interoperability in Product and Process Engineering*, Springer Series in Advanced Manufacturing,
DOI: 10.1007/978-1-4471-5073-2_6,

1.2 Function and Data Management of CAx

Advanced CAx tools merge many different aspects of product lifecycle management (PLM), including design, FEA, manufacturing, production planning, virtual product testing, product documentation, and product support. With the growing integration of these CAx tools, data and information management has become increasingly important to realizing those expected industrial benefits. Currently, the complex network of CAx systems and their various data cannot be handled without a product data management system (PDMS). PDMS was regarded as the backbone of modern product development and now is extended to support the whole product lifecycle. This new paradigm of coherent multistage and multi-view information management has led to a wave of research effort labeled PLM.

1.3 Main CAx Software Tools

CAx software tools have been produced since the 1970s for a variety of computer platforms. A landscape of the main CAx software tools is shown in Table 1. A kernel is the brain of the CAD application. A modeling kernel is a collection of classes and components that comprise mathematical functions performing specific modeling tasks [102]. Currently, in industry, CAD applications are usually generated from a commercially available kernel. AutoCAD, NX [64] and CATIA [13] use their own kernels, while most other applications use either ACIS from Dassault Systèmes [86] or Parasolid from Simence PLM [64].

Table 1 The main CAx software tools

CAD	AutoCAD
	Autodesk inventor
	NXCAD
	Solid edge
	Pro/ENGINEER
	CATIA
	Solid works
CAE	NX Nastran
	CATIAABAQUS
CAM	NXCAM
	CATIA
PDM	Team center
	Windchill

1.4 Customization

As CAx systems are so widely used in nearly every industry, deploying the right computer solution for each aspect of the engineering workflow demands exact data structure matching for information exchange and well-planned procedures to streamline the execution of computer functions. Industries that produce medical devices, machine tools, apparel, as well as those specialized engineering fields such as metrology and ship-building, are characterized by the need for special CAx software for specific functions. Rather than using the common "as-is" versions of CAx software tools, progressive companies often develop their own versions as a way to implement the required differentiation in the product development cycle. Such customized solutions can accelerate new process chains, and improve the final customer experience. Many software platforms like NX (see Fig. 1) and CATIA [13] offer customized solutions for specific CAx process chains.

The hierarchy within the CAx system is shown in Table 2, which illustrates the four levels of composition in the typical CAx application chain. Levels 1 and 2 are developed by various vendors as packaged commercial products. They are vendor-dependent and do not differ much according to customer requirements, with only limited customization features for user interfaces and user-defined templates. Advanced solutions are categorized as "extended application modules" in level 3 and "tailored solutions" in level 4.

In level 3, users select those extended application modules offered by the vendors according to more specific application areas. For example, MoldWizard will be selected for plastic injection mold design. There are hundreds of choices in this level: within the Siemens NX suite alone, there are many such modules offered, including machining application modules such as 3-Axis Machining, 5-Axis Machining, CAD for Numerical control (NC) Programming, Data Exchange, High Speed Machining, Machining Simulation, Multi-Function Machining, NCData Managament, Part Planning, Post Processing and Post Processor Library, Programming Automation, Resource Management, Shop Documentation, Wire EDM, and more [64].

Most platforms offer open API to support secondary development. The customization in level 4 therefore mainly comes from the connection with customized solutions. The development of customized solutions depends on the real needs of consumers and usually has an evident economic advantage. Further discussion on this topic can be found in Sect. 1.5.

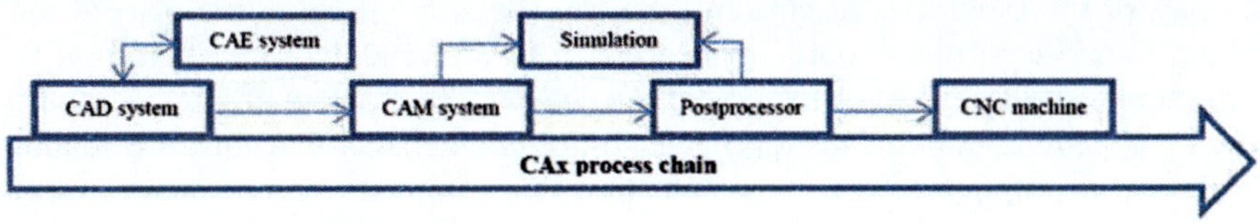

Fig. 1 Partial NX CAx process chain

Table 2 Hierarchies of a CAx system

Level	Composition	Description
4	Tailored solutions	Specific to end-user companies
3	Extended application modules	CAM for milling, mold wizard, etc.
2	CAD/CAM/CAE	Platforms such as NX, Pro-E
1	Kernel	Core kernels like parasolid and ACIS

1.5 Application Development

Application development based on the CAx system is a programming- and research-intensive process. Numerous applications have been developed and widely used in recent years, but this still cannot satisfy users' needs. Currently, most CAx software packages offer application programming interfaces (APIs) to satisfy application development needs [66]. An API is a source code-based specification intended to be used as an interface by software components to communicate with each other. An API may include specifications for routines, data structures, object classes, and variables. For example, CATIA and NX both offer their own open API for application development.

CATIA V6 can be adapted using Visual Basic and C ++ programming languages via component application architecture (CAA). CAA is Dassault Systèmes's comprehensive, open-development platform that enables developers to integrate their solutions. This collaboration expands Dassault Systèmes's system offerings and gives customers a larger set of CAx solutions to meet their specific industrial needs [14].

Pro/TOOLKIT is an API that allows Pro/ENGINEER functionality to be augmented and/or customized to meet the specific needs of PTC's customer base by using the "C" programming language. Specifically, Pro/TOOLKIT provides the ability to customize the standard Pro/ENGINEER user interface, automate processes involving repetitive steps, integrate proprietary or other external applications with Pro/ENGINEER, and develop customized end-user applications for model creation, design rule verification, and drawing automation [71].

2 Interoperability Among Systems

As part of the trend toward customizable and flexible product development tool suites, multiple software tools—even from different vendors—are often used for various phases of product development. As a result, the problem of interoperability among systems emerges. This reflects the unfortunate reality that software vendors tend to use proprietary data representation as a competitive advantage, which severely inhibits interoperability.

To solve the interoperability problem, it is necessary to identify two macro domains of application: horizontal data exchange and vertical data exchange [7]. Horizontal data exchange is taken to mean data exchange between different CAD systems, mainly focusing on geometric information. However, more important than just geometric information is the design intent, which is stored in design history and constraints. This makes the design intent reservation during data exchange something of a hot issue. At present, some commercial tools for CAD geometry data exchange are available, such as a conversion engine in CrossCAD [96], which can import, analyze, heal, and export models across CAD systems.

Currently, it is common practice to have a design geometry data model created from CAD systems translated into an intermediate data format like the STEP file format, and then imported into CAE and CAM applications. Conceptually, such data exchange is referred to as vertical because the data is transferred from the upstream application into downstream applications. In selection of the intermediate data format, there are two options: a proprietary data format or a neutral data format (NDF) [99]. Most commercial data exchange service providers tend to use a proprietary data format, which offers a competitive edge. For instance, NX and its PLM) solution Teamcenter are aimed at offering a complete solution from design to manufacture [64]. In contrast, an increasing number of industry companies have adopted neutral formats for data exchange; in the course of data exchange technology development history, several international standard data formats were proposed, such as IGES, PDES, and STEP [67]. Among these standards, STEP is the most advanced and complex, covering almost all the applications used in each product lifecycle phase. This topic is discussed in more detail in Sect. 3.

In contrast to file-based data exchange, recent research [7] has attempted to create a flat interapplication data service scheme that enables various engineering applications to share their models via the use of API functions. This approach is referred to as interface-based horizontal data exchange (see Fig. 2). Typically, client–server architecture is used for such a system: the CAD system provides its functions and data models via a coordination server, while the downstream applications receive services as subscribers. Bianconi et al. [7] summarize the advantages of such a system as data centralization, synchronization, and encapsulation.

2.1 Review of Interoperability and Related Technologies

The interoperability gaps among different computer-aided tools are well recognized across engineering domains, which call urgently for systematic integration to enhance interoperability and, hence, benefit the lifecycle. From the point of view of concurrent engineering, interoperability among applications can be enhanced on three levels: knowledge, information, and data [61]. As illustrated in Fig. 3, a NDF (such as IGES, STEP, or IDEF) provides standards to facilitate data sharing and

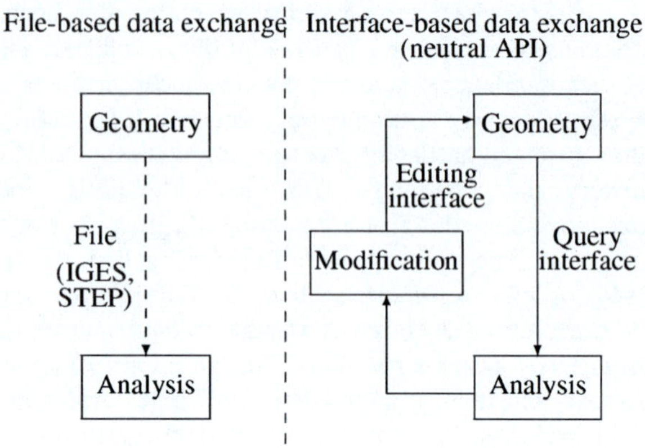

Fig. 2 Interface-based horizontal data exchange [7]

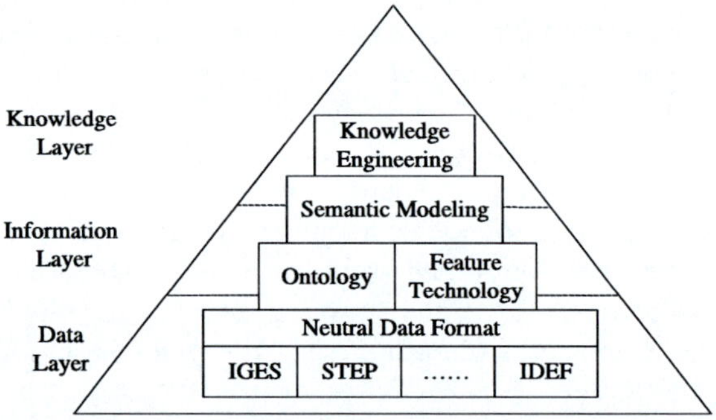

Fig. 3 Data representation pyramid for interoperability

exchange from the bottom data layer. Semantic modeling, a methodology that can effectively support knowledge engineering and feature knowledge, offers a flexible and scalable way to enhance interoperability.

2.2 Neutral Data Format

From the data layer, NDF provides a standardized intermediate data model to facilitate data exchange and sharing. An illustration of data transfer via NDF between computer applications among various domains is shown in Fig. 4. The purpose of NDF is to transfer data from all applications into a NDF, which requires the development of pre- and post-translators for each computer system involved to enable data transfer [67]. This significantly reduces the number of interfaces needed, as well as development effort and maintenance complexity. Based on this neutral data translation approach, any future potential development of more advanced application integration within a broader collaboration environment will be made feasible and efficient, as only the interface between NDF and the new

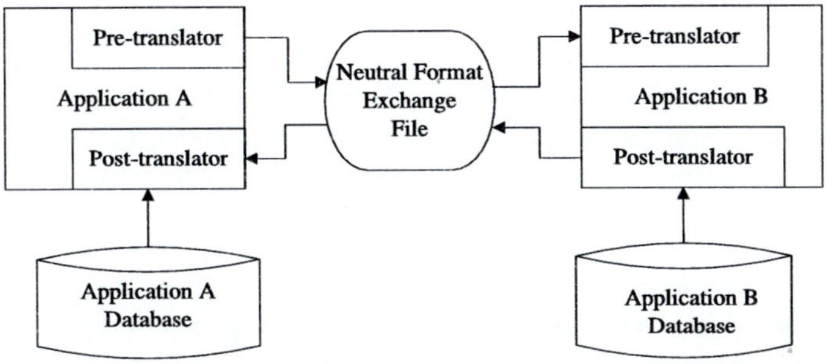

Fig. 4 Data transfer between computer-aided tools

application needs to be developed. Three foremost NDFs are IGES, STEP, and electronic design interchange format (EDIF), which are elaborated in the following subsections.

2.2.1 Initial Graphics Exchange Specification

The Initial graphics exchange specification (IGES), the foremost NDF, is a standard for graphics information exchange between CAx systems. It is designed to be independent of any computer systems but is capable of capturing all the information existing in the CAx applications, including binary information, start, global, directory entry, parameter data, and terminate sections [67].

Although efforts spent on improving capability with solid modeling has gained some results in IGES versions 4.0 and 5.0, the deficiency in solid modeling is still not significantly improved, which always leads to loss of information during the process of data exchange and sharing. The existence of the Standard for the exchange of product data model (STEP) reduced the urgency of further development and made IGES version 5.3 the last published standard in its series in 1996.

2.2.2 Standard for the Exchange of Product Data Models

STEP (ISO 10303), a standard for the representation and exchange of engineering product data, makes it possible to develop a complete and integrated product description in a NDF and, hence, to facilitate interoperability among different computer-aided systems throughout the product development lifecycle [40]. The standard is also known as the STEP. It is organized in a series of sections, or application parts (APs), covering the representation of product information (including components and assemblies) as well as the exchange of product data, which provides the capability of describing data throughout the lifecycle independently of any particular computer system.

STEP was developed as an alternative to IGES and boasts a more comprehensive set of definitions [87] for a set of neutral product information entities, especially for geometric ones. STEP provides a mechanism that describes a complete and unambiguous product definition throughout the lifecycle of a product, independent of any computer system. This international standard is accepted by most vendors, so it is quite suitable for use in realizing data interoperability. Users can implement proper APs to meet their product data exchange requirements [39]. Some APs are listed in Table 3, with their roles in integrating manufacturing activities shown in Fig. 5. Users can implement proper APs to meet their product data exchange requirements [39, 40].

APs across the disciplines of chemical, mechanical, and electrical engineering are illustrated in Fig. 6. AP 221, "Functional Data and Their Schematic Representation for Process Plants," specifies an exchange scheme that is applicable to chemical process projects [46]. The scheme describes the data structures used for communicating functional design and engineering specifications of system components, which can also be used for subsequent procurement and component manufacturing generally carried out by engineering, procurement, and construction (EPC) companies. Reference data, which comprises instanced templates in the form of library elements, is designed to be referenced together with the data structure specifications to facilitate collaborative system design across disciplines [3].

Another important application protocol, AP 227 ("Plant Spatial Configuration"), provides a standard for exchange among engineers from different disciplines as well as operation owners and EPC companies during the lifecycle of chemical process projects. In this AP, the information requirements for the exchange of design and layout models of a process plant are specified. It also specifies other integrated engineering resources, such as those models required for design, analysis, and fabrication of piping components and piping systems [44]. The exchange of functional characteristics of heating, ventilation, and air conditioning (HVAC), mechanical and piping components and systems, and schematic representation are also addressed. Similarly, it is specified in AP 231, "Process Engineering Data: Process Design and Process Specifications of Major Equipment," that the representation of process steps involved in a chemical process be

Table 3 STEP application protocols

AP	Title
AP204	Mechanical design using boundary representation
AP207	Sheet metal die planning and design
AP209	Composite and metallic structural analysis and related design
AP219	Managing dimensional inspection of solid parts or assemblies
AP223	Exchange of design and manufacturing product information for cast parts
AP224	Mechanical product definition for process planning using machining features
AP238	Application-interpreted model for computerized numeric controllers
AP240	Numerical control process plans for machined parts

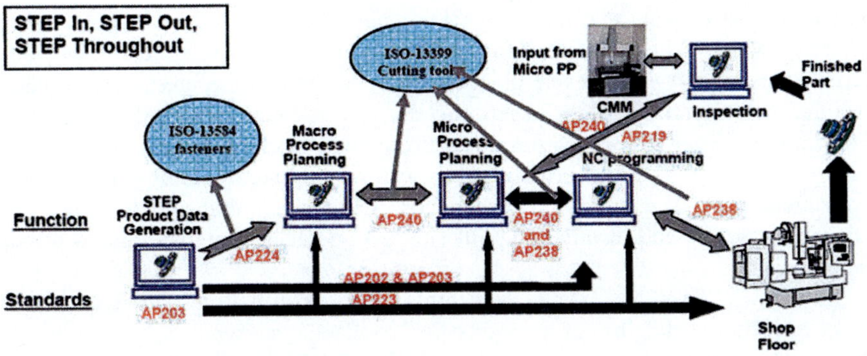

Fig. 5 Data exchange APs based on STEP [39]

Fig. 6 Application protocols providing lifecycle support

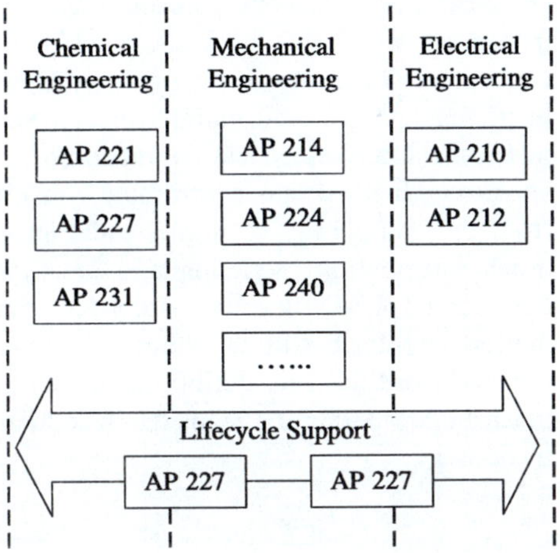

included, along with material and reaction data, process flow diagrams (PFDs), and detailed process and plant descriptions [3].

Within the mechanical engineering domain, there are even more APs defined in STEP, such as APs 203, 204, 214, 224, and 240. Application protocols for information representation and data sharing within mechanical engineering among CAx systems are addressed in these APs. This provides a neutral file exchange support independent of any particular computer-aided system.

Thanks to the collaboration between the International Organization for Standardization (ISO) and the International Electrotechnical Commission (IEC), STEP also defines such APs as AP 210 and AP 212, which are applicable to projects that span both mechanical, electrical, and electronic engineering domains. AP 210, "Electronic Assembly, Interconnect, and Packaging Design," specifies the transformation from detailed requirement data (such as functional descriptions of the device, manufacturing processes required and other design specifications) into data structures and formats that are analyzable and processable by the manufacturing systems, which will facilitate information-sharing between engineers from different domains [48]. However, this AP is not limited to the electronic domain. Another AP specified in the electronic domain, "Electrotechnical Design and Installation," (AP 212) specifies the information shared between the systems involved in the design and installation as well as the commissioning of electrical equipment [48]. The data specified in this AP, such as that which pertains to equipment design and installation, will effectively enhance interoperability across domains.

AP 232, "Technical Data Packaging Core Information and Exchange", addresses the packaging of products for exchange as well as the exchange requirements of product data groupings [42]. AP 239 is another member of the application protocol series specified for product lifecycle support, which can be applied to developing an integrated series of interfaces to provide data interchangeability within one domain or across disciplines [45]. These APs can be used to support the lifecycle of chemical process engineering projects (from process conceptual design, engineering analysis, and mechanical engineering and design) to implementation and maintenance, which involves the disciplines of chemical, mechanical, and electrical engineering.

Although STEP has contributed considerably to interoperability, it still suffers from the complexity of implementation in real applications, which requires too much information modeling and development [61]. Restricted by the APIs of commercial CAD systems, which are not designed for model exchange, information associated with the models, including the design intent, is very likely to be stripped from the data during the exchange process [52]. Hence, the stalled effort toward interoperability needs the introduction of a new technology, such as feature technology.

3 Current Standards' Limitations

Among the standards established for product information exchange, IGES and STEP, which have been set up and maintained by American National Standards Institute (ANSI) and ISO, respectively, are the most powerful and widely accepted. In this section, the advantages and limitations of both are described in detail.

IGES, a data format depicting product design and manufacturing information, mainly assists data exchange between CAD and CAM systems. IGES is independent of all CAD and CAM systems, and is thus an NDF [67]. As IGES was established in 1980, it has some (if limited) capability for the exchange of points, edges, and surface entities, but cannot handle solids at all; IGES, therefore, cannot support the full range of CAx entity chain processes. Ideally, data exchange should be supported across the product lifecycle [51]. Although IGES has obvious drawbacks, it still has 20 % of the usage level in the CAD data exchange field in

North America due to the simplicity of its implementation. In comparison, STEP only has about 15 % [37].

The best-developed application of STEP is the achievement of CAD/CAM and computer numerical control (CNC) integration. The relevant standard is ISO 6983 [38]. Recently, new standards have been developed, such as ISO 14649 [43] and ISO 10303 AP 238 [47]. They provide CAM and CNC vendors the opportunity to develop highly intelligent CNC controllers, which can realize bidirectional communication of standardized geometric and manufacturing data in the form of features [106].

For future research, new standards or new versions of existing standards should better support engineering semantics. Product ontology representation should be exchangeable for interoperability among information systems across the product lifecycle [90]. We can see that IGES and STEP have a common limitation in their inability to transfer design intent, such as construction history and constraints. Research efforts [52, 70] are increasingly devoted to this issue, but it has still not been adequately addressed.

4 Feature Technology

Feature-based product modeling was traditionally used for geometrical construction with certain predefined templates, and most CAD tools embraced this approach to facilitate interaction with designers. In this kind of CAD product model, geometric features are the basic components for building up the shape: a variety of features ultimately constitute a complete product in a hierarchical structural model. A *feature* encapsulates the engineering significance of a portion of the physical constitution of a part or assembly, and hence is important in product design and definition for a variety of computer-aided systems [77].

Generally, the feature-based approach uses a set of basic features as a starting point, then adds other advanced and user-defined features to enhance application-specific knowledge encapsulation and process automation. Moreover, if features are integrated with parameters and other features, they can be tracked. Numerous research efforts have been devoted to the feature-based design approach [108]. Monedero gives a basic definition of the integration of parametric design and modeling [63].

Ideally, when a feature's parameters change in a feature-based system, the other related features will be changed accordingly. This is the functional superiority of such an associative feature approach as compared to other procedural approaches. With the progress of feature technology, research has merged into two mainstream methodologies: Design by feature (DBF) and Feature recognition (FR) [11, 79].

DBF is a design modeling method used for pattern-based functionality and manufacturing geometric entities, in which the model is built in terms of features provided by an existing feature library [1]. One of the major challenges in applying the DBF method is that the limited and rigid definitions of available features constrain the creativity of designers. It is also impossible to predefine all design and manufacturing features.

FR is a method from an opposite perspective: instead of designing from features, it aims to recognize features from an existing geometry model. This method is mainly used for manufacturing purposes after CAD design models have been created and before CAM tool path generation is applied based on said models. FR is further elaborated in Sect. 5.1.

5 CAD/CAM Integration via Features

As yet there is still no consensus on a definition of a feature, but feature technology has been widely applied in the integration between CAx systems, such as CAD/CAM and CAD/CAPP [2, 109]. Feature-based CAD/CAM integration is a technology used to realize automatic transmission and conversion of product information among CAD, CAPP, and CAM systems [29, 78]. While it is true that the CAD/CAPP/CAM systems are at a maturation stage individually in their traditional functions; but because they have been developed separately, they emphasize their separate functionalities. They use different data models and formats, which severely inhibit product information exchange. (This problem is discussed in detail in Sects. 2 and 3). In this section, we will focus on the conversion of a design model to a manufacturing feature model using feature technology. A new trend of CAD/CAPP/CAM/CNC application integration is introduced in detail as well.

At present, most commercial CAD software tools support both solid modeling and feature-based modeling. A product model is usually constructed with the convenience of geometric construction, with features available in the CAD packages. When manufacturing processes are to be defined with the existing CAD models, the challenge of data reuse occurs. The majority of CAM tools on the market are feature-based, and certain specifically defined manufacturing features have to be used to define the processes that enable associativity with cutters, machine tools, jigs and fixtures, tool paths, and process conditions.

FR was the first challenge for feature applications in a domain based on CAD technologies, even though so far significant progress has been made. When a product is modeled simply in solids, FR is used to acquire engineering semantic features, such as manufacturing features, from CAD models. When the CAD model is created using a hybrid method of solid modeling and feature-based modeling, FR is still necessary to identify the manufacturing features. This is due to the fact that the design features will not be applicable in the manufacturing model because of the different definitions between design and manufacturing features and incomplete definition of geometry by design features. The interoperability problem between these design and manufacturing domains have been commonly recognized in industry due to historical CAD/CAM technology evolution. That is why FR plays a key role in feature-based CAM [34]. More elaboration follows in the next section.

However, inherited problems also exist in the FR approach, due to the restrictions of the hard-coded feature patterns to be recognized. To address the interoperability issue among computer-aided solutions at the feature level, other techniques have recently begun to emerge. In theory, with recent research progress in advanced feature-based modeling scenarios, if the CAD model is created with well-defined design features, there are two options. If the association between the CAD model and the manufacturing model is not required, then the FR approach can still be applied to generate manufacturing features from the resulting part solids. On the other hand, if the associativity is to be kept for updating future changes, then feature conversion is expected to map the design feature model to a manufacturing feature model. Feature conversion is also further discussed in Sect. 5.2.

5.1 *Feature Recognition*

FR is an interpretation of a geometric model to identify features [10], and can be achieved by the user interactively or automatically by algorithms. With the user-driven approach, the user can select certain entities in the parametric model to define a feature [108]. For example, a user will pick three imaginary faces and two real faces to define a notch in the boundary representation (B-rep).

However, to realize complete CAD/CAM integration, automatic feature recognition (AFR) is necessary. AFR is the process of matching the parametric model with the predefined generic features to identify features. Babic et al. [1] specified three interrelated tasks which are necessary for AFR: geometric feature extraction, part representation formation suitable for form identification, and form feature matching. The specific tasks in this process are searching the database to match geometric patterns; extracting recognized features from the database; determining feature parameters; and completing the feature geometric model [77].

5.1.1 Rule-Based Methods

Rule-based methods) use production rules to depict features. The rules show the necessary conditions for the elements in the model such as convexity, perpendicularity, or adjacency. The expert system then uses these rules to perform the FR [10, 34]. Rule-based methods) were the initial ideas for FR. However, they have such obvious drawbacks that writing rules for all the features is a huge project and recognition will consequently be very slow.

5.1.2 Graph-Based Methods

A common data structure for B-rep models is the graph [10], especially the face-edge graph as shown in Fig. 7. Nodes represent the faces in the model and links of

Fig. 7 Face-edge graph for feature recognition

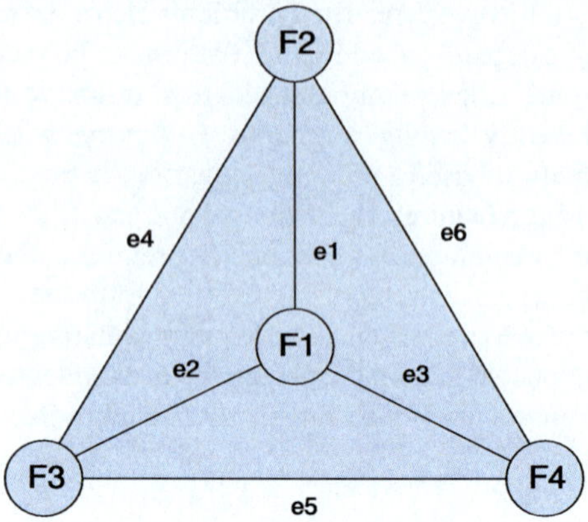

nodes represent the edges between the faces. The properties of the links represent the adjacency relations between the faces. In this way, the graph-matching method realizes FR [34]. Graph-based methods are currently the most frequently used FR technique, largely due to their efficiency. A variety of approaches with respect to each task are classified in Table 4.

A comparison between DBF and FR is illustrated in Table 5. Although great research effort has been spent in this field with a certain amount of progress, there are still some limitations associated with feature technology. Especially among computer-aided systems, a multiple view for engineers across different disciplines is required to support interoperability. To further extend the capability of feature technology in the enhancement of interoperability among different computer-aided systems, some new technologies (such as associative feature, unified feature,

Table 4 Classification of AFR approaches [1 CiI]

Form feature extraction		Pattern recognition
Geometric feature extraction	Form feature identification	
1. External approach 2. Internal approach	1. Syntactic pattern recognition 2. State transition diagrams and automata 3. Logic (if–then) rules and expert systems 4. Graph-based approach 5. Convex-hull volumetric decomposition 6. Cell-based volumetric decomposition 7. Hint-based approach 8. Hybrid approach	Logic rules
	1. Graph-based approach 2. Face coding 3. Contour-syntactic approach 4. Volume decomposition	Artificial neural networks

Table 5 Comparison between design by feature and feature recognition in manufacturing

	Advantage	Disadvantage
Design by feature [77]	• Rich information associated with models • Real-time manufacturability evaluation • Concurrent design and process planning	• Restricted design creativity by the feature availability • Limited and specific features • Complex and undefined features resulting from feature interactions
Feature recognition [53]	• No need for input feature information • No constraint to design creativity • Possible automatic recognition	• Complex recognition algorithms • Limited features to be recognized • Input required for associated manufacturing information

multiple view feature, and semantic feature modeling) have been introduced, all of which are covered in the scope of semantic modeling [8, 11, 19, 61].

5.2 Feature Conversion

Feature-based design is a relatively new approach for CAD/CAM integration [74]. For FR, it is a process that transfers low-level parametric models into high-level features. If we set up the product model with features initially, the extraction of features will be quite easy. There are several requirements for a feature modeling system, as follows:

1. The system must be interactive and graphical, as this is the best way to support the modeling system.
2. There must be a library for the storage of generic descriptions of features, and a mechanism to create instances of features by specifying the features.
3. Constraints must be represented and maintained consistently to guarantee the validity of features.

Design features usually consist of form features coupled with functions, design intents,, and other design-related information. As mentioned above, manufacturing features consist of special form features coupled with distinctive machining operations and other manufacturing-related information [29, 30]. Such different feature domains are supposed to be associative, to cater to constant change throughout the product lifecycle; hence, after design modeling, there is a need to convert the CAD feature model into a CAM feature model. Much research has been dedicated to this conversion process [74], which can be divided into three parts: form feature mapping, dimension mapping, and the mapping of other attributes such as tolerance or surface finish; more details can also be found in the work of Gao et al. [29].

5.3 Feature Interaction

Most research in this area has focused on simple FR and conversion, while feature interaction is often encountered in practice and causes difficulties in FR and conversion. Some researchers have tried to recognize composite features as a combination of simple features, but have not resolved the issue completely [69, 94]. Lee [55] concludes that are based on nine kinds of simple features: step, blind step, slot, blind slot, pocket, hole, wedge, fillet, and sector, and brings out the projective FR algorithm to recognize composite features. Gao et al. [30] propose a mathematical description of the feature mapping process to solve the feature interaction problem. However, all these works can only partially address the problem, and feature interaction remains an extremely challenging research issue.

5.4 CAD/CAPP/CAM/CNC Integration

At the end of the twentieth century, most research effort had been put into the integration of CAD/CAPP/CAM with the neutral file standard and feature technology. However, for a complete manufacturing process chain, the critical interoperable connection problem from CAD to CNC is still not fully solved.

In traditional CNC manufacturing, control is based on axis-movement description programming techniques (G and M codes) [38] which cannot perfectly support the advancement of CNC machines. Hence, machine manufacturers add their proprietary instructions to the standard [105]. Consequently, there is a need for specific post-processing programs for different configurations of CAM tools and CNC machines, which is a big obstacle for the interoperability of CAM/CNC applications. Furthermore, CAM systems add manufacturing-related information to the design, such as machine processes, tools used, and operations. However, after post-processing, the output files are NC programs, which are only machine-interpretable by CNC machines. The result is that the date translation is a one-way process, involving huge information loss [99].

STEP greatly helps the CAD/CAPP/CAM integration process, and STEP-NC has been developed with the aim of better CAM/CNC integration. STEP-NC is a standard specifically for NC programming, helping to achieve the goal of a standardized CNC controller and NC code generation facility. This standard has two notable advantages: first, STEP-NC is vendor-independent: if the vendors accept this standard, then a neutral data is achieved for exchange; second, STEP-NC files have the data regarding "what to do" instead of "how to do," which are easily accepted by different intelligent CNC controllers [99]. There are three types of STEP-compliant CNC: (1) conventional control; (2) new control; and (3) intelligent control [92, 105].

For conventional control, STEP-NC translators read the STEP-NC file and output an NC file, which is similar to post processing. It has achieved partial interoperability, as different configurations can be connected using the same neutral file. With this method, the CNC machines do not need to be retrofitted [92]. The new CNC controllers can process the STEP-NC file inside the CNC machine and then convert it into NC programs which consist of G and M codes. At present, this method does not have many intelligent functions, and most researchers are building the CAM/CNC integration in this way [91, 106]. Xu and Wang [104] proposed a G code-free machining procedure. In their system, the CNC controller will make full use of the information stored in the STEP-NC file, such as the work plan, work step, machining features, and cutting tools, to work out the NC codes using its own programmable control language.

Intelligent control is the most promising type for STEP-compliant CNC. As both design and process plan information are stored in the STEP-NC file, many intelligent functions can be achieved by the CNC controller [105]. Suh et al. [91, 92] have developed a new CNC controller called STEP-CNC. It uses the STEP-NC file as input and can realize intelligent machining control functions, such as decision making for unexpected changes, program validation at the time of execution, monitoring, and recovering.

6 CAD/CAE Integration

CAD systems are commonly used for modeling the geometry of a product with a variety of tools; CAD geometry is used as input for FEA in CAE. CAD and CAE data models are always different from one another, due to the nature of the operations they carry out. In order to decrease the length of the product development cycle, the integration of CAD and CAE is in high demand; numerous efforts have been made in recent decades. Ideally, the integration of CAD and CAE will efficiently decrease the design cycle time, reduce cost, and simplify the fine-tuning process for the product; Gabbert and Wehner did a feasibility study on CAD/FEA integration as early as 1993 [28], and many researchers continue to work toward a seamless integration of CAD and CAE systems, without yet to achieving a satisfactory result. Gordon [31] summarized CAD/CAE integration into three approaches: geometry conversion, CAD-centric geometric modeling, and CAE-centric geometric modeling.

In the geometry conversion approach, CAD geometry is used and then converted into simulation mesh geometry. In this approach, the same geometric source is used in both design and analysis. However, this type of integration can only be used with simple parts, such as pipelines.

The second approach uses a CAD-centric geometric model. In this type of integration, the CAD solid models contain too many details that are not suitable for the CAE-required abstracted models. There thus needs to be an idealization process that includes detail removal and dimensional reduction. Currently, the idealization process is a major obstacle for CAD/CAE practice. However, due to the ease of access to the modern 3D feature-based CAD technology, researchers tend to use this type of integration.

The third approach can be classified as a CAE-centric geometric model. The simulation model is built first, and is based on the design concept and analysis method. After analysis, verification, and modification of the simulation feature model, designers are to work out full details and manufacturing features to support downstream process planning. This type of integration is recommended by Gordon but requires analysts to know upfront about the product's function details.

The subsections below are intended to introduce the key technologies and remaining problems related to integration. First, data interoperability issues between CAD and CAE are discussed, followed by an introduction to geometry transformation practice. Recent feature-based integration research is also reviewed in detail. The basic concepts of three specific methods—the multimodeling method, common data model method, and analysis feature method—are introduced along with the visual framework structures. The benefits, technological improvements, and limitations of these three methods are discussed as well. Though the feature-based product method for CAD/CAE integration is in development, there are still some gaps to be filled.

6.1 Data Interoperability Between CAD and CAE Systems

Many researchers have tried to build an integrated CAD/CAE data model. Remondini et al. [73] developed a unified data model supporting both design and structure analysis activities, which can build the bridge between the CAD model and CAE analysis. However, this model is restricted to the treatment of linear analysis. For interoperability, Foucault et al. [26] recommended using a polyhedral model as a transitional model between CAD and the finite element method (FEM). However, this method has to modify and update the product design model repeatedly during the product development evolution.

Semantically, data is the lowest level of information, which is used in software to represent different kinds of information in product design. STEP standard AP 209 provides a means to build an integrated model, including nominal geometry (CAD), various idealized CAE geometries, and associated FEM analysis models and results, along with PDM and separate version control [48]. Users can customize the combination as needed. Liang et al. [58] raise the idea of using an integrated product data-sharing environment (IPDE) based on STEP, to allow CAD/CAM/CAE programs from different vendors to share data conveniently. This model is mainly based on several STEP application protocols: AP 203, 209, 214, and 224. Among these, AP 203 specifies data structure definitions for the configuration-controlled 3D designs of mechanical parts and assemblies, while AP

209 supports the design elements through analysis of composite and metallic structure. However, implementation of STEP in CAD/CAE integration is still limited, because STEP has mainly been developed for CAD/CAM integration.

6.2 Geometry Transformation for CAD/CAE Integration

A common platform that can contain information from both the CAD and CAE sides has been a popular research topic since the early 1980s. Early research mainly focused on the idealization of CAD models and automatic mesh generation. A workflow of the CAD/CAE integrated modeling method has been suggested by Li et al. [57], as shown in Fig. 8. CAD models are usually set up to satisfy requirements for design, process planning, and manufacturing [23] and therefore usually contain too many complex details for CAE analysis. Hence, the models need to be idealized before they can be subjected to CAE application. In most cases, the idealization process will not affect the accuracy of the analysis but can reduce analysis time significantly. There are two main methods for CAD model idealization: CAD detail feature simplification and dimension reduction [5, 23].

6.2.1 CAD Detail Feature Simplification

Detail simplification is the process of removing those unnecessary detail design features that do not affect analytic accuracy or mesh quality but do increase analysis time. Detailed features refer to the small shape features of the product model, such as small fillets and minor local ribs. Usually, these small features have little influence on the analysis result compared with the overall parameters of the model. Ji et al. [49] group the detailed features needing to be removed into a number of types, including chamfer features, edge blending features, thread features, groove features, holes features, pad and boss features, and slot features.

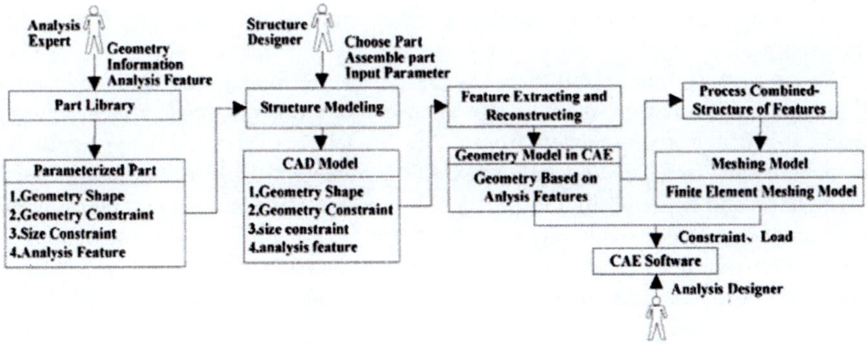

Fig. 8 CAD/CAE integrated modeling scheme [57]

To remove these features automatically, a simplification processing module must be built, for which functions include unwanted detail FR, selection, removal, and recovery. The process involves three steps: searching all the features in the model, determining the parametric information of the features by feature recognition, and carrying out rule-based simplification.

6.2.2 Dimension Reduction

In CAE systems, wireframes are used to represent beams and sheets for plates and shells. This kind of representation requires a process of abstraction. The mid-surface approach is suggested for this abstraction [24, 72], which usually involves three steps:

- *Face judgment*. Judge whether the two planes of the model can be faces. If yes, then the mid-surface is created between the faces.
- *Mid-surface modification*. This step modifies the mid-surfaces by removing those small facets that have little effect on the analysis results.
- *Extend and seam*. The mid-surface model is completed as a whole with extension and seaming of faces.

Currently, the idealization process still requires human interaction, while automatic mesh generation is realized by most commercial CAE tools [23, 27].

6.3 Feature-Based CAD/CAE Integration

As mentioned above, all three of the methods proposed by Gordon [31] require two separate models for one product, to the severe detriment of efficiency. Moreover, there are only geometric connections between these two models. The connections are thus a one-way process, and some semantic information needed for CAE analysis is lost [23]. Recently, researchers have been trying to develop a unified data model and concurrent modeling environment for seamless CAD/CAE integration.

Kao et al. [50] recognized that most of the feature models can be used in both CAD and CAE software, including geometric and non-geometric features. In their work, the changes propagated from CAD model to CAE model appear to be automatic. In fact, all the related CAD parameters had to be recalculated beforehand and exported into a spreadsheet, and the corresponding changes in the CAE model had to be made interactively. Chen et al. [16, 17] discuss semantics in information entities, relations, and constraints in each phase, and generalized common entities in order to develop a consistent product information model. In the course of their work, they then created a conceptual framework by applying the unified feature concept for CAD and CAx model integration [17].

Features, a form of well-defined data structure expressing engineering patterns associated to geometric entities and relations, are recognized as the basic and essential entities for product model design and interoperability between different types of software, such as CAD and CAE software. However, the feature association of design models between CAD and CAE is considered to be the main area of difficulty in terms of integration. For example, design form features used in CAD are usually represented geometrically, while features used in FEM have mesh and material data, which are derived from the imported CAD geometry without considering design features. These different types of features are easily confounded can cause mistakes in the design updating phase.

Multi-model technology (MMT) introduces object-oriented technology (OO) into the product modeling process and combines OO with feature-based modeling technology. It utilizes OO to create the object model of a product and uses feature-based modeling technology to build the model of an object. In this way, it can sustain system-level modeling along with design-level modeling. Because the object model consists of multi-models, it is called MMT [107]. The object model of the product is a multi-model structure (MMS), consisting of a finished part model level (assembly model level), a rough part model level (part model level), a function model level, and a basic model level [107]. Every model in the MMS is created by feature-based technology in the design process. With the help of MMS, CAD engineers and CAE engineers are expected to work concurrently, and the integration of CAD and CAE can be achieved. Figure 9 shows a visual structure of CAD and CAE interaction processes.

Lee [56] contributed to a feature-based multi-resolution and multi-abstraction modeling approach. This technology is realized using techniques such as

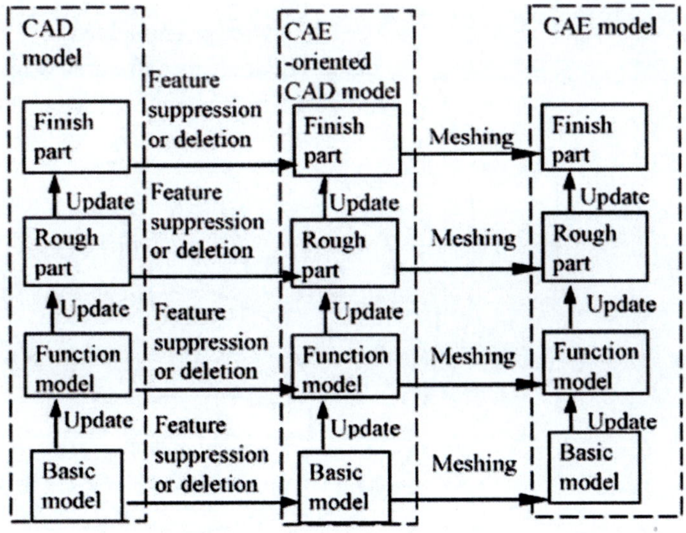

Fig. 9 The integration of CAD/CAE in MMT [107]

design-by-feature, non-manifold topological (NMT) modeling, multi-resolution solid modeling, and multi-abstraction NMT modeling. The CAD and FEA models are built up simultaneously into a unified master model, in which design and analysis features are embedded; this research supports the buildup of CAD and FEA models at multiple levels. There is a drawback to this method, however: boundary conditions such as load and displacement conditions cannot be transferred from CAD models into CAE models automatically.

With feature-based technology, if a product needs to be modified, a synchronization mechanism can be developed to update the FEM model with persistent connections between the CAD model and the CAE model. This could mean significant benefits for the product development process. However, this feature-based approach also requires higher knowledge and skill competency for the analysis engineers, who must initially extract useful analysis features from the CAD model to create the associated CAE model. The CAE engineer needs to give specific working condition definitions for such analysis features in the early stage.

Following the development of feature-based methodology, Gujarathi and Ma [33] tried to integrate CAD and CAE models using a joint data structure called a common data model (CDM). This CDM consists of semantic design parameters used in three ways: building the CAD model, building the FE analysis mesh model, and performing engineering analysis functions with the assistance of knowledge-based algorithms and software APIs. Figure 10 shows the basic concept of CDM [33].

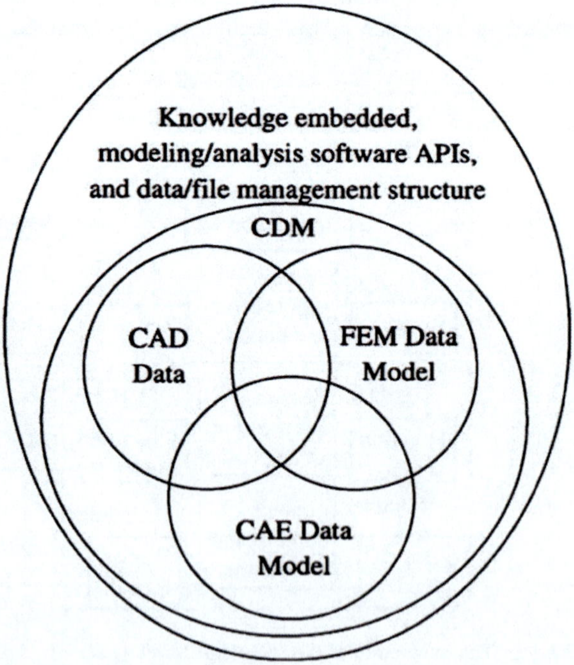

Fig. 10 General working aspects of CDM [33]

In the proposed method, the CDM is initially generated by an engineering concept calculation module which works out the key driving parameters within the engineering project scope. The CDM is used to store the initial conditions and final results of the engineering conceptualization result in parameters and their constraints. In this conceptualization procedure, basic physics/chemical principles are implemented and verified.

A structural designer then constructed the CAD models by retrieving templates from a part and assembly library; the templates provided input for design parameters, and defined the assembly relations among parts.

Third, product's FEM geometry information, based on the analysis feature information that was already embedded in the CAD templates, was constructed automatically in the CAE system. Since 1990s, a feature-centric CAD/CAE integration approach has been developed by a number of researchers [56, 57, 82, 110]. In an early effort [82], a part library with built-in analysis features was first established by an expert CAE engineer. In the proposed method [32, 33], geometrical CAE meshes are generated in sequence using an automatic meshing technique. The meshes of features are ultimately combined into a complete mesh model for the product, which is guaranteed by a structure-combining algorithm. Finally, CAE analysis is automatically carried out and the results are also recorded in the CDM.

This method offers centralized design parameters and data for CAD/CAE. The CDM method for CAD/CAE integration has two specific advantages. First, it supports parametric design and analysis in the integrated CAD/CAE environment. Second, CDM updates its parametric data dynamically over the processes involved in design consolidation. The content parameters in CDM largely belong to three general categories: geometric, non-geometric/functional, and intermediate design parameters. The structure of the CDM is shown in Fig. 11 [32].

The approach can also be extended to include engineering rules used in various models. These rules can then be consistently applied to multiple domains. Such a centralized "control board" enables an enhanced control mechanism that offers the flexibility of adopting and changing a variety of design codes, standards, and expertise in the cycle of design procedures. As a kind of parametric data model, CDM data can also be further customized to incorporate manufacturing requirements.

Although quite flexible, the proposed method has some limitations. The most difficult task in the initial phase of parametric CAD modeling is the associative relation model development. The initial identification of parameters and the logics of different kinds of relationships require considerable programming and set-up effort for automatic model generation and updating.

Further, the design procedure in the terms of computer system operations has to be fully defined beforehand in consideration of building the CAD and CAE analysis models with logical constraints. Therefore, the proposed method offers long-term efficiency only for those well-established, generic, and set design problems [33]. Efficiently or expeditiously conducting simulation for the testing of candidate solutions and demonstrating design scenarios for provisional customers

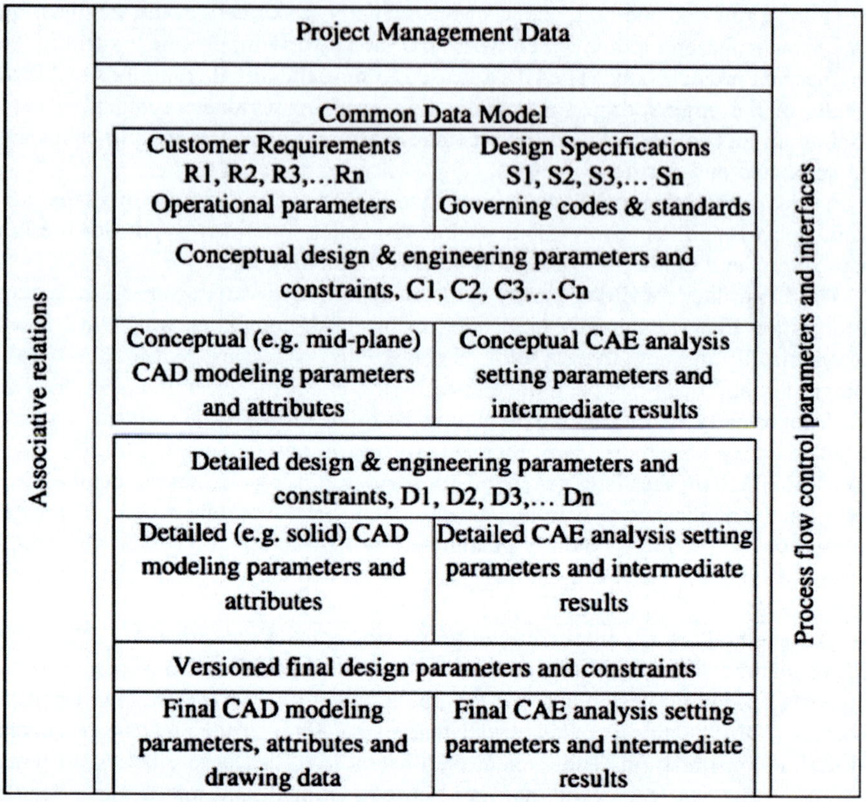

Fig. 11 Structure of CDM with the progress of the design process [32]

can be problematic because of the ad hoc procedures involved and the short response time required.

The authors believe the CDM integration approach [32, 33] has two contributions. First, the method abandons an oft-required feature reformation process (from CAD to CAE) that is still technologically immature, and instead binds the analysis features within the parameterized CAD model [82, 110]. More progress has been made to improve the reliability of obtaining the analysis feature from the CAD model [57]. Second, automatic mesh generation based on analysis features is a straightforward technique. The meshing method improves the quality of the mesh model. The improved method leads to better analysis results and shorter simulation time. In addition, more complicated component shapes and assemblies can be managed.

This common data model involves only the preliminary design and is limited to sizing and the essential operational concepts. Refinement of the model by adding more design features has to be carried out before it can be usable in day-to-day industrial applications.

7 Toward Feature-Based Integration and Interoperability in Chemical Process Engineering

The complexity of chemical process design and engineering requires engineers to work collaboratively across disciplines. Existing software tools have allowed engineers from disparate domains to deal with the complexity embedded in each specific domain. However, interoperability of the heterogeneous data generated by different tools remains a problem [62]. This reality highlights the urgent need to integrate the software tools involved in any given chemical process project to enhance interoperability, not only on the levels of syntax and structure, but also on the semantic level. As should be clear from the technology reviews in previous chapters, semantic modeling and feature technology have been adopted by researchers to construct a variety of integration frameworks. This section proposes one such framework, under which the semantic feature associations between two domains are analyzed and a new, more efficient design process is proposed based on the concept of collaborative engineering. A case study further demonstrates how the framework functions, and allows engineers from different domains to work collaboratively within it.

7.1 Integrated System Architecture for Chemical Process Engineering

The common project engineering practice in chemical process development involves multiple disciplines, such as chemical process engineering and mechanical engineering. Ideally, the engineering design efforts should be coordinated coherently, with close interactions among relevant disciplines due to their heavy dependency on one another. Traditional discipline-centric engineering practice and the relevant engineering software tools are becoming outdated, because networked computer information technology has made interdisciplinary collaboration much easier. To address this kind of industrial demand, the authors have proposed a system integration architecture [103] based on a common framework of semantic modeling and feature technology, and consisting of disciplinary modules such as mechanical and process design modules. The centrally unified feature management system (a common base module) consists of a product feature module and a process feature module, which are built on top of a networked federation of data repositories representing different disciplinary domains.

The improvement of the proposed semantic integration framework over the individual disciplinary engineering approach is that it incorporates semantic interoperability. The feature information will be retrieved from the files or databases generated by different domain software tools and mapped onto a central database according to the relevant semantic schema and a generic mapping mechanism. Based on the data collected, the central unified feature management system will generate a view according to domain-specific schema and display it to the domain-specific engineers through their respective user interfaces. An ontology library and a knowledge library have been developed to support semantic feature mapping. As shown in Fig. 12, Module 6, the central unified feature management is the core of the system, maintaining and validating all the features according to a unified scheme with the related mapping mechanism, and managing view generation and updates. In addition, a standard feature library can be established using the generic data feature structure specified by Xie et al. [103], which will facilitate semantic feature mapping and also reduce the modeling workload of mechanical engineers.

Due to space constraints, the framework provided here lists only a few of the software tools involved in chemical process engineering. There are more software tools used in industry, however, and these vary from company to company. Extensive software tools have been developed and applied in chemical process engineering as well. Within this framework, whenever a new version or a completely new software tool is created, only one new "translator" needs to be added into the shared interface library. This leads to a considerable decrease in the development efforts needed, as compared to merging different modeling schema into an integrated schema [4].

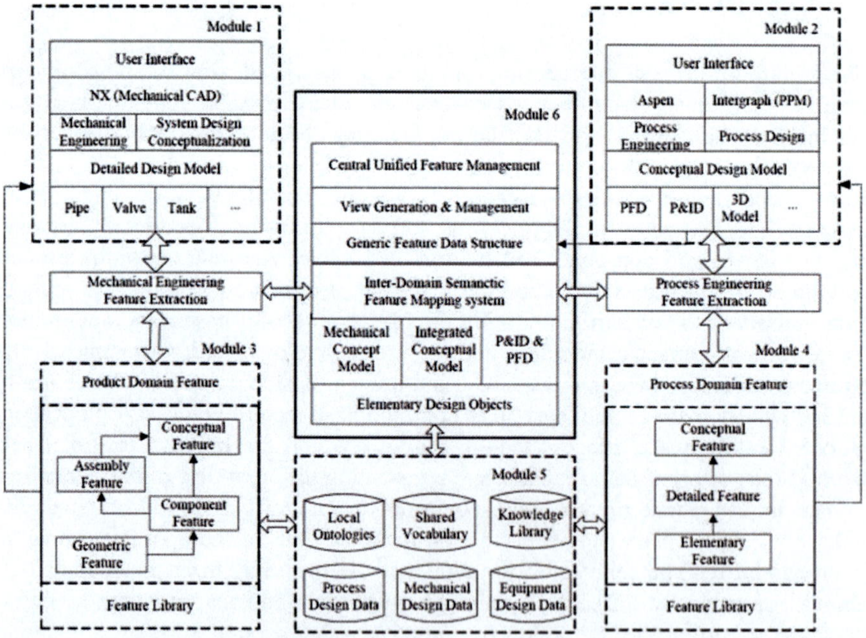

Fig. 12 Integrated system architecture [103]

7.2 Semantic Feature Associations Between Process Conceptual Design and Mechanical Detail Design

The first challenge in facilitating semantic integration in chemical process engineering is to identify the semantic feature association among the phases of the lifecycle [60]. The activities involved in chemical as well as mechanical process engineering and design lead to the generation of domain-specific features. These features are classified, as suggested by Han [35], into two categories: chemical process conceptual design features (CPCDFs) and mechanical detail design features (MDDFs), as shown in Fig. 13. CPCDFs are the features created in the chemical process conceptual design and engineering phases, which are designed to satisfy the requirements of the chemical process projects, as in, for example, the capacity of a chemical plant under construction. The parameters involved in CPCDFs can be mapped to the constraints, which will influence the downstream mechanical detail design [60]. MDDFs are the features created in the mechanical engineering and design phase to satisfy the requirements and be subject to the constraints stated in the process conceptual design. The specification of the equipment design will in turn place constraints on the process conceptual design. These mappings are implemented by knowledge-based reasoning, as shown in Fig. 13.

The semantic associations between feature parameters that are associated with equipment engineering and design are shown in Fig. 14. The corrosion allowance constraint (CAC), temperature (T), pressure (P), and residence time (RT) can be considered by the input and output parameters, which are specified based on the project requirements during the conceptualization phase [33]. Further, the shell thickness constraint (STC) is derived by T and P. The flow rate (Fr) is calculated based on the capacity (Cap) requirement and diameter of piping (DP), which will

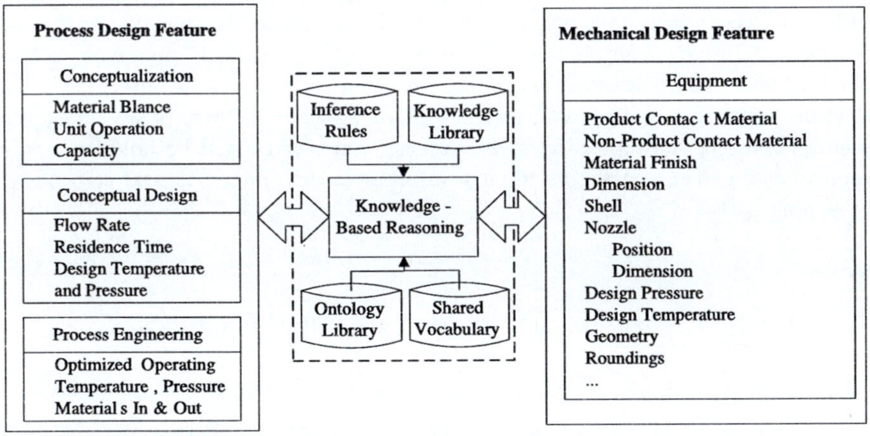

Fig. 13 Semantic feature associations between process design and mechanical design

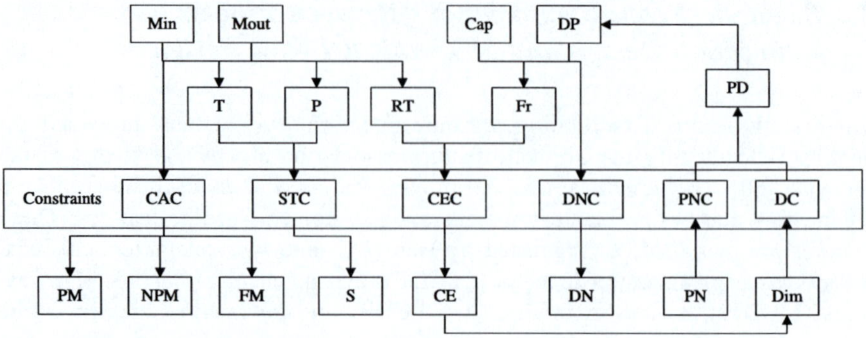

Fig. 14 Semantic associations of feature parameters and constraints

further determine the capacity-of-equipment constraint (CEC). And with DP alone, the diameter of nozzle constraint (DNC) is identified and will further determine that diameter of nozzle (DN) should be equal to the DNC. Meanwhile, the equipment's shell thickness (ST) and capacity-of-equipment (CE) should be larger than the STC and the CEC, respectively, while product contact material (PCM), non-product contact material (NPCM), and finish material (FM) are identified by the CAC. Further, mechanical engineers will work out the dimension (D) and geometry of the equipment, which determines the dimension constraint (DC). Similarly, DCs are used to describe the relationship between the position of nozzle (PN) and those position-of-nozzle constraints (PNC). Lastly, the DP is identified based on the piping design (PD), which is influenced by the DC and the PNC.

The mechanical design features associated with the chemical process conceptual design features should be kept consistent by implementing an active checking mechanism known as the "association" [33]. This has to be implemented in two ways. Each CPCDF and its properties are mapped to constraints specified for mechanical design, and any later changes in the CPCDF will be reflected in the update of the constraints, which will further influence the parameters in the MDDFs. Similarly, further design changes by mechanical engineers within MDDFs will update those related constraints that are mapped from the CPCDFs; then the constraint changes will trigger CPCDF updates. If there is any constraint conflict emerges during the updating process, the change will be hold up, and a report generated for engineers' review in order to determine the next step of the reasoning path.

7.3 *Proposed Workflow Under the Integrated Framework*

Conventionally, chemical process design and mechanical design work are sequentially coupled with verbal consultations between engineers. After the process design engineer works out the process specifications, mechanical detail design begins. However, this second stage is often delayed by several iterations of the process conceptual design modification. The specifications of the mechanical design may then require that the process conceptual design again be changed, especially when "non-standard" equipment is applied. However, changes to the process design will also lead to adjustments in the specifications of equipment and hence of the mechanical design [19]. Several iterations of both phases are usually needed before the process design and mechanical design are finalized. The work associated with these iterations is tedious, time-consuming, and error-prone. It is difficult to maintain consistency, as engineers' work during iterations will often conflict with those constraints defined in earlier cycles, without the engineers noticing [4, 21]. An example is shown in Fig. 15 [4].

To reduce the amount of time spent on interdisciplinary collaboration during design phases, a new design process is proposed here by the authors based on collaborative engineering principles under an integrated framework, as shown in Fig. 16. For example, given a capacity expansion project in chemical engineering, the project scope and reusability of knowledge are first determined; the material balance and operation capacity are identified in the conceptualization based on the project requirements. Instead of working sequentially in the conventional design process, conceptual design, process engineering, and mechanical design are now implemented in a concurrent and collaborative environment. The associations involved are supported by a systematic knowledge-based reasoning procedure [101]. Meanwhile, engineering constraint checking should be implemented to keep designs originating from different domains consistent. If the design change is rejected by the constraint checking module, the engineer can retain the most recent valid model while trying another round of iterations to reach a new solution. In this collaborative work environment, the likelihood of redesign, as well as workload, is greatly reduced. Furthermore, knowledge will be extracted from the complete project case and added to the knowledge library for future reuse after the project is complete.

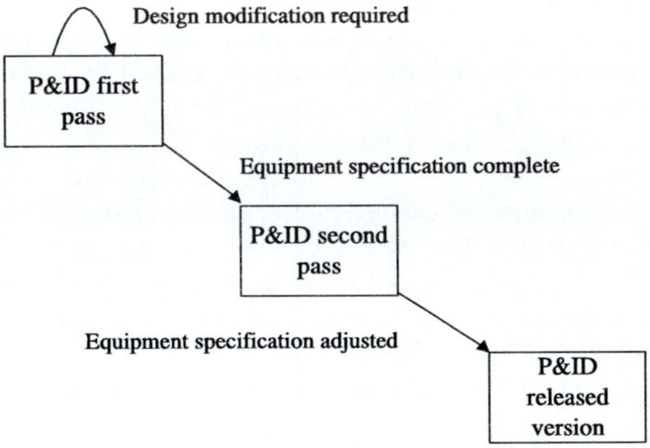

Fig. 15 Iteration of modifications in the conventional design process [4]

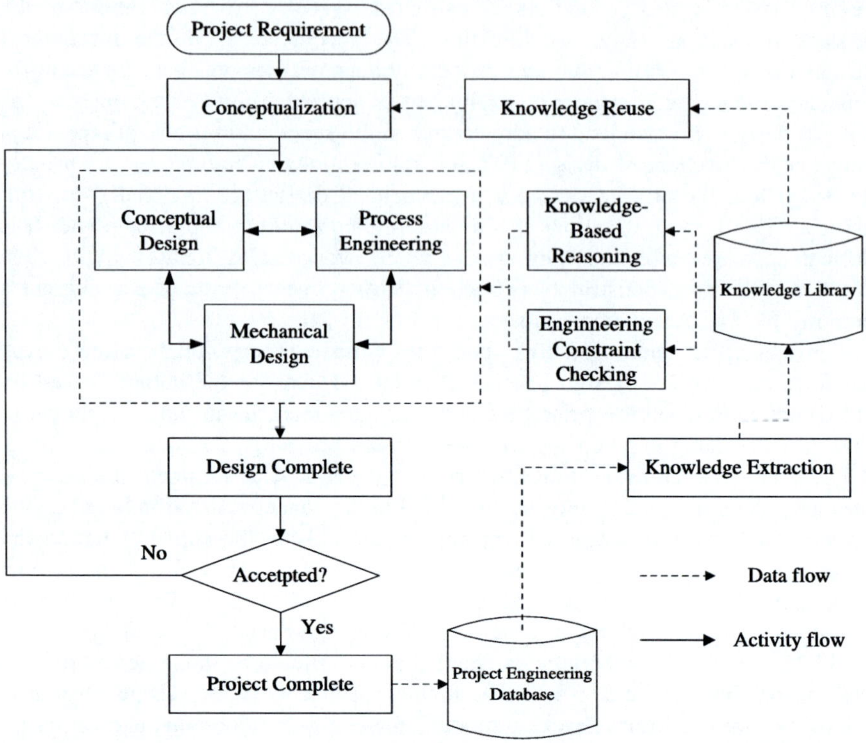

Fig. 16 Proposed design process flow under an integrated framework

7.4 Case Study

An example of the integrated system is shown in Fig. 17. Figure 17a shows the process and instrumentation diagram (P&ID) of the process based on its process flow diagram (PFD). Both the PFD and the P&ID are created with the conceptualization of design intent which can be expressed with a set of associated attributes. Such Attributes can be retrieved from a central database according to the project requirement analysis. The P&ID specifies the equipment, instrument, key piping, process control schema, and so on. For the downstream mechanical detail design, there is too much irrelevant and incomplete information, as only those equipment characteristics that are significant from the process point of view are specified. However, all of this information will be mapped onto the central database. With this information, as well as other information mapped from, for example, the PFD, the "standard" equipment can be selected from the equipment library, as shown in Fig. 17b. However, sometimes custom-designed equipment will be needed. In this case, the mechanical engineer can work on similar equipment and just make some minor modifications. During this process, the knowledge library will provide data support.

However, there is again too much detailed and mechanically relevant information included in the solid model shown in Fig. 17b, such as small chamfering or filleting parameters embedded in the design features. Should the full solid model be transferred to process engineers or added to the process design model directly, it would cause a network burden with unnecessary information being transferred, and would also confuse the process engineers. Instead, a process engineering and design view, which has been tailored to include only the tank's process features, is generated according to the view definitions. This process view presents only process engineering feature properties, such as operating pressure, capacity, key dimensions, and other process-related attributes, and they will be further referenced in the process model as external data resources. Thus, the tank generated in Siemens NX shown in Fig. 17b is mapped to the tank in the pink wire frame in SmartPlant 3D, as shown in Fig. 17c.

8 System Architecture for Interoperable Network-Based Engineering Systems

Competition in global marketing forces companies to develop products in the shortest time with the highest quality. Collaborative engineering aims to shorten the length of the product development process. In collaborative engineering, tasks can be performed by engineers who are both spatially and temporally distributed. Two critical technologies help to realize collaborative engineering: web technology and agent technology.

8.1 Web Technology

Web technology enables centralized information integration through a shared web server and a central database [81]. It usually uses the client–server architecture to realize communication between those servers and distributed developing teams. Ideally, the collaborative engineering system will be web-based, semantics-enabled, comprehensive, and agent-based [36, 81].

Critical issues about web-based engineering systems need to be tackled. Expert systems tend to use a variety of software tools and computer systems. The system should therefore have the ability to support heterogeneous computer applications and data sharing. Distributed object technology such as the Common object request broker architecture (CORBA) and DCOM/ActiveX can solve these problems [68, 83]; the details of data exchange have been discussed in earlier sections.

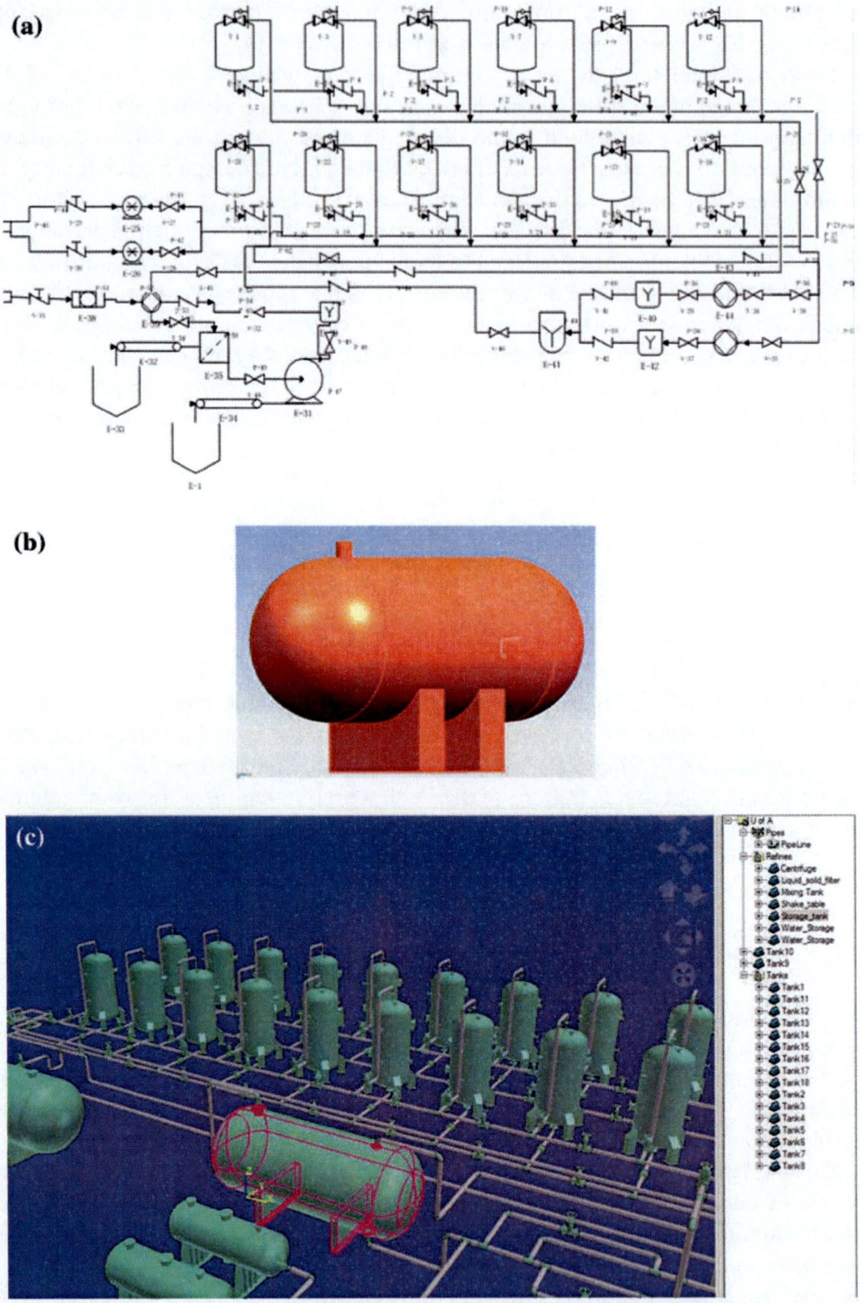

Fig. 17 An integrated system: **a** P&ID, **b** the 3D solid model generated in Siemens NX, and **c** the 3D model generated in SmartPlant

For collaborative design, multiple teams often work with the same model for different disciplinary purposes; hence, conflicts occur frequently. There has to be some rules for decision making. For instance, the system should notify those engineers in charge about the conflicting constraints and to the engineers make decisions via negotiation to resolve conflicts; sometimes multiple solutions exist simultaneously.

Web technology can only satisfy the data communication and exchange requirements of collaborative engineering systems. There are interwoven intellectual exchanges of opinion, consultations, and compromises that need complete, accurate, and sustainable information models instead of web technology alone. Further, the product-related data should be complete and can be translated into different application models; designers must have access to the complete design model if required to visualize, manipulate, and retrieve all the geometry and the semantics of the design, and negotiate modifications. It is also better for the collaborative engineering system to have flexible and modular architecture, and agent technology is useful in facilitating automatic process flow management requirements. Thus, ideally, the collaborative engineering system will be web-based, semantics-enabled, comprehensive, and agent-based [36, 81].

8.2 Agent Technology

Agents are programs acting for a user or another program under predefined conditions. The aim of Agent technology is to integrate heterogeneous, distributed, and semiautonomous knowledge-based software tools into a collaborative application [54]. There have been numerous efforts to develop agent-based collaborative engineering systems. Palo Alto collaborative test bed (PACT) is one of the earliest CE web-oriented platforms satisfying multiple sites and various disciplines. PACT is agent-based and allows agents working on different aspects of design to share and exchange information with one another [22]. Agent interaction relies on three elements [22]: shared concepts and terminology for communicating knowledge across disciplines, an interlingua for transferring knowledge among agents, and a communication and control language that enables agents to request information and services. Shen and Barthes [80] have developed a prototype of a distributed intelligent design environment (DIDE) in which the internal structure of an agent and the inter-agent communication mechanism are illustrated in detail. The internal structure of an agent is shown in Fig. 18.

8.3 Multi-Agent Systems

A web-based interoperable engineering system is composed of various engineering software tools that rely on different principles. In such systems, multi-agent

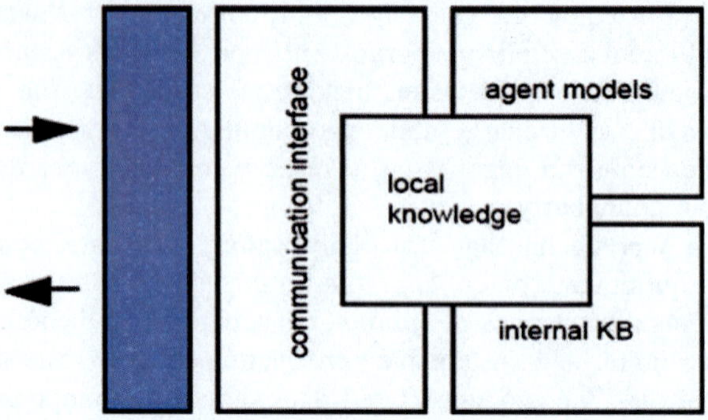

Fig. 18 Internal structure of an agent [80]

technology is more suitable and makes the systems more flexible. As each agent is coupled with a certain function and a well-defined application program interface, the engineering system can change its configuration based on practical requirements. At the same time, agent modules can be reused in different systems [54].

Wang et al. [100] have developed a distributed multidisciplinary design optimization (MDO) environment that supports seamless interaction between designers, agents, and servers. The architectural framework for MDO environments is shown in Fig. 19. Hao et al. [36] have developed a lightweight agent framework for mechanical product design by applying intelligent software agents, Web, workflow, and database technologies. The framework developed is compliant to Foundation for Intelligent Physical Agents (FIPA) standard, called autonomous agent development environment (AADE).

8.4 System Architecture

For the agent- and web-based interoperable engineering system, the target is to use software agents to reduce reliance on large, complex, centralized systems and to efficiently facilitate collaboration.

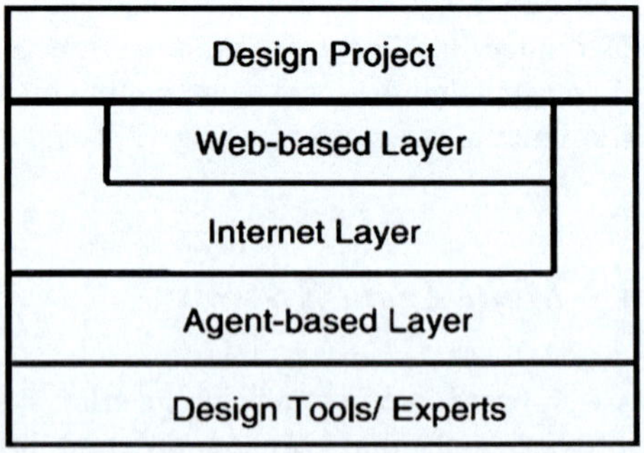

Fig. 19 Architectural framework of integrating the internet, web, and agent technologies for MDP environments [100]

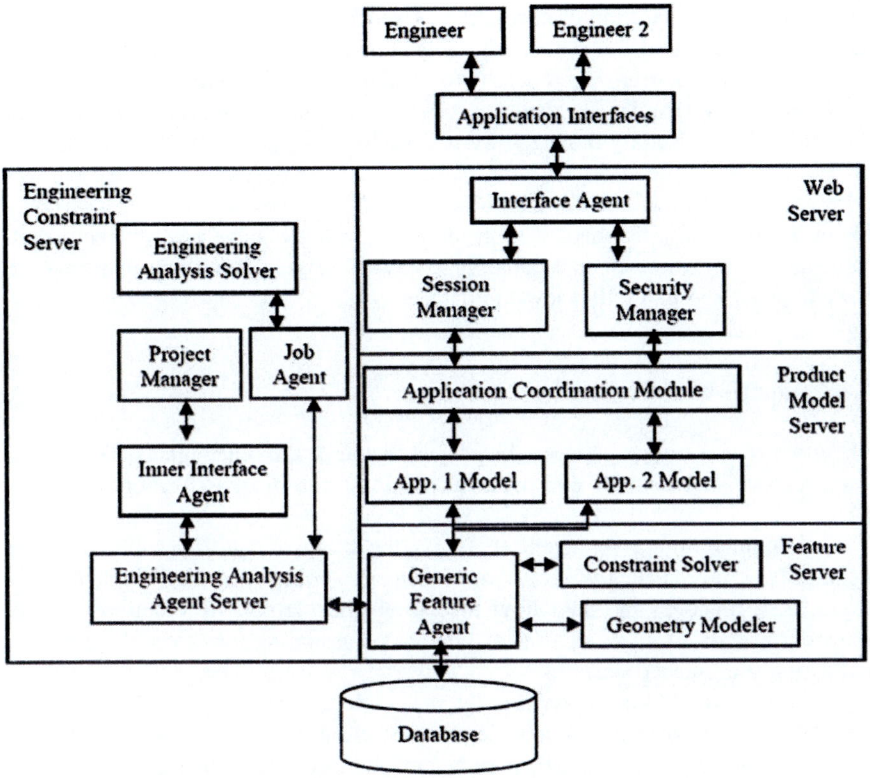

Fig. 20 A system design for agent-based engineering collaborations

Ulieru et al. [95] describe three layers of the system architecture: a low-level inter-networking communication support layer, a coordination layer (managing inter-agent cooperation through intelligent conversation/communication mechanisms), and an agent layer consisting of five categories of agents: interface, collaboration, knowledge management, application, and resource agents.

In this section, a unified feature model is applied as the system modeling basis to realize interoperability. A simplified architecture is proposed, as shown in Fig. 20.

8.4.1 Web Server

The web server contains an interface agent (IA), security manager, and session manager. The IA provides shared access for multiple users. It can instantiate different data for different users according to the users' requirements. The security manager is used to check whether the user has the right to access the product model data, and what kind of access it is. Users are separated into several groups, each with different access rights. All management data are stored in the database.

The system supports shared access for multiple users, so a comprehensive data management system is designed for maximum concurrent access to the data. Access to data is mainly managed by the session manager. If multiple users with different priorities require the same data, the user with highest priority will get access to the data while other users have to wait (i.e., can only view the data) until his or her session is finished. If multiple users with the same priority require the same data, it follows the "first come, first served" principle: other users remain on a waiting list and can only view the data [9, 93].

8.4.2 Agent-Based Design

The inner IA is a bridge between the project manager and other agents. The project manager assigns new jobs and manages on-going job progress through the inner IA.

The engineering server agent (ESA) is the brain of the Agent-based design system. It communicates with the job manager to accept jobs and manages messages, then manages the data flow to operate the system. Its functions include transferring data files to and from the database, assigning jobs to job agents, and validating the finished design.

The job agent (JA) is responsible for automatic task arrangement. In the Agent-based design system, a design job is composed of job ID, task ID (which is formed following the task sequence), job parameters, and task files. When a new job enters the JA, the JA can automatically distribute tasks to the appropriate available designers following the task sequence.

8.4.3 Working Procedure for Agent-Based Design

1. The project manager gets access to the inner IA to give a new job to the system.
2. The ESA gets the message from the project manager and starts the function "starting job#." It then reads the job data from the database and transfers it to the JA.
3. The JA receives the imported data and assigns specific tasks to various designers. Tasks will be arranged following predefined sequences.
4. When a job is finished, the finished design files are sent back to the ESA. The JA sends the design to the problem solver to be validated. If there is no failure, the design data is stored in the database and the job status changes to "finishing-job#." However, if there is a defect, the process will go back to step 3 [36].

8.4.4 Downstream Application Management

To realize collaborative engineering, the collaborative design system needs support from many collaborating functional modules, which can provide services through agents as well. For example, features are managed by a feature agent. Other downstream applications can also be consolidated by an administrative downstream application management agent for their services or interactions.

The feature agent can provide feature objects for application packages and separate application packages for discrete users, so that users can use specific feature models for certain downstream applications. The feature agent has the functions of FR, feature extraction, and feature modification. The feature agent can also receive feedback from users and process it. Every time the feature agent modifies the feature model, it will call the constraint solver and the geometry modeler to validate the modified feature model. The constraint solver can check the validation of all constraints, which are part of the feature definition. The geometry modeler can validate feature geometry. Finally, the unified feature model in the database will be updated.

8.4.5 Database

The database provides physical storage for all kinds of data, including product model data and security management data. Geometric data and the unified feature models are stored within the database. The unified feature models are composed of various generic feature models, which are stored as data elements across tables [18, 19]. In this manner, the database manager can reorganize these data elements for flexible use by different applications [93].

9 Information Views, Granularity, and Knowledge-Driven Engineering

9.1 Information Granularity

Granularity is the extent to which information is broken down into small components of computer system entities. Coarse-grained information model consists of fewer and larger components than fine-grained model. Granularity becomes an important issue for data modeling when trying to represent levels of information with data structures across systems or databases [25]. In practice, information can be granulated into four parts, as shown in Fig. 21.

Data. The most-granulated information type is data. Data as an abstract concept can be viewed as the lowest level of abstraction from which information and then

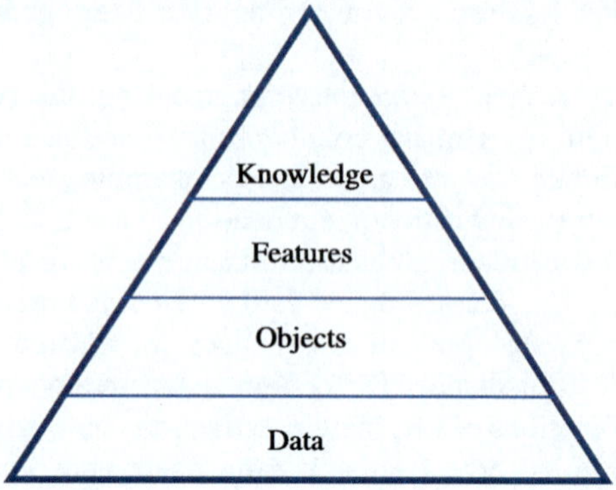

Fig. 21 Information granularity

knowledge are derived. Data on its own carries no meaning. For data to become information, it must be interpreted and take on a meaning.

Object. Objects can be thought of as wrapping their data within a set of functions designed to ensure that the data are used appropriately, and to assist in that use. The object's methods will typically include checks and safeguards that are specific to the types of data the object contains. An object can also offer simple-to-use, standardized methods for performing particular operations on its data.

Feature. A generic feature representation in a database can be expressed as shown in Fig. 22 [93]. A feature has *feature_id*, *product_id*, and *domain* as its attributes. The *feature_id* attribute is an object identifier, which can uniquely identify a feature object in database. *Product_id* specifies which product a particular feature belongs to. *Domain* is a predefined data type, which can be instantiated for *design*, *manufacturing*, or *analysis*, and their relevant setting parameters are stored in a domain table. A *feature* will also contain a list of referenced entities, a list of constraints, and a list of parameters. Dimensions and tolerances are regarded as subtypes of *constraint* bounded to certain geometrical entities.

Knowledge. Knowledge-based engineering (KBE) is OO and rule-based, where knowledge is the object and rules are the operations. Features can be a part of the knowledge, and rules are responsible for the reasoning and mapping of the features.

9.2 Information View

An information view is a selective set of information that is specially filtered for a purpose. As shown in Fig. 23, the functional view can be designed or implemented with a specific purpose and scope as well.

The design of user-specific and need-based information views plays a significant role in the integration of CAx applications. These views are context-dependent interpretations of self-contained subsets of information about the entire product model (EPM). With STEP technology, all of the functional views can be expressed

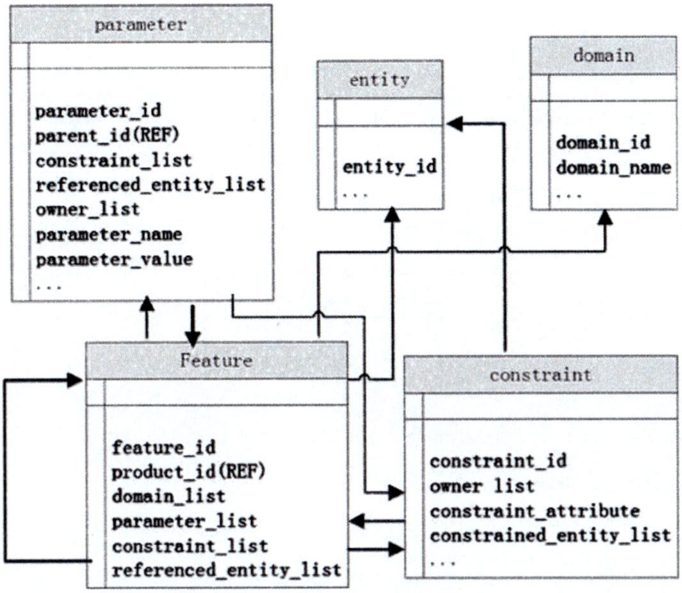

Fig. 22 Generic feature representation in a database [93]

with the same language, EXPRESS, and an arbitrary view can easily be translated into other views. Building a common product model representation is crucial to achieving different functional views. EPM describes information across applications, and contains the domain classification ontology and metadata. In practice, application feature sub-models can provide specific views of the EPM [93].

9.3 Introduction to Knowledge-Based Engineering

Knowledge-based engineering (KBE) is a special type of knowledge-based system (KBS) that focuses on product development activities such as design, analysis, process planning, and manufacturing. Stokes [89] defines KBE as "the use of advanced software techniques to capture and re-use product and process knowledge in an integrated way."

There are many advantages of KBE. With the product and process knowledge stored in the system, KBE can reduce the time spent on routine work and also save time for innovation. As the expertise is stored in the database, companies will be less affected staff turnover. A major drawback, however, is that it takes time to develop and update the KBE system.

KBE is currently widely used in industry. KBE systems are usually developed by individual companies to generate product concepts using captured product and process knowledge, and can later be used to help prepare for FEA [12, 15].

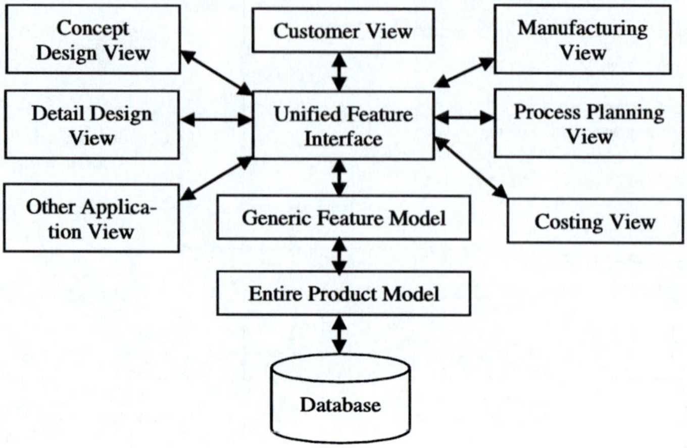

Fig. 23 Functional information views

A recent challenge is to integrate manufacturing-related knowledge into KBE systems to aid in manufacturing; this often takes the form of assessment of manufacturability and process planning [6, 85].

9.4 Foundations of KBE

Knowledge is central to KBE, and it interacts with information about how a certain product should be developed. Considering specific knowledge-based product engineering applications, knowledge can be managed with reference to three categories of information: geometry, configurations, and engineering rules. The rules are complex and powerful expressions composed with mathematical formulae and conditional statements [59, 75].

Geometry. Most product-oriented KBE systems have limited capability of CAD functions and hence are usually integrated with CAD packages. Very often, the output of the KBE system is a CAD model.

Configuration. This refers to a functional product model that is an assembly of a set of existing modular components. At present, vast KBE applications for configuration design are used with many real cases. A typical example is Toyota's configuration management system [88].

Engineering rules. This refers to specific domain-related or analysis-related knowledge that consists of well-organized logical rules and assists decision making based on the input conditions of engineering constraints imposed in product development processes. Figure 24 shows the framework of the KBE system.

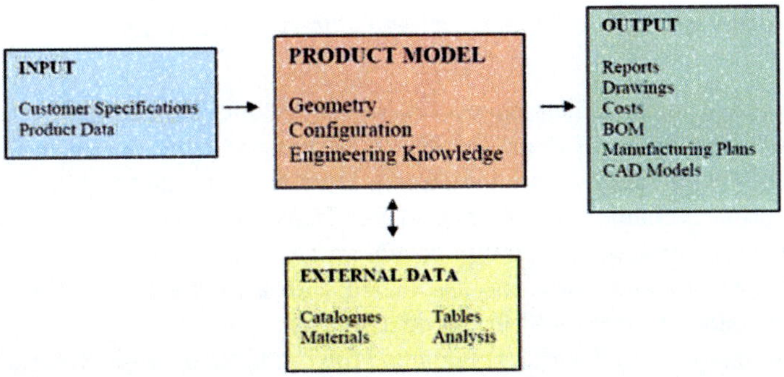

Fig. 24 The KBE system [75]

9.5 Methodology to Develop KBE Systems

The existing KBE methodologies mostly focus on systems (KBSs). For example, CommonKADS, a widely known methodology for knowledge engineering and knowledge management, is powerful but also difficult to learn and complex to use [20]. However, it was not appropriately developed for KBE applications [59]. Methodology- and tools-oriented knowledge-based engineering applications (MOKA) is another project aimed at developing a methodology to form the basis of the international standard for KBE. MOKA is based on eight KBE lifecycle steps: *identify, justify, capture, formalize, package, distribute, introduce,* and *use*; however, it mainly focused on the *capture* and *formalize* steps [59, 89].

Knowledge capture is the first KBE process, intended to elicit knowledge from experts or extract it from other sources (such as documentation). This collected knowledge can be structured in an informal model. Correctness and completeness are checked, and the ambiguity of language expression is eliminated. Only after such post processing will the collected knowledge become understandable and usable. There are several ways to elicit knowledge, depending on the knowledge source. The most widely used method of extracting knowledge from experts is to interview them. Another common method is to use data mining technology, which originates from the artificial intelligence domain, to extract knowledge from documents. For example, MOKA elicits knowledge from both experts and documents with its engineering domain ontology, which enables the identification of a large number of knowledge objects [89].

Knowledge formalize is the process of transforming knowledge into a neutral and formal model that can be embedded in any KBE applications. This is a process to convert knowledge into a computer-interpretable representation that facilitates encoding into a computer program. In MOKA, knowledge is represented with the MOKA modeling language (MML), which is adapted from the UML.

9.6 Implementation of KBE in Industrial Practices

Product configuration management (PCM) in product development is a good example of KBE application areas [88]. PCM provides the tools to translate the engineering specifications and validation logic for option-oriented, customer-specified product lines into a centralized PCM knowledge base. The PCM knowledge base is made up of multiple rule types, data tables, and algorithms that are maintained independently and associated with a product line, allowing PCM logic to be shared across multiple product lines as required.

Take the example of Toyota. It launched the brand Scion with two models, XA and XB, and more than 40 types of accessories for customization. Customers can refer to detailed information offered online or from dealers to customize the configuration (color, transmission, exterior, interior, wheels, and sound). Once the order is finished, the customized car will be ready for pickup. The same KBE technique has been applied to other Toyota products, such as Camry and Corolla [88].

As to commercially available implementation platforms, Siemens NX knowledge fusion [65] is a typical and fully integrated knowledge-based engineering tool that permits knowledge-based extension of NX by the end user. Compared to traditional KBE technologies, the tight integration of Knowledge Fusion into the NX digital product development system provides a significant industrial advantage. Knowledge fusion permits the creation of powerful applications that take advantage of engineering knowledge. It supports the capture and reuse of design intent and user intelligence to increase design speed and productivity, while intelligently controlling change propagation [84].

As illustrated in Fig. 25, companies can customize NX to include a set of features common to their particular design practices with user-defined features. They can be used to streamline the design process, promote reuse, and ensure that product designs follow common methods and utilize standard design components. User-defined features can take advantage of a robust set of additional knowledge fusion capabilities, which is a built-in easy-to-use design scheme that allows the designer to create rules to capture the design intent and the rationale behind design decisions. This capability is made possible by allowing rules to be attached to the user-defined feature. These rules can be used to alter the geometry, location, and even the selection of the appropriate user-defined features based on model conditions.

9.7 Future Research Issues

Knowledge technology and Artificial intelligence have a long history of development, but KBE applications specific to product engineering are still new and have yet to mature. For example, KBE systems are only used by big companies like Boeing and Toyota; tools for building KBE applications such as the methodology for

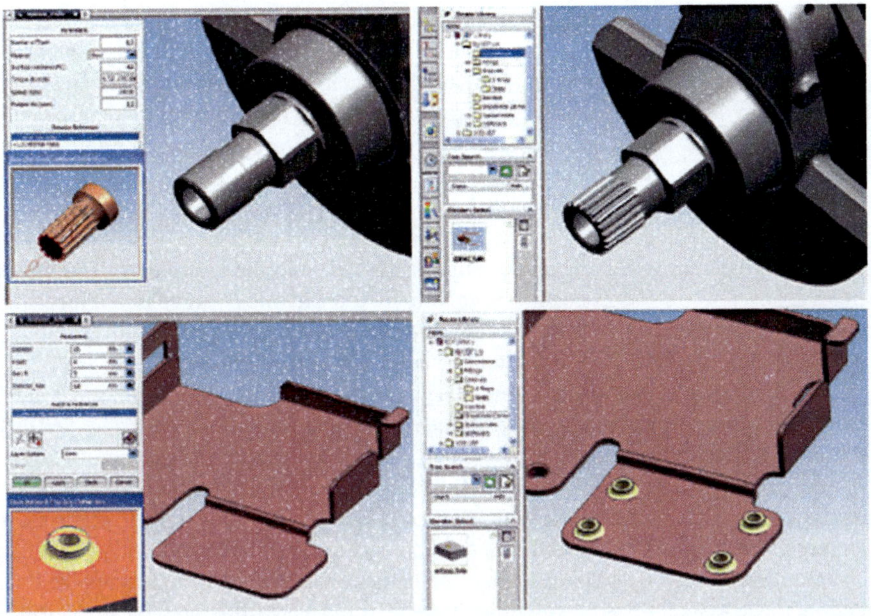

Fig. 25 User-defined features supported by knowledge fusion [84]

knowledge-based engineering applications (MOKA) [89] are similarly geared toward big companies. Little research effort has been spent on building KBE systems for small and medium-sized enterprises [59].

Transparency of reasoning procedure in KBE applications is also in need of improvement. At present, most KBE applications operate in a black box: nobody knows what the justifications are for deriving the results. If the built-in reasoning logics are faulty, the results will be very misleading. A more user-friendly, flexible, and adaptable knowledge base is needed.

10 Summary

Given the growing industrial demand for engineering information integration, there should be a systematic and scalable approach to developing a uniform implementation platform for informatics solutions that can deal with real world diversity and complexity. This chapter presented a set of challenges that require a new paradigm to address them. Among the many challenges, in-house knowledge representation and implementation, associative information-sharing and management, and cross-domain data and semantics integration (such as CAD and CAE integration) by consistent referencing and constraint management all offer new grounds for research and development.

This chapter tried to address these challenges by investigating their complex requirements and exploring some initial conceptual solutions. It seems that, theoretically, extended feature technology offers the requisite flexibility in associating entities from different domains with different levels of granularity. One of the application industries is oil and gas, where chemical process engineering is the leading field. A special section was dedicated to discussing informatics solutions in this field. Every chemical process engineering project is a complicated task, requiring the collaboration of engineers from different domains, such as chemical process engineering and mechanical engineering. Due to the complexity and close associations among the activities involved in the chemical process engineering project, interoperability is a major issue. A NDF (such as STEP) and traditional feature technology can only deal with structural heterogeneity.

To enhance interoperability, especially on the semantic level, an innovated integration framework was proposed based on semantic modeling and feature technology, and was designed to support semantics, patterns, association, and change propagation in the chemical process engineering project. This unified semantic integration framework is an open architecture and is designed to integrate a number of software tools with a common system infrastructure of consistent information referencing and updating mechanisms via a cluster of features. In that section, those associations between semantic features were illustrated by an example of equipment design, which provided a generic semantic representation of the associations across multiple design stages in the chemical process project lifecycle.

Compared to conventional design, ideally the lifecycle of the design phase can be shortened and design consistency can be more easily maintained with a significant decrease in modeling effort. In the future, constraint management and the consistency mechanism need to be enhanced, and knowledge extraction and management need further research effort.

References

1. Babic BR, Nesic N, Miljkovic Z (2008) A review of automated feature recognition with rule-based pattern recognition. Comput in Ind 59:321–337
2. Babic BR, Nesic N, Miljkovic Z (2011) Automatic feature recognition using artificial neural networks to integrate design and manufacturing: review of automatic feature recognition system. Artif Intell for Eng Des, Anal Manuf 25:289–304
3. Bayer B, Marquardt W (2003) A comparison of data models in chemical engineering. Concurrent Eng 12:129–138
4. Bayer B, Marquardt W (2004) Towards integrated information models for data and documents. Comput Chem Eng 28:1249–1266
5. Belaziz M, Bouras A, Brun JM (2000) Morphological analysis for product design. Comput Aided Des 32:377–388
6. Bernard A, Deglin A (2000) Knowledge-based environment for the generation of rapid product development processes. J Manuf Sci Prod 3:167–174

7. Bianconi F, Conti P, Diangelo L (2006) Interoperability among CAD/CAM/CAE systems: a review of current research. In: Proceedings of the conference on geometric modeling and imaging: new trends, Washington, D.C
8. Bidarra R, Bronsvoort WF (2000) Semantic feature modeling. Comput Aided Des 32:201–225
9. Bidarra R, Berg E, Bronsvoort WF (2001) Collaborative modelling with features. In: Proceedings of ASME design engineering technical conference, Pittsburgh
10. Bronsvoort WF, Jansen FW (1993) Feature modeling and conversion: key concepts to concurrent engineering. Comput Ind 21:61–86
11. Bronsvoort WF, Noort A (2004) Multiple-view feature modeling for integral product development. Comput Aided Des 36:929–946
12. Bylund N (2012) Modes, methods and tools for car body development. Licentiate thesis, Lulea University of Technology
13. CATIA (2012) http://www.3ds.com/products/catia/. Accessed 19 Aug 2012
14. CATIA/CAA (2012) http://www.3ds.com/plm-glossary/caa-v5/. Accessed 19 Aug 2012
15. Chapman CB, Pinfold M (2001) The application of a knowledge based engineering approach to the rapid design and analysis of an automotive structure. Adv Eng Softw 32:903–912
16. Chen G, Ma YS, Thimm G, Tang SH (2004a) Unified feature based integration of design and process planning. In: Hinduja S (ed) Proceedings of the international MATADOR conference. Springer, London
17. Chen G, Ma YS, Thimm G, Tang SH (2004) Unified feature modeling scheme for the integration of CAD and Cax. Comput Aided Des Appl 1:595–601
18. Chen G, Ma YS, Thimm G, Tang SH (2005) Knowledge-based reasoning in a unified feature modeling scheme. Comput Aided Des Appl 2:173–182
19. Chen G, Ma YS, Thimm G, Tang SH (2006) Associations in a unified feature modeling scheme. ASME Trans J Comput Inf Sci Eng 6:114–126
20. CommonKADS (2012) http://www.commonkads.uva.nl/. Accessed 1 Oct 2012
21. Couper JR, Penney WR, Fair JR et al (2004) Chemical process equipment: selection and design, 2nd edn. Elsevier, Burlington
22. Cutkosky MR et al (2006) PACT: an experiment in integrating concurrent engineering systems. IEEE Comput 26:28–37
23. Deng YM, Britton GA, Lam YC, Tor SB, Ma YS (2002) Feature-based CAD-CAE integration model for injection-moulded product design. Int J Prod Res 40:3737–3750
24. Fischer A, Wang KK (1997) A method for extracting and thickening a mid-surface of a 3D thin object represented in NURBS. J Manuf Sci Eng Trans ASME 119:706–712
25. Fonseca F, Egenhofer M, Davis C, Câmara G (2002) Semantic granularity in ontology-driven geographic information systems. AMAI Annals Math Artif Intell 36:121–151
26. Foucault G, Cuilliere JC, Francois C, Leon JC, Maranzana R (2008) Adaptation of CAD model topology for finite element analysis. Comput Aided Des 40:176–196
27. Francois V, Cuilliere JC, Gueury M (1999) Automatic meshing and remeshing in the simultaneous engineering context. Res Eng Des 11:55–66
28. Gabbert U, Wehner P (1993) Steps towards CAD–FEA integration. Eng Comput 9:17–26
29. Gao J, Zheng DT, Gindy N (2004) Extraction of machining features for CAD/CAM integration. Int J Adv Manuf Technol 24:573–581
30. Gao J, Zheng DT, Gindy N (2004) Mathematical representation of feature conversion for CAD/CAM system integration. Robot Comput Integr Manuf 20:457–467
31. Gordon S (2001) An analyst's view: STEP-enabled CAD-CAE integration. In: Proceedings of presentation materials of NASA's STEP for aerospace workshop, jet propulsion laboratory, Pasadena
32. Gujarathi GP, Ma YS (2010) Generative CAD and CAE integration using common data model. In: Proceedings of the 6th annual IEEE conference on automation science and engineering, Toronto

33. Gujarathi GP, Ma YS (2011) Parametric CAD/CAE integration using a common data model. J Manuf Syst 30:118–132
34. Han JH, Pratt M, Regli WC (2000) Manufacturing feature recognition from solid modes: a status report. IEEE Trans Robot Autom 16:782–796
35. Han S, Lee T (1999) Information sharing between process and engineering design activity in CAD environment. Comput Chem Eng Suppl:S573–S576
36. Hao Q, Shen WM et al (2006) Agent-based collaborative product design engineering: an industrial case study. Comput Ind 57:26–38
37. Hu KM, Wang B, Liu Y, Huang J, Yong JH (2012) An extended schema and its production rule-based algorithms for assembly data exchange using IGES. Int J Adv Manuf Technol 58:1155–1170
38. ISO 6983-1 (1982) Numerical control of machines-program format and definition of address words—part 1: data format for positioning, line motion and contouring control systems. International Organization for Standardization, Geneva
39. ISO 10303-42 (1994) Industrial automation systems and integration: product data representation and exchange—part 42: integrated generic resources: geometric and topological representation. International Organization for Standardization, Geneva
40. ISO 10303-1 (1995) Industrial automation systems and integration: product data representation and exchange—part 1: overview and fundamental principles. International Organization for Standardization, Geneva
41. ISO 10303-212 (2001) Industrial automation systems and integration: product data representation and exchange—part 212: application protocol: electrotechnical design and installation. International Organization for Standardization, Geneva
42. ISO 10303-232 (2002) Industrial automation systems and integration: product data representation and exchange—part 232: application protocol: technical data packaging core information and exchange. International Organization for Standardization, Geneva
43. ISO 14649-1 (2003) Industrial automation systems and integration: physical device control—data model for computerized numerical controllers—Part 1: overview and fundamental principles. International Organization for Standardization, Geneva
44. ISO 10303-227 (2005) Industrial automation systems and integration: product data representation and exchange—Part 227: application protocol: plant spatial configuration, 2nd edn. International Organization for Standardization, Geneva
45. ISO 10303-239 (2005) Application protocol for product life cycle support: HTML document and electronic inserts (includes access to additional content). International Organization for Standardization, Geneva
46. ISO 10303-221 (2007) Industrial automation systems and integration: product data representation and exchange—Part 221: application protocol: functional data and their schematic representation for process plants. International Organization for Standardization, Geneva
47. ISO 10303-238 (2007) Industrial automation systems and integration: product data representation and exchange—Part 238: application protocol: application interpreted model for computerized numerical controllers. International Organization for Standardization, Geneva
48. ISO 10303-210 (2011) Industrial automation systems and integration: product data representation and exchange—Part 210: application protocol: electronic assembly, interconnect, and packaging design (2nd edn). International Organization for Standardization, Geneva
49. Ji AM, Zhu K, Huang JC, Dong YP (2011) CAD/CAE integration system of mechanical parts. Adv Mater Res 338:272–276
50. Kao YC, Cheng HY, She CH (2006) Development of an integrate CAD/CAE/CAM system on taper-tipped thread-rolling die-plates. J Mater Proc Technol 177:98–103
51. Kemmerer S (1999) STEP: the grand experience. National Institute of Standards and Technology Special Publication, USA

52. Kim J, Pratt M, Iyer RG, Sriram RD (2008) Standardized data exchange of CAD models with design intent. Comput Aided Des 40:760–777
53. Lam SM, Wong TN (2000) Recognition of machining features: a hybrid approach. Int J Prod Res 38:4301–4316
54. Lander SE (1998) Issues in multi-agent design systems. IEEE Expert Intell Syst Appl 12:18–26
55. Lee HC, Jhee WC, Park HS (2007) Generative CAPP through projective feature recognition. Comput Ind Eng 53:241–246
56. Lee SH (2005) A CAD-CAE integration approach using feature-based multi-resolution and multi-abstraction modeling techniques. Comput Aided Des 37:941–955
57. Li SH, Shao XD, Ge XB (2010) A kind of CAD/CAE integrated modeling technology based on feature. Adv Mater Res 97–101:3436–3442
58. Liang J, Shah JJ, D'Souza R et al (1999) Synthesis of consolidated data schema for engineering analysis from multiple STEP application protocols. Comput Aided Des 31:429–447
59. Lovett PJ, Ingram A, Bancroft CN (2000) Knowledge-based engineering for SMEs: a methodology. J Mater Process Technol 107:384–389
60. Ma YS (2009) Towards semantic interoperability of collaborative engineering in oil production industry. Concurrent Eng 17:111–119
61. Ma YS, Tong T (2003) Associative feature modeling for concurrent engineering integration. Comput Ind 51:51–71
62. Marquardt W, Nagl M (2004) Workflow and information centered support of design processes: the IMPROVE perspective. Comput Chem Eng 29:65–82
63. Monedero J (2000) Parametric design: a review and some experiences. Autom Constr 9:369–377
64. NX (2012) http://www.plm.automation.siemens.com/en_us/products/nx/. Accessed 19 Aug 2012
65. NX knowledge fusion (2010) http://www.plm.automation.siemens.com/en_us/products/nx/. Accessed 19 Aug 2012
66. NX programming and customization (2010) http://www.plm.automation.siemens.com/zh_cn/Images/4988_tcm78-4564.pdf. Accessed 19 Aug 2012
67. Owen J, Bloor MS (1987) Neutral formats for product data exchange: the current situation. Comput Aided Des 19:436–442
68. Pahng GDF, Bae S, Wallace D (1998) Web-based collaborative design modeling and decision support. In: Proceedings of the ASME design engineering technical conferences, Atlanta
69. Perng DB, Cheng CF (1997) Resolving feature interactions in 3D part editing. Comput Aided Des 29:687–700
70. Pratt MJ, Anderson BD, Ranger T (2005) Towards the standardized exchange of parameterized feature-based CA models. Comput Aided Des 37:1251–1265
71. Pro/TOOLKIT (2012) http://www.ptc.com/product/creo/toolkit. Accessed 19 Aug 2012
72. Razayat M (1996) Midsurface abstraction from 3D solid models: general theory and applications. Comput Aided Des 28:905–915
73. Remondini L, Leon JC, Trompette P (1996) Generic data structures dedicated to integrated structural design. Finite Elem Anal Des 22:281–303
74. Salomons OW, van Houten FJAM, Kais HJJ (1993) Review of research in feature-based design. J Manuf Syst 12:113–132
75. Sandberg M (2003) Knowledge based engineering in product development. Technical report, Lulea University of Technology
76. Sequin CH (2005) CAD tools for aesthetic engineering. Comput Aided Des 37:737–750
77. Shah JJ (1991) Assessment of feature technology. Comput Aided Des 23:331–343
78. Shah JJ, Mantyla M (1995) Parametric and feature based CAD/CAM: concepts, techniques and applications. Wiley, New York

79. Shahin TMM (2008) Feature-based design: an overview. Comput Aided Des Appl 5:639–653
80. Shen WM, Barthes JP (1997) An experimental environment for exchanging engineering design knowledge by cognitive agents. In: Mantyla M, Finger S, Tomiyama T (eds) Knowledge intensive CAD-2. Chapman & Hall, London
81. Shen WM, Hao Q et al (2010) System integration and collaboration in architecture, engineering, construction, and facilities management: a review. Adv Eng Inform 24:196–207
82. ShuMing Gao (1998) A survey of automatic feature recognition, Chin J Comput 21:281–288. In Chinese
83. Siegel J (1996) CORBA: fundamentals and programming. Wiley, New York
84. Siemens PLM Software (2012) http://www.plm.automation.siemens.com/en_us/. Accessed 1 Oct 2012
85. Singh N et al (1997) A knowledge engineering framework for rapid design. Comput Ind Eng 33:345–348
86. Spatial (2012) http://www.spatial.com/products/3d-acis-modeling. Accessed 12 Sep 2012
87. Srinivasan V (2008) Standardizing the specification, verification, and exchange of product geometry: research, status and tends. Comput Aided Des 40:738–749
88. Stanford Graduate School of Business (2005) Toyota: demand chain management, global supply chain management forum. Case: GS-42
89. Stokes M (ed) (2001) Managing engineering knowledge: MOKA—methodology for knowledge based engineering applications. Wiley, New York
90. Sudarsan R, Fenves SJ, Sriram RD, Wang F (2005) A product information modeling framework for product lifecycle management. Comput Aided Des 37:1399–1411
91. Suh SH, Cho JH, Hong HD (2002) On the architecture of intelligent STEP-compliant CNC. Int J Comput Integr Manuf 15:168–177
92. Suh SH, Lee BE, Chung DH, Cheon SU (2003) Architecture and implementation of a shop-floor programming system for STEP-compliant CNC. Comput Aided Des 35:1069–1083
93. Tang SH, Ma YS, Chen G (2004) A feature-oriented framework for web-based CAx applications. Comput Aided Des Appl 1:117–126
94. Tseng YJ, Joshi SB (1994) Recognizing multiple interpretations of interacting machining features. Comput Aided Des 26:667–688
95. Ulieru M, Norrie D, Kremer R, Shen WM (2000) A multi-resolution collaborative architecture for web-centric global manufacturing. Inf Sci 127:3–21
96. UnifiedCAD (2012) http://www.unifiedcad.com/. Accessed 19 Aug 2012
97. van Dijk CGC (1995) New insights in computer-aided conceptual design. Des Stud 16:62–80
98. Wang H (1995) An approach to computer-aided styling. Des Stud 16:50–61
99. Wang XV, Xu XW (2012) DIMP: an interoperable solution for software integration and product data exchange. Enterp Inf Syst 6:291–314
100. Wang YD, Shen WM, Ghenniwa H (2003) Webblow: a web/agent-based multidisciplinary design optimization environment. Comput Ind 52:17–28
101. Wiesner A, Morbach J, Marquardt W (2011) Information integration in chemical process engineering based on semantic technologies. Comput Chem Eng 35:692–708
102. Wittenborn D (2004) CAD interoperability. http://www2.tech.purdue.edu/cimt/courses/cimt311/cad_interop.pdf. Accessed 13 Aug 2012
103. Xie Y, Wei J, Ma YS (2012) Multi-view feature model representation to support integration of chemical process and mechanical design. Comput Aided Des Appl: accepted
104. Xu XW, Wang J (2004) Development of a G-code free, STEP-compliant CNC lathe. In: Proceedings of 2004 ASME international mechanical engineering congress and exposition, Anaheim
105. Xu XW, Newman ST (2006) Making CNC machine tools more open, interoperable and intelligent: a review of the technologies. Comput Ind 57:141–152

106. Xu XW et al (2005) STEP-compliant NC research: the search for intelligent CAD/CAPP/CAM/CNC integration. Int J Prod Res 43:3703–3743
107. Xu XY, Wang YY (2002) Multi-model technology and its application in the integration of CAD/CAM/CAE. J Mater Proc Technol 129:563–567
108. Yin CG, Ma YS (2012) Parametric feature constraint modeling and mapping in product development. Adv Eng. doi:10.1016/j.aei.2012.02.010
109. Zhou X, Qiu Y, Hua G, Wang H, Ruan X (2007) A feasible approach to the integration of CAD and CAPP. Comput Aided Des 39:324–338
110. Zhu GuoWang (1994) An overview of feature technology on CAD/CAM, Chin J Comput Appl 8–12. In Chinese

Data Representation and Modeling for Process Planning

Chulho Chung and Qingjin Peng

Notations

AOR	Annual operational requirement (h)
γ	Annual preventive maintenance cost including servicing, inspection, calibration
C_o	Machine capital cost
C_m	Normalized machine cost based on machine capital cost and maintenance cost
d_{ij}	Dimensional tolerance between manufacturing faces f_i and f_j
EIDL	End item design life
ET_j	Elapsed time of the *j*th preventive maintenance task for the *i*th failure mode
F_i, F_j	*i*th and *j*th manufacturing features
f_i, f_j	*i*th and *j*th manufacturing faces
m	Number of relative tolerances
MTBF	Mean time between failures
MTTR	Mean time to repair for corrective maintenance
n	Number of manufacturing faces in a part to be machined
P_{cm}	Corrective maintenance cost rate per hour, including payment and spare parts
P_{pm}	Preventive maintenance cost rate per hour, including payment and supplies
q	Number of failure modes in a machine
R_a	Roughness average
r	Number of preventive maintenance tasks in a failure mode
s_i	Surface roughness given to a manufacturing face f_i
TF_j	Task frequency of the *j*th preventive maintenance task for the *i*th failure mode

C. Chung · Q. Peng (✉)
Department of Mechanical and Manufacturing Engineering,
University of Manitoba, Manitoba, Canada
e-mail: pengq@cc.umanitoba.ca

Y.-S. Ma (ed.), *Semantic Modeling and Interoperability in Product and Process Engineering*, Springer Series in Advanced Manufacturing,
DOI: 10.1007/978-1-4471-5073-2_7,

1 Introduction

Process planning designs the details required for the manufacture of a product according to its design specifications and available manufacturing resources. It includes a variety of activities such as interpretation of product design, selecting, and sequencing of machining processes, selection of machine tools and cutting tools, determination of cutting parameters, choice of jigs and fixtures, and calculation of production time and cost. Process planning is recognized as a bridge between design and manufacturing. Computer-aided process planning (CAPP) plays an important role in an integrated CAD/CAM system [11].

CAPP systems can not only generate consistent and reasonable process plans, but also provide a user interface with CAD systems in a computer-integrated manufacturing (CIM) environment. Investigation shows that an efficient CAPP system can result in a total reduction of production costs by up to 30 %, and the length of a manufacturing cycle can also be reduced by up to 50 % [1].

As a feasible process plan has to include resources available in a specific industry where a product is produced, process planning can be a very complex process. The selection of machine tools, cutters, process parameters, and material removal processes is based on a variety of factors. Process planning relies on a good understanding of manufacturing environments. Data representation and process modeling are important parts of a feasible plan, as CAPP systems depend greatly on available data, data presentation, and communication between planners and computer systems.

A great deal of research has been done in the area of CAPP, in large part devoted to addressing the needs of dynamic processes, integration of design and manufacturing, and data collection and management in process planning. Artificially intelligent (AI) technologies have been employed in CAPP systems for the recognition and representation of part features, for machining operation and tool selection, and in operation sequence planning [2, 7–10, 15, 16].

There are two major traditional approaches to CAPP systems: variant and generative methods [14]. The variant CAPP system uses an information retrieval procedure based on group technology (GT). All the parts involved in this system are grouped into component families according to their similarity in design and manufacture. This method is useful for relatively well-established companies, as they are likely to have stable products. In contrast, the generative CAPP system uses manufacturing knowledge to generate process plans. This has been advantageous for new product development, but it relies on the development of AI technologies. Most CAPP systems have various functions to aid process planning activities, including selection of machining processes, sequencing of machining operations, choice of machine tools and cutting tools, decision of machining parameters, and estimation of time and cost. A conventional CAPP procedure is shown in Fig. 1. This process is followed a "serial" approach from the input of part information to the output of the process plan.

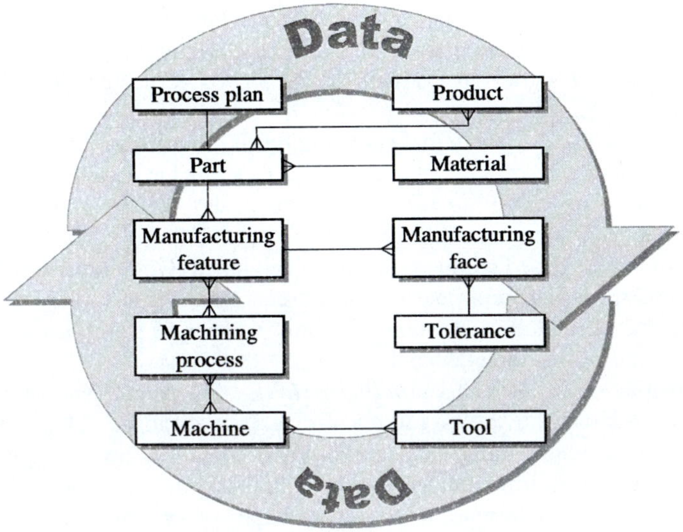

Fig. 1 Data and relationships in process planning

This chapter introduces process planning for methods in manufacturing feature recognition, set-up planning and sequencing, machining process determination, and the selection of tools and machines in a dynamic manufacturing environment. Database (DB) search algorithms and knowledge are introduced to efficiently determine tool and machine alternatives. Rotational parts are used as examples of process planning in the discussion.

2 Data Structure and Database for Process Planning

Figure 1 shows entities and data relationships in process planning. Each entity represents data from manufacturing environments associated with others in the environments. For instance, product data are based on a hierarchical structure, such as the bill of materials (BOM), which may be associated with a set of part data. A part consists of a number of manufacturing features that may require different machining processes. A manufacturing feature can be divided into a set of manufacturing faces based on tolerance assignment. A suitable process with available tools and machines can be decided based on the tolerance information and a chosen material type for a manufacturing feature. Therefore, data interaction is critical to efficiently support process planning for a manufacturing part.

The data can be defined and managed in a database management system with machine DB, tool DB, and material DB. A DB is defined with tables and relationships such as one-to-one, one-to-many, many-to-many, one-to-one recursive, and one-to-many recursive relationships [17]. For instance, 36 tables with several types of relationships (Appendix 1) are used to define the tool and machine DBs in the system introduced in this chapter. In particular, the tool DB is modeled via ISO standards. This DB is generic, so that it can be commonly used in a variety of industries. Figure 2 shows an example of the ISO code designation of external insert holders.

As shown in Fig. 2, an external insert holder is represented by ten pre-designated codes, and the description of each code is described in detail. Figure 3 shows an example of data modeling for the external insert holders shown in Fig. 2. A table is defined via the data definition language (DDL) syntax of SQL, which is presented in Fig. 3. Based on this syntax, a table named *tblExternalInsertHolder* is defined, consisting of appropriate fields and data types. Every piece of external insert holder data, including the unit price and number of quantities on the shop floor, are stored in this table as well.

As shown in Fig. 3, the *tblExternalInsertHolder* table is modeled using one-to-many relationships with tables for ten pre-designated codes. This structure is advantageous for maintaining data integrity and efficiently managing data: each table stores its own information, and this information is neither kept in other tables nor managed by them. Instead, relationships based on primary and foreign keys efficiently make connections among the tables. Moreover, primary and foreign key interactions enable a reduction in searching time to find or form specified information, allowing an index-based searching scheme.

3 Manufacturing Features and Recognition

Manufacturing features are defined here as elements that can be formed in a single manufacturing process. The feature parameters consist of design geometric data, surface roughness, dimensional and geometric tolerances, and methods to form the manufacturing feature. A feature is not directly available from a product design, such as a CAD file. Therefore, feature recognition is required to convert the design information into manufacturing features used to generate the process plan.

Several data exchange standards have been established to mediate between heterogeneous CAD systems. Among these standard data exchange formats, the mostly widely used are data exchange file (DXF), initial graphics exchange specification (IGES), and the standard for the exchange of product data (STEP). Owing to simplicity of format and ease of access, the DXF format has been selected as the data input of the CAPP system for manufacturing feature recognition in the discussion that follows here. This selection is also based on the wide use of AutoCAD systems in industries where a DXF-formatted part drawing is usually given to a planner for process planning. A feature-management module is developed here for this purpose, which supports the feature-based modeling of a part for process planning. A part design is represented by geometric entities such as lines, curves, and surfaces. From a DXF data file, the feature recognition starts with obtaining geometric entities used to form manufacturing features. Figure 4 shows the brief steps involved in converting the DXF representations into the

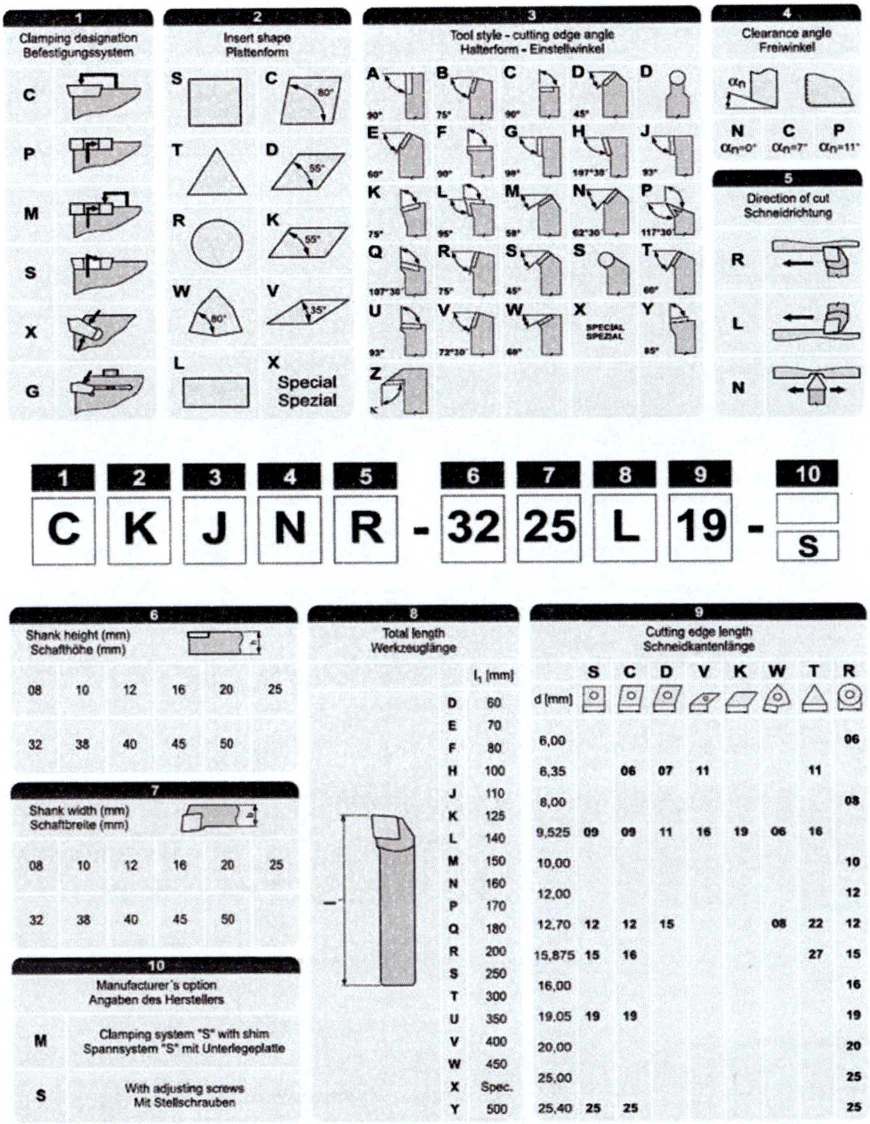

Fig. 2 Designation-external insert holders [12, 13]

geometric entities defined in this method. The detailed flowchart of these steps is illustrated in Appendix 2.

As can be seen in Fig. 4, a line entity is converted from the DXF representations such as *AcDbLine* or *AcDbPolyline*, and stored in a defined entity format (*Line_type*, *Point1*, *Point2*). *Line_type* defines either a straight line or a tangential line. In particular, if a line entity from a DXF includes the number 42, it will need

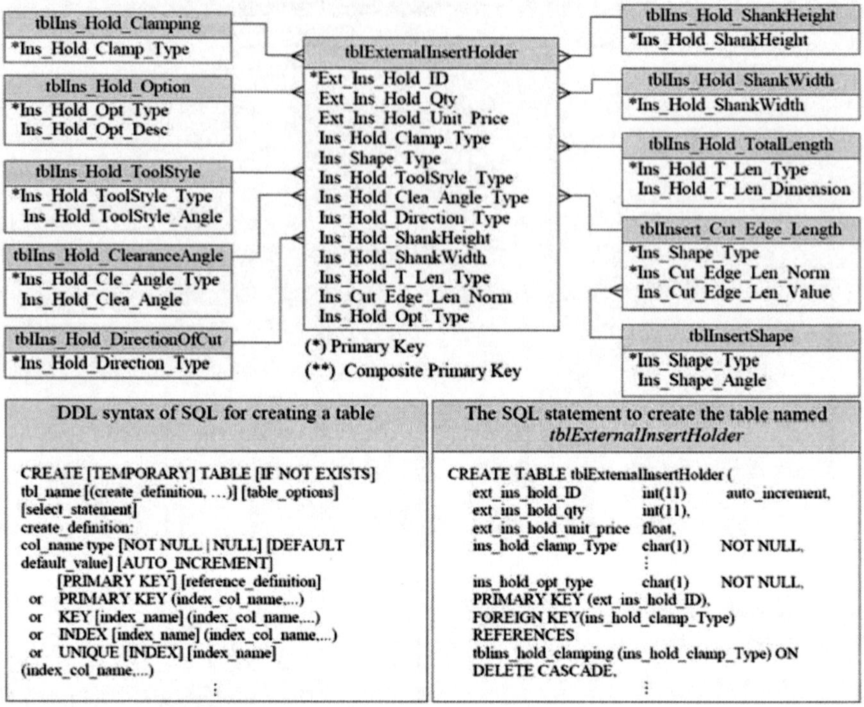

Fig. 3 An example of data modeling and table definition for the external insert holders shown in Fig. 2

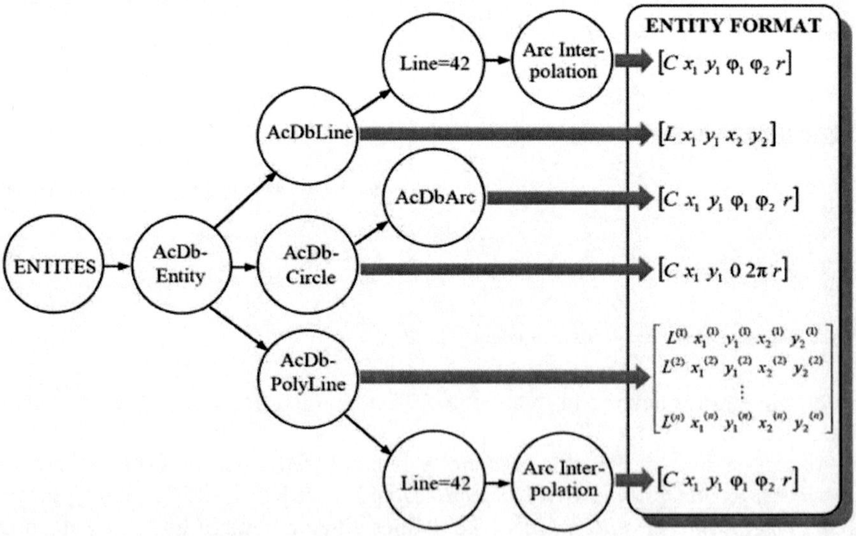

Fig. 4 Steps of converting geometric entities from DXF representations

a circular interpolation in the following line, because the DXF format does not provide information for the circle between two lines. The circular entity interpolated here is also stored in a defined circular entity format (*Circular_direction, Centre_point, Start_angle, End_angle, Radius*). In the above entity format, *Circular_direction* is defined as one of the two directions: counter-clockwise (CCW) and clockwise (CW). Figure 5 and Eqs. (1), (2), and (3) are used to interpolate a circle between two lines from a DXF.

Determination of the center point x_c and y_c is

$$x_c = \left(\frac{b_2 - b_1}{a_1 - a_2}\right), \quad y_c = a_1 x_c + b_1 \tag{1}$$

where, $a_1 = -\left(\frac{x_1' - x_0'}{y_1' - y_0'}\right)$, $b_1 = y_1' - a_1 x_1'$ and $a_2 = -\left(\frac{x_1'' - x_0''}{y_1'' - y_0''}\right)$, $b_2 = y_0'' - a_2 x_0''$

Determination of the radius r:

$$r = \sqrt{(x_c - x_1')^2 + (y_c - y_1')^2} \tag{2}$$

Determination of two angles ϕ_1 and ϕ_2:

$$\phi_1 = \left|\tan^{-1}\left(\frac{y_c - y_1'}{x_c - x_1'}\right) \cdot \frac{180}{\pi}\right| \text{ and } \phi_2 = \left|\tan^{-1}\left(\frac{y_c - y_0''}{x_c - x_0''}\right) \cdot \frac{180}{\pi}\right| \tag{3}$$

Once geometric entities are obtained and stored, the next procedure converts these entities into manufacturing features. This process uses decision-making diagrams to determine the type of feature. There are three types of decision-making diagrams (Appendix 3) developed in this research, one of them is illustrated in Fig. 6.

In Fig. 6, if the second layer entity is the same line entity as the first, it should be an *L_Void* or *E_Thread* feature depending on the type of the next layer entity. In this case, the type of the next layer entity (i.e., the third layer entity) is examined to determine the filling type of the feature for the second layer entity. Based on these rules, a feature can be formed from obtained geometric entities. In particular, an object-oriented manufacturing feature (OOMF) scheme based on object-oriented programming (OOP) is introduced here to efficiently store and manipulate the defined feature. Figure 7 shows a hierarchical class diagram of the OOMFs

Fig. 5 Circle interpolation between two arbitrary lines

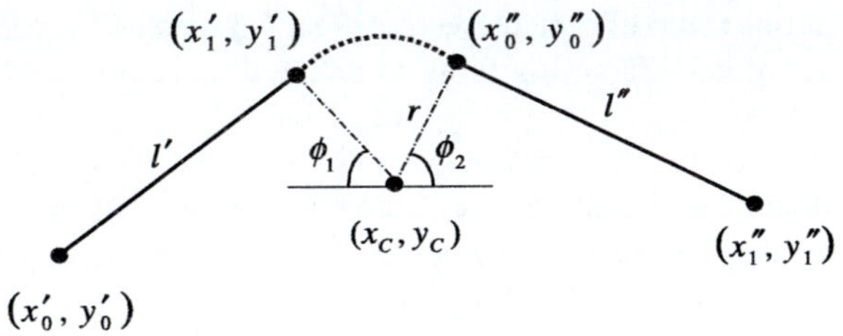

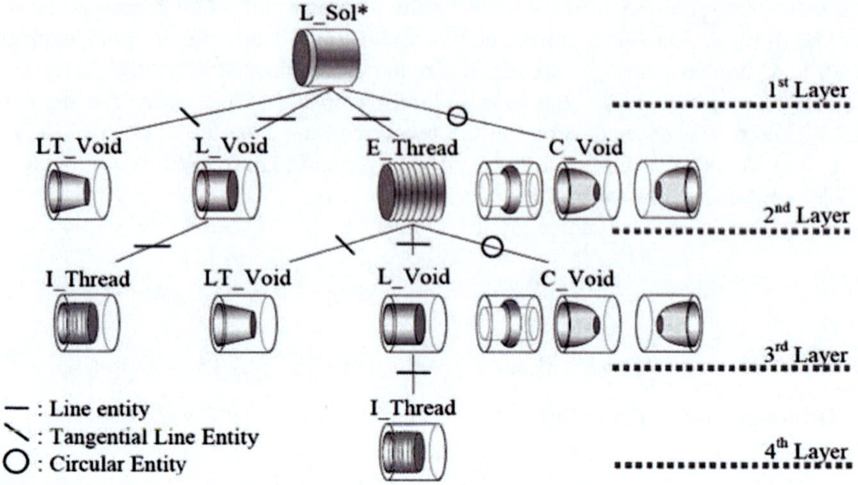

Fig. 6 A decision-making diagram for feature definition based on obtained geometric entities

representing a groove, a tap, an internal and an external cylinder, and a thread, which are inherited from a super class called the *ManufFeature* class.

The *ManufFeature* Java class is described in Appendix 4. This class is interfaced with the *FeatureManagement* class, which imports the DXF of a part, and converts the obtained geometric entities to OOMFs via methods presented earlier in this section. As shown in Appendix 4 in particular, a defined feature has the information about its manufacturing faces used for tolerancing, or setup planning and sequencing. The manufacturing faces here are simply calculated from defined features. For instance, each feature initially generates four manufacturing faces: right, left, upper, and lower faces. Then, the related manufacturing faces in adjacent features are merged or redefined via simple rules. These rules are formulated by using the normal vectors ($\pm X$, $\pm Y$) of adjacent manufacturing faces, which are examined by such properties as *prevFeatureID* and *ingredientList*.

4 Part Setup Planning and Process Sequencing

Setup planning and sequencing define a set of part faces that can be manufactured at one setup of a machine tool. The features defined in Fig. 7 can be divided into a set of manufacturing faces to define dimensional and geometric tolerances. These tolerances are categorized into local tolerances (e.g., tolerance of a diameter of a hole) and relative tolerances (e.g., a geometric tolerance such as concentricity, or a tolerance of a dimension between two faces). Depending on the number of faces used to define a tolerance, the notion of this categorization is applied for calculating tolerance factors proposed by Huang [5].

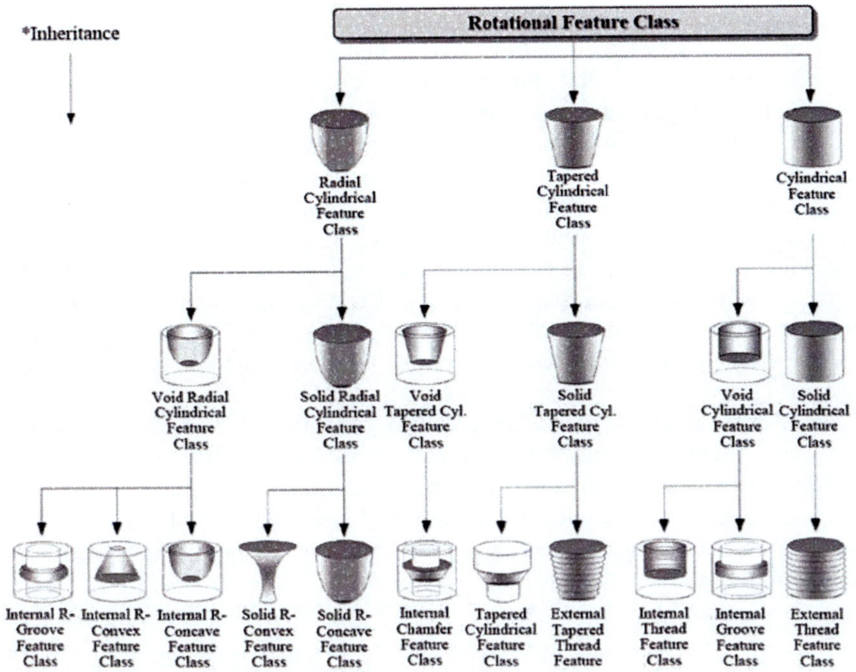

Fig. 7 Hierarchical class diagram of the OOMFs

Using this method, the setup planning and sequencing algorithm via the tolerance factors is adopted to minimize the number of setup sequences. This algorithm was suggested by Huang [5] and Huang and Zhang [6]. A tolerance defined in single or multiple manufacturing faces can be converted into a tolerance factor when dividing the tolerance value by the representative length. This is relative tolerance information, representing which manufacturing face combination has a tighter tolerance relationship in a part to be manufactured. Several examples of calculating tolerance factors are included in Appendix 5.

The setup planning and sequencing algorithm is summarized in Appendix 6. This algorithm begins by grouping manufacturing faces with tight tolerance relationships, which should be machined at the same setup based on tool approach directions. The selection of setup data is then executed to facilitate tolerance control, namely, manufacturing faces that have tight tolerance relationships to be mutually clamped and referenced. These setup data consist of a datum plane and a clamping position in a lathe machine. Subsequently, the machining procedure that determines a setup sequence is checked with the available setup data (they are the base faces for setting up machining processes) and those faces to be machined with tight tolerance relationships.

The machining process of a manufacturing face is determined by two factors: the surface roughness obtained from machining processes [3] and the ISO tolerance number (*IT-number* or ISO tolerance grade) assigned for different machining processes [18]. The former is applied when surface roughness is given to a manufacturing face while the latter is applied for a manufacturing face with a dimensional and geometric tolerance. Moreover, the most precise factor will be chosen if two or more tolerances given to a manufacturing face lead to different machining processes. Figure 8 shows a decision-making diagram to determine a machining process when surface roughness is given to a manufacturing face. In Fig. 8, a turning process based on the given surface roughness is chosen among the machining processes such as roughing, semi-roughing, finishing, fine turning, and external and internal threading. Grinding and honing are not included. Note that each referenced roughness value (i.e., 63μ in Fig. 8) refers to the roughness average (R_a) that is defined as the average deviation from the mean surface or arithmetical average.

For a dimensional or geometric tolerance given to a manufacturing face, it is necessary to determine the *IT-number* that is based on both the basic size of the face and its tolerance limitation. For instance, a manufacturing face shown in Fig. 9 is assigned a dimensional tolerance of $d = 0.2$ mm. The basic size of this face is $D = 124.9$ mm. The tolerance coefficient for this face can then be calculated as follows [18]:

$$\text{tolerance coefficient} = \frac{d \cdot 10^3}{\left(0.45D^{\frac{1}{3}} + 0.001D\right)} \tag{4}$$

Using Eq. (4), the resulting coefficient ≈ 84, and it corresponds to the *IT-number* 10 (i.e., IT10) via the tolerance-grade conversion table (refer to Appendix 7). With this *IT-number*, the semi-roughing turning process is subsequently chosen in the machining process allocation table in Appendix 7.

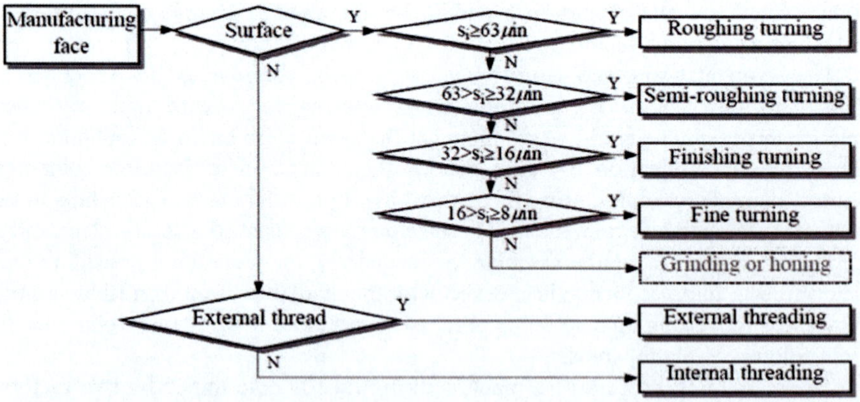

Fig. 8 A decision-making diagram to determine a turning process for a given surface roughness

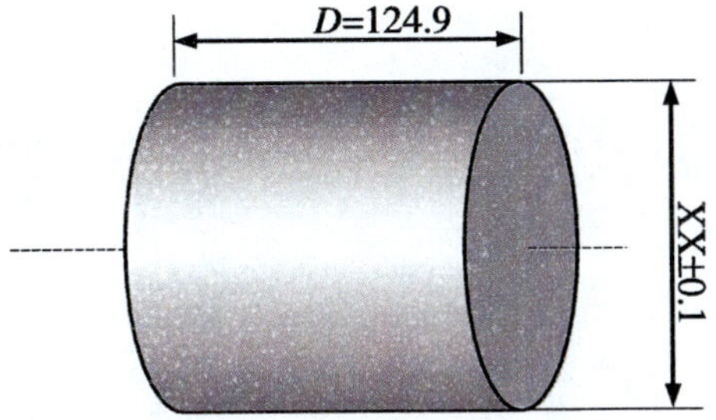

Fig. 9 A manufacturing face with a dimensional tolerance to calculate the *IT-number*

5 Tools and Machines Selection

Tools and machines can be selected in both static and dynamic manufacturing environments. Manufacturing resources (including tools, machines, and personnel) at a shop floor are always available in a static environment, which is assumed in traditional CAPP systems. In a dynamic manufacturing environment, however, it is essential to consider resource availability and time schedules for a feasible process plan [4]. The following process for selecting tools and machines will enable process planning to generate timely alternative plans for workable and economical production.

5.1 Tools Selection

5.1.1 Tool Alternative Retrieval from Tool DB

For tools selection, it is important to initially determine all possible tools that will be needed to machine the faces of a part. These tools are retrieved from the shop floor's tool DB. However, the selected tools and their combinations may create a huge amount of choices that would appear to be required to machine the whole part. In the method proposed here, two approaches are taken to reduce the number of tools retrieved from the tool DB. The first is to reduce the number of transactions with the DB, and the second is to reduce the amount of data to be retrieved, which responds to user requests. An efficient DB search strategy is required in order to satisfy these considerations.

DB search algorithms are thus developed to efficiently search the tool DB in retrieving tool alternatives for manufacturing faces of a part. These algorithms are based on dynamic SQL queries and searching criteria. Figure 10 shows a DB search algorithm to retrieve the tool alternatives for external manufacturing faces of a part. As shown in Fig. 10, this algorithm starts with forming a manufacturing face direction (i.e., *manufFaceDirection* in Fig. 10) based on a set of external manufacturing faces in a part, which are expected to process with an external

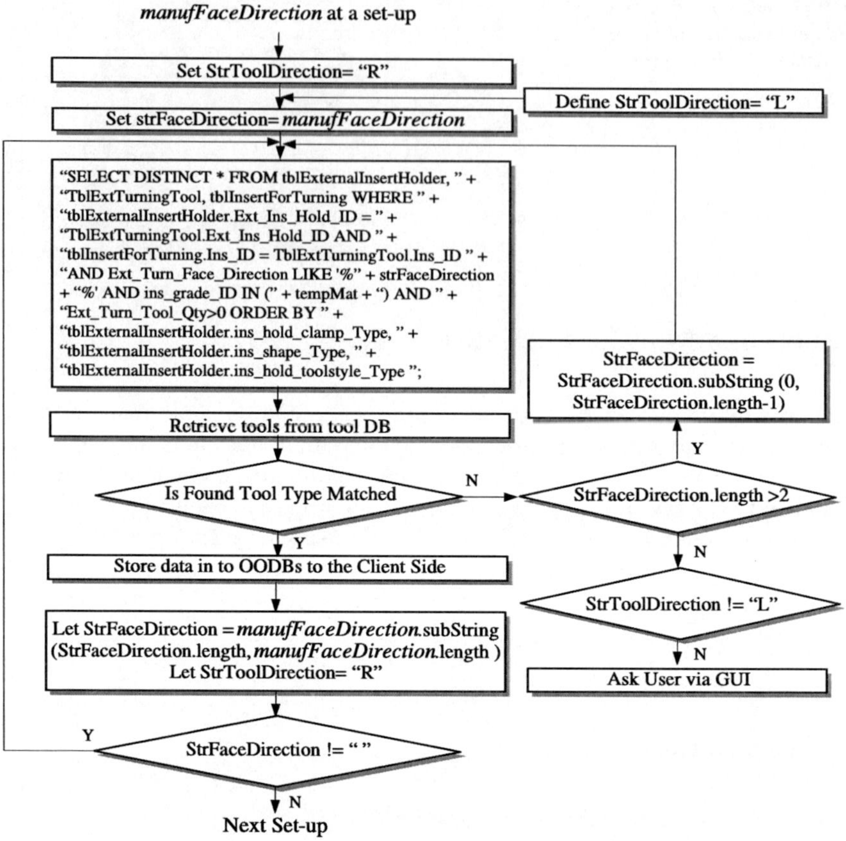

Fig. 10 DB search algorithm to retrieve possible tool alternatives for external manufacturing faces of a part

turning tool at one setup. A formed manufacturing face direction is a set of direction vectors, and is represented by a set of predefined numerals. Moreover, this set of geometry-based vectors can be easily converted into the required directional constraints of an external turning tool to machine the set of external manufacturing faces.

To retrieve the least number of tool alternatives from the tool DB, the manufacturing face direction is initially formed via all the external manufacturing faces of a part at one setup. This manufacturing face direction is then inserted into a dynamic SQL query with other search criteria, including required tool material types and tool availability. The determination of the required tool material types is based on the material type of the work-piece, hardness, and required machining processes. The material type of the work-piece and hardness are based on the part's design requirements, while machining processes are determined by given tolerances (as discussed in the previous section). To efficiently support the determination process of a tool material, a tool manufacturer's handbook [12, 13] based on ISO 513 is used to collect related knowledge. A summary of knowledge tables is illustrated in Appendix 8.

The availability of a tool is examined by checking the remaining quantity of the tool in the tool DB. As shown in Fig. 10, a searching criterion such as the *Ext_Turn_Tool_Qty>0* is simply inserted for this check. When a tool is reserved for a part to be manufactured, its quantity information in the tool DB can be reduced by one. The DML (Data manipulation language) used for defining the dynamic SQL query shown in Fig. 10 is also used for this update process (e.g., UPDATE *TblExtTurningTool* SET *Ext_Turn_Tool_Qty=Ext_Turn_Tool_Qty*-1 WHERE *Ext_Turn_Tool_ID=…*). Moreover, this process can increase the quantity of a tool in the tool DB when the tool returns to an available state.

The dynamically generated SQL query shown in Fig. 10 is sent to the DB, and is used to search appropriate external tools. If there is no tool matched in terms of the tool direction vector, the manufacturing face direction represented by numerals is reduced by a string operation. In Fig. 10, this string operation is described by the Java expression *StrFaceDirection.subString (0, StrFaceDirection. length-1)*. This reduction process is repeatedly executed until a matched tool is found. In this method, tool direction vectors can be represented as a set of directions with which the tool moves in a machine. They are also pre-stored in the knowledge DB with constraints. The used tool direction vectors and constraints are based on tool holder types and insert shapes according to ISO. Therefore, all tools designated by the ISO standard can use this knowledge without any further modification.

Figure 11 shows a type of an internal turning tool with tool direction vectors and constraints in the knowledge DB. As shown in Fig. 11, the tool type is *SCXC*, as classified by ISO. It has the tool direction vectors of *454565* and *121818*, depending on the insert-holder direction types. Four constraints for this tool type are also listed in Fig. 11. For instance, if a formed manufacturing face direction is represented as *4545*, it will be matched with the tool direction vectors of *454565* (i.e., $4545 \subset 454565$). A checking process to avoid a tool collision is then executed for the four constraints of this tool, as shown in Fig. 11, namely, the *Tool_Height* (i.e., $a + 0.5b$) < *Hole_Radius*, θ_4, $\theta_6 = 40°$, and $l_6 < c\sin 40°$.

This method is expected to reduce the number of transactions as well as overwhelming numbers of possible tool alternatives retrieved from the tool DB. It can also protect a possible tool-path from interference from adjacent features, which may occur when a tool is recommended from only one manufacturing feature. Subsequently, the tools recommended for machining a whole part consisting of manufacturing faces are stored in an object-oriented database (OODB) , called *TurningToolOODB*. In particular, the *TurningToolOODB* (see Appendix 9) stores tool information and manufacturing faces to be machined at a setup. These tool alternatives are further used to create complete tool sets to machine a whole

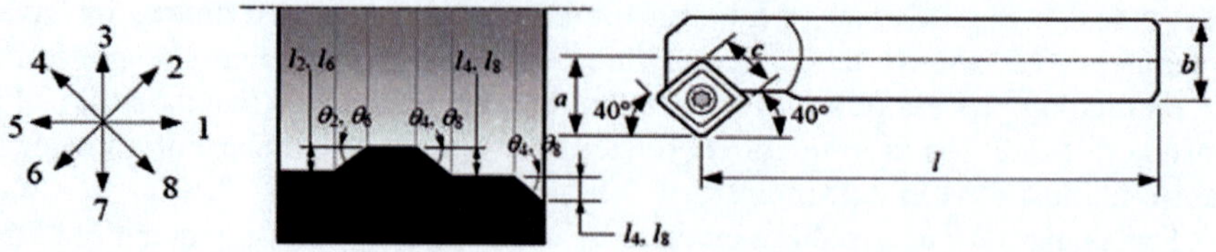

Tool type	Tool direction vectors		Constraints
	R-type	L-type	
SCXC	454565	121818	$\begin{cases} \frac{\text{hole Dia.}}{2} > \left(a + \frac{b}{2}\right) \\ \theta_\xi, \theta_\zeta = 40^\circ;\ \sum l_\zeta < c \sin 40^\circ \end{cases}$ $\text{Where, } (\xi, \zeta) = \begin{cases} (4,6) \text{ for } R-\text{type} \\ (2,8) \text{ for } L-\text{type} \end{cases}$

Fig. 11 An internal turning tool type with face direction patterns and constraints

part. Thus, the complete tool sets form a basis for performing the tool- and machine-selection process by weighting the factors of production time and cost, which will be further discussed later in this chapter.

5.1.2 Creation of Complete Tool Sets Based on Tool Alternatives

The creation of possible complete tool sets is based on all tool alternatives retrieved for a whole part. These tool alternatives are stored in the *TurningToolOODB*. Because a huge amount of possible tool sets can be created, an algorithm is developed to efficiently generate complete tool sets using the *TurningToolOODB*. For instance, Fig. 12 shows manufacturing faces to be machined at one setup. As shown in Fig. 12, the manufacturing faces are grouped into *p*-sets of manufacturing faces. Each face set (i.e., $i = 1, 2, \ldots, p$) is machined by a tool type but it can have a number of tool alternatives of different types. In Fig. 12, the number of tool alternatives for each face set is represented as *Nt*(*i*). The total number of possible tool sets created is $Nt(1) \times Nt(2) \times \cdots \times Nt(p)$.

Algorithm 1 is used to create possible complete tool sets that are based on the retrieved tool alternatives for manufacturing faces in one setup. In Algorithm 1, the *TurningToolOODB* is an array of the *TurningToolOODB* classes to store tools, and possible tool sets (i.e., *toolset* array) store indexes of the *TurningToolOODB* array. For the rest of the setups, this algorithm is repeatedly applied in the order of the setup sequences discussed in Sect. 4. The *toolset* array subsequently stores all the possible tool sets for a part.

Algorithm 1: Creation of possible tool sets

Input $TurningToolOODB[p][t_m]$, where $t_m = \max \forall Nt(i), i = 1, 2, \ldots, p$
Let $t_{all} = Nt(1) \times Nt(2) \times \cdots \times Nt(p)$
Let $toolSet\ [t_{all}][p]$
For $i = 0, \ldots, t_{all} - 1 \Rightarrow$
  Let $k = i$
  For $j = 0, \ldots, p - 1 \Rightarrow$
    $toolSet\ [i+1][j+1] = k\ \mathbf{mod}\ Nt(j+1) + 1$ /* mod : Integer Modulus */
    $k = k \setminus Nt(j+1)$ /* \ : Integer Division */

5.1.3 Tool Selection

Once possible tool sets (i.e., *toolSet*[][] in Algorithm 1) are created, the next step is to rank each tool set with weighting factors related to production time and cost. An optimized equation (see Appendix 10) is used for ranking, as suggested by Usher and Fernandes [16]. In Appendix 10, Eqs. (7) and (8) are used to calculate production time and cost, respectively. The best tool set to machine a whole part will be the set with the minimum score after calculating with Eq. (9) from the Appendix 10. The equation determines the minimum value of combining production time and cost.

A total cutting length (L) is required to calculate machining time [i.e. $T_m = L/(f_r \times N)$] in Eq. (7). The total cutting length is determined as a machining volume (MV) divided by the required cutting depth for a machining process. In this method, several algorithms are developed to build 2D MVs. As shown in Fig. 13a, there is a set of manufacturing faces to be machined at a setup.

By contouring the geometry defined by the manufacturing faces, an MV is first generated for a fine or finishing process as shown in Fig. 13b. The depth (ε) in

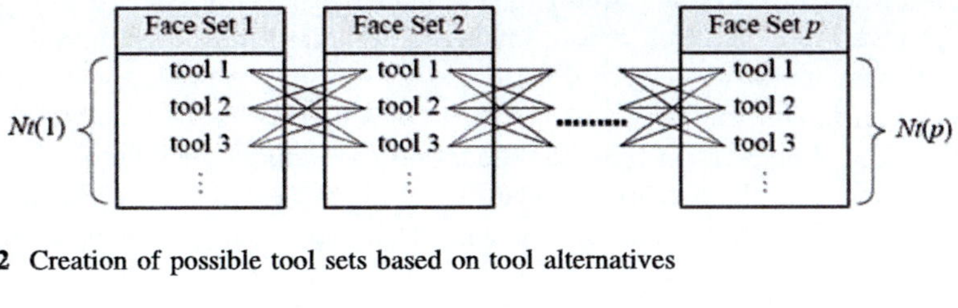

Fig. 12 Creation of possible tool sets based on tool alternatives

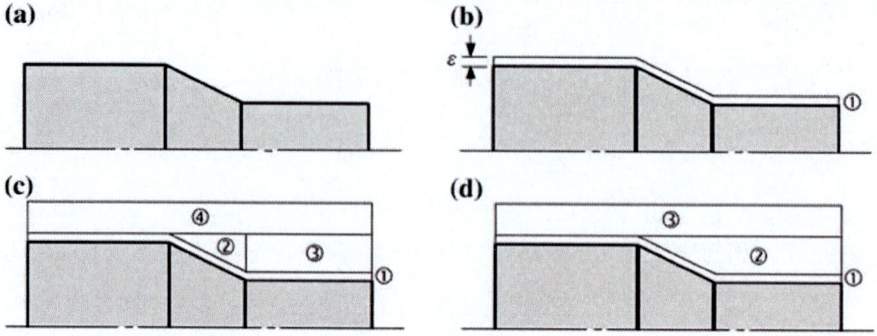

Fig. 13 Machining volume generation in this research

Fig. 13b is determined here to allow three cuts for the generated MV. As shown in Figs. 13c and d, other MVs for rough cutting are subsequently generated and merged via the algorithms suggested in this chapter.

In calculating the machining time using Eq. (7) (see Appendix 10), the most rigid tool in a recommended tool set is first assigned to machine the generated MVs for rough cutting. Then, each tool in the tool set is used to machine the MV for the finishing process. The machining parameters such as depth of cut, feed rate, cutting speed, and tool life are collected from tool manufacturers' catalogs [12] and stored in the knowledge DB. Appendix 8 summarizes knowledge tables used for calculating machining parameters, and presents an equation to calculate the optimized cutting speed using the knowledge tables.

5.2 Machine Selection

The machine selection for a particular part is similar to the process of tool selection. A DB search algorithm is presented to efficiently search the machine DB using a dynamic SQL query based on defined searching criteria. It is also advantageous to reduce the number of available machines retrieved from the machine DB. Constraints used to select a machine include machine availability, maximum turning diameter and length, maximum spindle speed, maximum motor power, and tail stock stroke. These constraints are inserted into a dynamic SQL query to search appropriate machine alternatives for a tool set. Figure 14 shows the developed DB search algorithm for the machine selection.

For a given tool set, such constraints as the required maximum spindle speed and maximum motor power are determined by consulting the knowledge tables and equations in Appendixes 8 and 11. The knowledge tables and equation are also summarized in Appendix 11 based on the tool manufacturer's handbook [13]. As a searching criterion, the machine availability is simply examined by checking the two time-related fields used in the machine DB, such as *lm_machine_start_date*

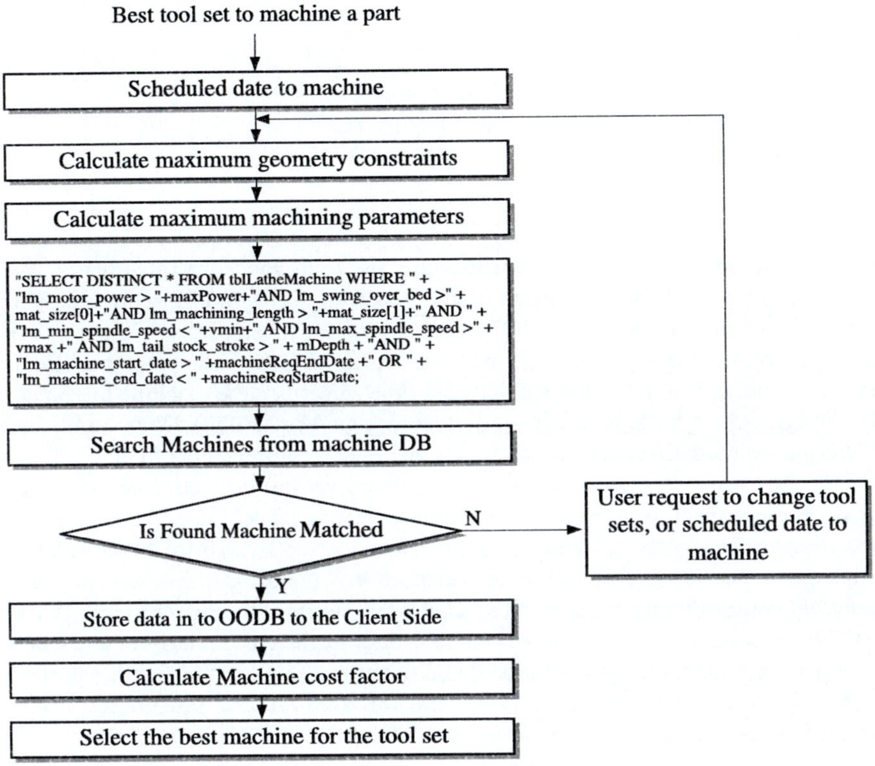

Fig. 14 DB search flow chart to retrieve possible machine alternatives

and *lm_machine_end_date* of the *tblLatheMachine* table (see Appendix 1). These two fields are updated when a lathe machine is reserved for a part to be manufactured. The DML of the SQL is used for this update process (i.e., an UPDATE statement).

As shown in Fig. 14, the best tool set ranking method that has been described in Sect. 5.1.3 is first applied in the procedure for retrieving possible machine alternatives. If there is no machine matched, then there are two options: the user can change the time schedule of machine usage, or the second-best tool set can be applied by the user. When the process is complete, machine alternatives selected for a tool set are stored in OODBs, named *MachineOODB*s (see Appendix 12).

To select the best machine in *MachineOODB*s for a tool set, a machine cost model is developed. The machine cost model consists of two parts: machine capital cost and maintenance cost. Thus, the developed cost model for a single machine selection is described by Eq. (5).[1]

[1] The symbols used in Eq. (5) can be found in the Notations list at the beginning of this chapter.

$$
\begin{aligned}
C_m &= \sum_i^q \left(\left(\frac{C_o}{\text{EIDL}} + \frac{\text{MTTR}}{\text{MTBF}} \times P_{\text{cm}} \right) \cdot \text{AOR} \right)_i + \sum_i^q \left(\sum_j^r (\text{TF}_j \cdot \text{ET}_j) \times P_{\text{pm}} \right)_i \\
&= \sum_i^q \left(\left(\frac{C_o}{\text{EIDL}} + \frac{\text{MTTR}}{\text{MTBF}} \times P_{\text{cm}} \right) \cdot \text{AOR} + \gamma \right)_i \\
&\cong \left(\frac{C_o}{\text{EIDL}} + \frac{\text{MTTR}}{\text{MTBF}} \times P_{\text{cm}} \right) \cdot \text{AOR} + \gamma
\end{aligned} \tag{5}
$$

Two system effectiveness measures are used in this equation: mean time between failures (MTBF) and mean time to repair (MTTR). MTBF represents how often a machine is failing, while MTTR represents the mean of the distributions of the time-intervals needed to repair an item. In Eq. (5), the value of a machine is reduced annually, and it will eventually become zero when the machine life equals its design life, which is called end item design life (EIDL). This value can therefore be normalized based on annual operation requirements (AOR).

The maintenance cost is divided into preventive and corrective maintenance cost. With respect to scheduled preventive maintenance intervals, preventive maintenance cost can be defined as the annual cost to perform tasks such as adjustment, servicing, calibration, and inspection without including any machine failure. On the other hand, corrective maintenance cost (with respect to MTBF and MTTR for corrective maintenance) can be defined as the cost to perform tasks such as repair, replacement, and operational and functional tests, in addition to the same tasks of preventive maintenance done with machine failure. Equation (5) is the result normalized by AOR to give the two categories the same dimension.

6 Processing Simulation

6.1 Graphical User Interface

As shown in Fig. 15, a graphical user interface (GUI) is implemented for simulation of process planning. The main GUI in Fig. 15 consists of five components: (1) an applet for server selector, (2) an applet for feature and face managements, (3) an applet for Java 2D, (4) an applet for part and machining information, and (5) a VRML browser. Three sub-GUIs based on the Java *Frame* class support users in three activities: (1) selecting a part DXF and choosing its material, (2) providing input parameters used in tools and machines selection, and (3) showing a generated process plan that is used for machining simulation.

6.2 Data Communication for Simulation

Data communication between the simulation and process plan is supported by the interaction between Java and the virtual reality modeling language external

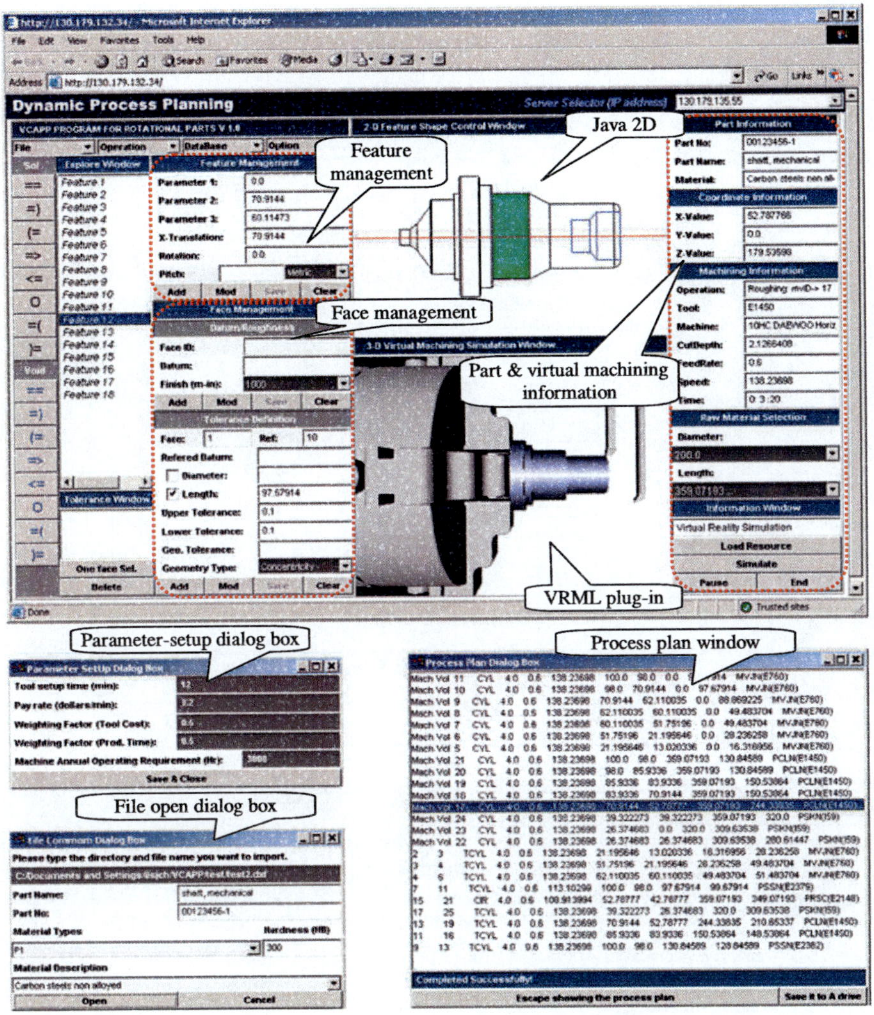

Fig. 15 GUI for process planning

authoring interface (VRML-EAI). Figure 16 represents the implementation of communication between a Java applet and a VRML plug-in. With the external authoring interface (EAI), it is possible for an applet method in a standard Web browser to access the scene graph of an embedded VRML plug-in. As shown in Fig. 16, the EAI provides two important services: enabling external programs to read and change the scene graph, and enabling external programs to register some of their functions as callbacks. Every callback is bound to an event. Whenever this event is generated in the VRML scene, the browser invokes the associated callback function and passes the current value of the event as an argument.

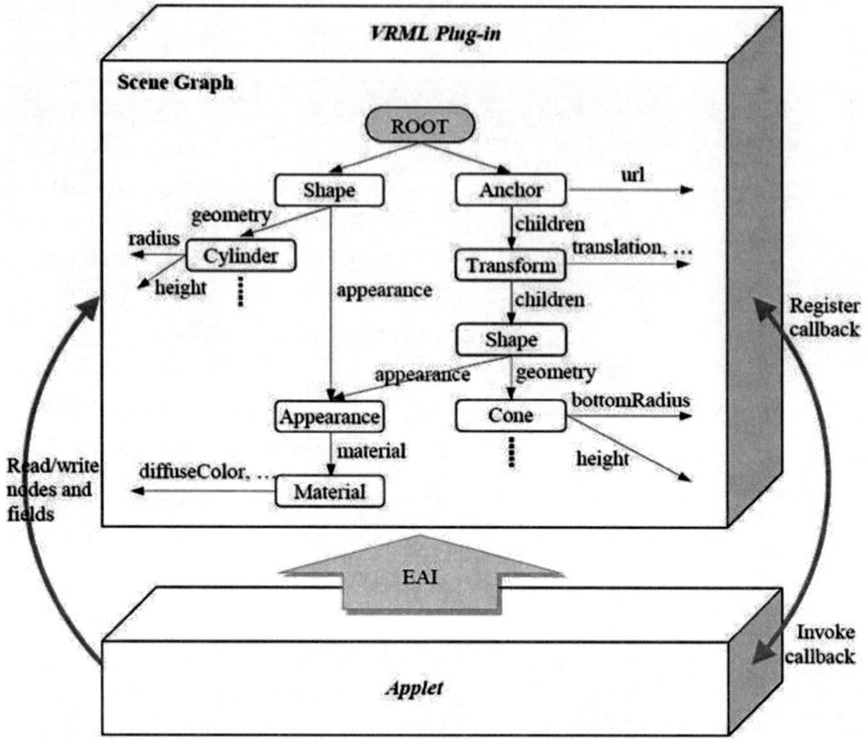

Fig. 16 Implementation of communication between a Java applet and a VRML plug-in in a Web-based client application

The following is a part of the program description:

```
import java.awt.*;
import java.awt.event.*;
import java.applet.*;
import vrml.eai.*;
import vrml.eai.event.*;
import vrml.eai.field.*;
public class VR_EAI_Applet1 extends Applet implements VrmlEventListener{
        Browser browser0;
                    ⋮
        private EventOutSFBool [] a_TouchSens_changed=new EventOutSFBool
                [nTotalNode];
                /* Total number of nodes consisting of parts and fasteners */
        private EventOutSFVec3f[] a_translation_changed=new EventOutSFVec3f
                [nTotalNode];
                    ⋮
}
```

This program first imports a Java package called *vrml.eai*, provided by the Cortona VRML plug-in. The package contains classes and sub-packages of the EAI. An EAI interface called *VrmlEventListener* is implemented to allow a VRML browser to invoke a callback occurring at a VRML object; subsequently, the applet receives an event message from the VRML browser. Inside the program, a VRML browser is declared with EAI variables including *EventOutSFBool* and *EventOutSFVec3f*. In particular, these variables are assigned to parts in a retrieved product from a DB, and used to receive event messages from the VRML browser. If an object in the VRML browser is touched by a user, for instance, the value of its *EventOutSFBool* variable changes and this result is signalled to the applet.

The following are Java program codes written in a Java method called *start()*, in which VRML browsers, VRML nodes, and accessible VRML variables are sequentially defined. This program shows how to define the touch sensors of components in a Java applet. The touch sensors are efficiently used for an interaction between a user and VRML objects.

```
public void start(){
    try {
            int i, j;
            browser0=BrowserFactory.getBrowser(this,null,0);
            if(browser0!=null){System.out.println("Accessing Browser0");}
                                    ⋮
            for (i=0, j=0;i<nTotalNode;i++) {
                    Node a;
                    if (i<nTotalPartNode) /* total number of parts retrieved */
                                    a=browser0.getNode("tPart"+i);
                    else
                                    a=browser0.getNode("tFastener"+j++);
                    a_TouchSens_changed[i]=(EventOutSFBool)a.getEventOut
                                    ("isActive");
                    a_TouchSens_changed[i].setUserData(new Integer(i));
                    a_TouchSens_changed[i].addVrmlEventListener(this);
                                    ⋮
            }
                                    ⋮
        }
    catch(VrmlException e){System.out.println("error occurred : "+e);}
}
```

For instance, a VRML browser, *browser0*, is obtained within a Java applet by calling up the *BrowserFactory.getBrowser()* method. Then, its nodes are defined by invoking the *getNode()* method. The nodes shown here are VRML touch sensor nodes, each of which is attached to a component. An *eventOut* variable (e.g., *isActive*) of the defined VRML node is assigned to an *a_TouchSens_changed* variable declared in a Java applet. Subsequently, this *a_TouchSens_changed* variable interfaces with *VrmlEventListener* to await an event message. Moreover,

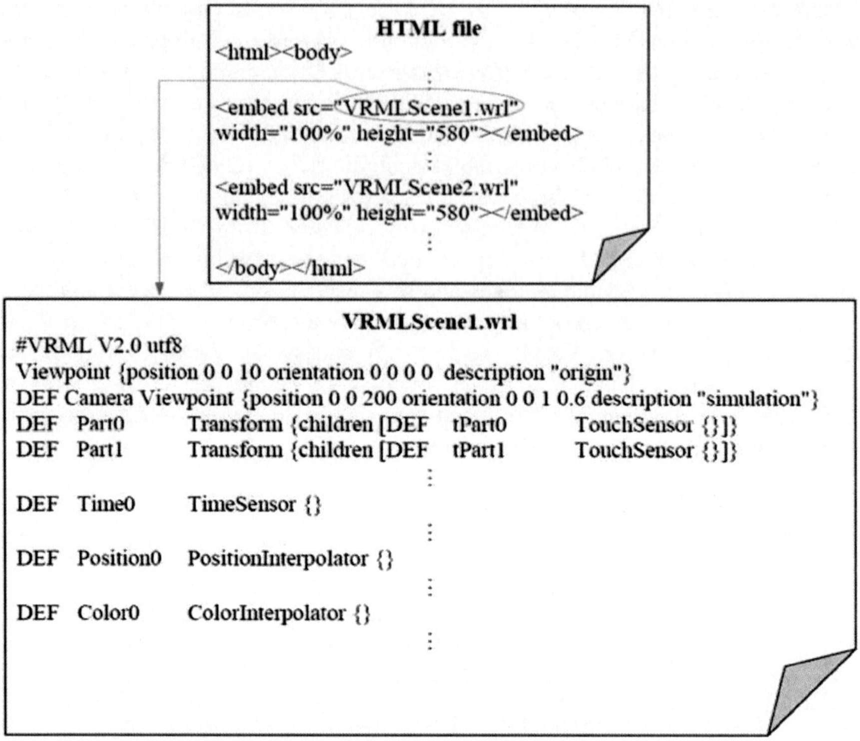

Fig. 17 VRML file integrated into an HTML file

each *a_TouchSens_changed* variable is given a serial number by invoking the *setUserData()* method. This allows quick identification of a VRML object generating an event message during user-VRML interactions.

In the above program code, a defined touch sensor node is the one bound with *DEF* to the node named $tPart_i$ in a VRML file. *DEF* in VRML is a VRML keyword and enables a VRML node to have a specific name which enables other nodes to call it by the unique name. Node names such as $tPart_i$ and $tFastener_i$ should be predefined within an empty *Transform* node in a VRML file (**.wrl*). Figure 17 shows a predefined VRML file, which is integrated into an HTML file with the HTML tag *<embed>*.

As shown in Fig. 17, a VRML node bound with *DEF* can be dynamically controlled by a Java program. For instance, *Part0* node in the *VRMLScene1.wrl* is used to represent part geometry, appearance, and transformations, while its embedded touch sensor called *tPart0* is used to generate an event message during user-VRML interactions. Moreover, other empty nodes in Fig. 17 are predefined in the VRML file, including time sensors, position interpolators, and color interpolators. These nodes are used to implement 3D process planning simulation.

Once VRML browsers, VRML nodes, and accessible VRML variables are defined, the Java program waits for an event message generated in a VRML browser. A VRML-EAI method named *eventOutChanged()* accomplishes this function in the Java program. The following Java program code describes how this method is applied:

```
VRMLGeometryData[] vr;
VRMLModelHandler vh=new VRMLModelHandler();
                        ⋮
public void eventOutChanged(VrmlEvent event){
    ItemEvent a=null;
    String nodeName;
    Integer infor=(Integer)event.getData();
    int inf=infor.intValue();
    float[] transVal=new float[3];
    if (inf<nTotalNode) {
        lstPartList_itemStateChanged(a);
        lstFastList_itemStateChanged(a);
        this.lstPartList.select(inf);
        nodeName=lstPartList.getSelectedItem();
        if (funcKey==0) vh.removeComponent(browser0, vr, nodeName);
        if (funcKey==1) vh.LoadComponent(browser0, vr, nodeName);
        if (funcKey==2) vh.highLightOneComponent (browser0, vr, nodeName);
        if (funcKey==3) vh.hideComponent (browser0, vr, nodeName);
        if (funcKey==4) vh.showComponent (browser0, vr, nodeName);
        if (funcKey==5) {
                vh.moveComponent (browser0, nodeName);
                transVal=a_translation_changed[inf].getValue();
                vh.movingAngleCone(transVal);
                }
                        ⋮
        }
                        ⋮
}
```

In this program, an event message is delivered from a VRML browser to *eventOutChanged()* method in the Java applet program. The program then finds the assigned number of the VRML object that generates the event. Based on this number and a function key triggered by a user, the program invokes various interactions including loading, hiding, deleting, showing, highlighting, and moving a VRML object. The following sections describe two major interactions implemented for static and dynamic VRML-EAI-Java interactions.

6.3 Static Interaction

The static interaction manipulates three properties of a simulation object: its geometry, its appearance (e.g., colors, lighting, and, textures), and the transformations of its position, orientation, and scaling. A *getEventIn()* node is used to allow an external program to change the properties of the simulation object.

Geometry represented by VRML can be defined and deleted by manipulating the *addChildren* and *deleteChildren* property of a target VRML node. *EventIn* typed variables in VRML are used to set a value into the VRML node. For instance, the following Java program codes describe how to load a VRML model into a VRML browser. In the program code, the *EventInMFNode* typed variable called *set_ROOT_addChildren* is declared to handle the *addChildren* property of the target node. Then, the *Node[]* typed variable called *node* initially stores the actual VRML code of a component to be loaded. This VRML code is text based and dynamically generated by Java. The defined *node* variable is transferred to the VRML node via the *setValue()* method.

```
public void LoadComponent(Browser b, VRMLGeometryData[] vr, String nodeName)
{
    browser=b;
    int arrayNo;
    /* Find the array number of the VRMLGeometryData with the nodeName */

    Node root=browser.getNode(nodeName);
    EventInMFNode set_ROOT_addChildren=(EventInMFNode)root.getEventIn
                ("addChildren");
    Node[] node=browser.createVrmlFromString ("Transform { \n" +
                " translation 0 0 0\n" + " children [\n" + vr[i].getVRMLCode()
+"]}");
    set_ROOT_addChildren.setValue(node);
}
```

Similarly, the appearance and transformations of a node are also set or changed by manipulating node properties with *getEventIn()* and *setValue()* methods. For instance, node appearance (including color, lighting, and texture) changes through the manipulation of such properties as *transparency*, *diffuseColor*, *emissiveColor*, *shininess*, and *specularColor*. The method applies this appearance manipulation for several user-interactive effects, including selecting, hiding, and showing a VRML object. The following Java program codes describe how to highlight a VRML object with user selection:

```
public void highLightOneComponent (Browser b, VRMLGeometryData[] v, String
                                   nodeName) {
        int i;
        browser=b;
        EventInSFColor set_HightLight_Color;
        float[] fColor=v[0].getSelectionColor();
        for (i=0;i<v[0].nTVRML;i++) {
                Node root=browser.getNode("m"+v[i].name);

        set_HightLight_Color=(EventInSFColor)root.getEventIn("diffuseColor");
                set_HightLight_Color.setValue(v[i].getOrigColor());
        }
        Node root=browser.getNode("m"+nodeName);
        set_HightLight_Color=(EventInSFColor)root.getEventIn("diffuseColor");
        set_HightLight_Color.setValue(fColor);
}
```

In this program code, all VRML objects in a VRML browser are changed to their original colors by setting the *diffuseColor* property via the *getEventIn()* and *setValue()* methods. Then, a pre-defined color is assigned to a selected VRML object by *nodeName*.

6.4 Dynamic Interaction

Dynamic interactive simulation is implemented by VRML interpolators including *PositionInterpolator* and *ColorInterpolator*. These VRML interpolators create a time-dependent simulation with a VRML object. The basic concept of these interpolators is illustrated in Fig. 18. As shown in the figure, a clock based on *TimeSensor* starts generating events immediately after it is created. The value of its *eventOut fraction_changed* is a type of *SFFloat* (single float data type) and its value is in the interval key value [0, …, 1]. Subsequent events have increasing values. By default, if the end of the interval is reached, no more events are generated. If the field loop has true value, then, when reaching the end of the interval, the clock starts over at the beginning of the interval. As shown in Fig. 18, the event *fraction_changed* is routed to the *event_fraction* for either *PositionInterpolator* or *ColorInterpolator*. This node generates the *eventOut value_changed* of type *SFVec3f* (e.g., positions in 3D space or RGB color codes). These are sent to such *eventIn* properties as *set_translation*, *set_scale*, and *set_diffuseColor*.

Figure 19 shows the concept of the interaction process in brief. OOMFs and a generated process plan are imported into the VRML simulation module. The OOMFs are used to define a number of cylinders for simulating a lathe machining process. The radius of each VRML cylinder node is the same size as that of a chosen work-piece, and its thickness is assigned a default value, 0.5 mm. There would thus be 480 VRML cylinders defined when selecting an initial work-piece of 240 mm.

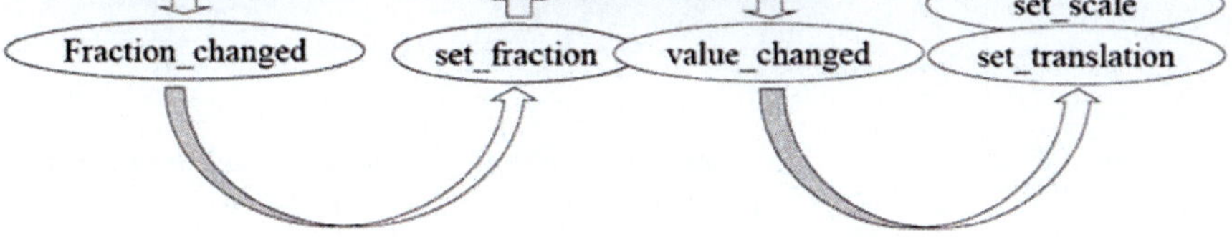

Fig. 18 Basic animation concept of VRML interpolators

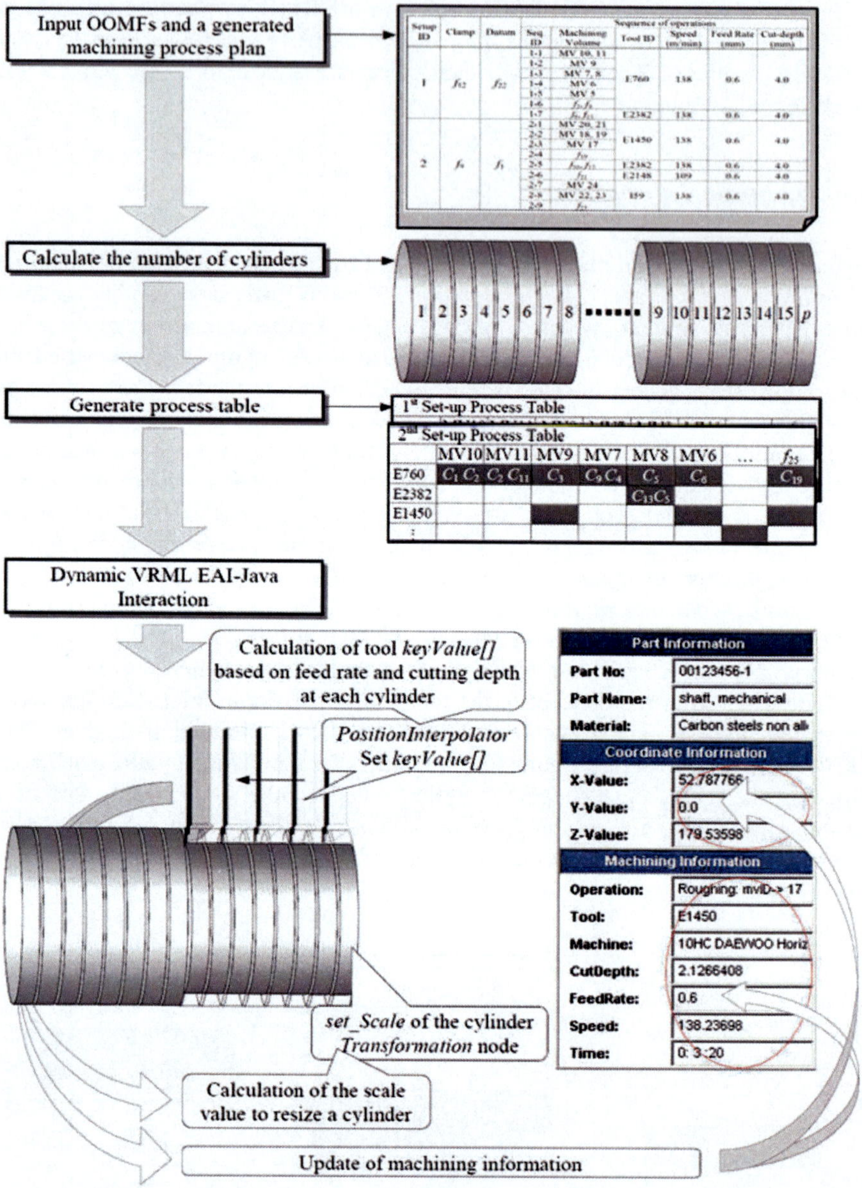

Fig. 19 VRML-EAI-Java interaction for the machining process machining process

Process tables in Fig. 19 are formed by the defined cylinders and the generated process plan. In a sequence, each MV consists of a series of VRML cylinders. When a tool moves along an MV, its cylinders are resized by manipulating the *set_scale* property of the *Transformation* node. Once process tables are completely formed, a dynamic VRML-EAI-Java interaction is executed as shown in Fig. 19. At each cylinder that belongs to an MV to be machined, the following processes are repeatedly executed: (1) calculation of *keyValue [Current_Tool_Position, Next_Tool_Position]*, (2) initiation of a *PositionInterpolator*, (3) transmission of the *keyValue[]* to the *PositionInterpolator*, (4) a tool move and stop, and (5) an update of coordinate and machining information including accumulated machining time.

The following Java program code describes how to implement tool position interpolation and resize a VRML cylinder according to the process shown in Fig. 19.

```
boolean isContinue;
Browser b=browser0
float[] fkey={0.0f,1.0f};
float[][] keyValue=new float[2][3];           /* (x_s, y_s, z_s) and (x_e, y_e, z_e) */
Node tool = b.getNode("Tool0");               /* Tool */
Node timer = b.getNode("Timer0");             /* Timer node */
Node position=b.getNode("Position0"); /* PositionInterpolator node */
                    ⋮
EventInSFTime interval = (EventInSFTime) timer.getEventIn("cycleInterval");
isActive= (EventOutSFBool) timer.getEventOut("isActive");
EventInSFBool loop=(EventInSFBool)timer.getEventIn("loop");
EventInMFFloat key=(EventInMFFloat) position.getEventIn("key");
key.setValue(fkey);
EventInMFVec3f keyVal=(EventInMFVec3f) position.getEventIn("keyValue");
b.addRoute(timer,"fraction_changed",position,"set_fraction");
b.addRoute(position,"value_changed",tool,"set_translation");
                    ⋮
/* Calculation of keyValue[] */
keyVal.setValue(keyValue); /* Transmission of KeyValue to PositionInterpolator */
interval.setValue(speed);    /* Determination of animation speed */
loop.setValue(true);                 /* Tool activation (Loop mode) */
loop.setValue(false);                /* Turn off the loop mode */
do {
        isContinue=isActive.getValue();
} while (isContinue);
                    ⋮
/* Find a cylinder k to be resized */
/* Calculate the new radius of the cylinder: newRadius=current radius-cutting depth */
Node cylinder = b.getNode("Cyl"+k);
EventInSFVec3f t_scale = (EventInSFVec3f) cylinder.getEventIn("set_scale");
s_xyz[0]=newRadius/radius;
s_xyz[1]=1f;
s_xyz[2]=newRadius/radius;
t_scale.setValue(s_xyz);
                    ⋮
b.deleteRoute(timer,"fraction_changed",position,"set_fraction");
b.deleteRoute(position,"value_changed",tool,"set_translation");
```

7 Examples

To evaluate the proposed method, two examples of rotational parts are tested. Tooling data for process planning are collected from product catalogs [12] and stored in the tool DB. The tool DB contains 4,200 tools, including external, internal, grooving, and threading tools. External turning tools consisting of various external holders and inserts are listed in the MySQL query test window shown in Fig. 20. The transaction result shown is generated via a SELECT query based on the DML of SQL. From the tool DB, this query retrieves only the tool information satisfying its selection criteria. The data from five computer numerical control (CNC) machines are stored in the machine DB. However, most information related to the machines is assumed in Table 1.

As shown in Table 1, an EIDL of 60,000 operation hours is given to all the machines equally, and machine cost is assumed in accordance with the complexity of the machines. The mean time between failures (MTBF) is used to measure the machine reliability. Based on this notion, it is assumed that more complex machines are more reliable in accomplishing their missions, although they come with a much higher maintenance cost. This assumption is supported by the fundamental concepts of existing engineering design, such as redundancy, fail-safe, or damage tolerance design.

A similar assumption is used for mean time to repair (MTTR) for corrective maintenance. The more expensive machines are normally design for easy repair so that the MTTR can be reduced. The self-diagnosis function, the built-in test (BIT), and modularization for easy replacement, can support this assertion. In spite of lower MTTR, the maintenance cost rate of the more expensive machines is much higher. The maintenance cost, including preventive and corrective maintenance costs, consists of the costs associated with diagnosis, spare parts, and paying skilled personnel.

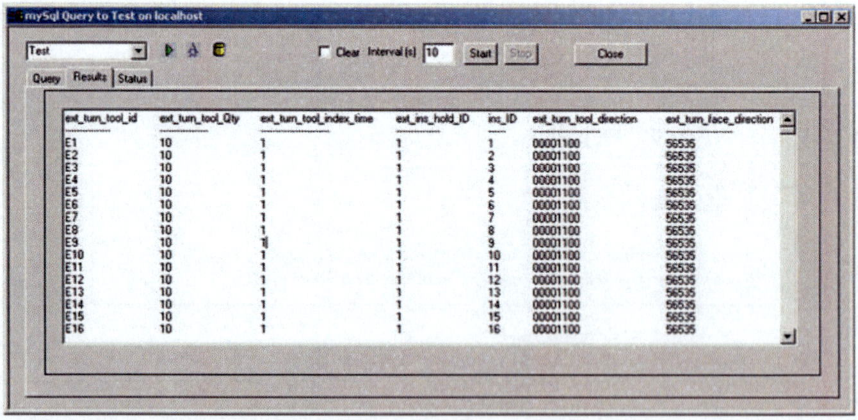

Fig. 20 Transacted result of external turning tools in the tool DB

Table 1 Assumed machine data

Machine model	Machine cost (C$)	EIDL (h)	MTBF (h)	MTTR (h)	Preventive maint. cost per Year (C$/Year)	Corrective maint. cost rate per hour (C$/h)
10HC DAEWOO Horizontal Lathe	80,000	60,000	100,000	20	400	800
TUR-630 M	30,000	60,000	30,000	85	200	400
CJK0620 NC Lathe	40,000	60,000	20,000	100	200	400
CJK0632 NC Lathe	50,000	60,000	20,000	100	200	500
CJK6125 NC Lathe	55,000	60,000	20,000	100	200	550

7.1 Example 1

As shown in Fig. 21, a part is tested to evaluate the approach developed here. The part is made of the heat- and creep-resistant steel that is classified in the M1 group (see Appendix 8) based on ISO 513, and has a hardness of 300 HB (Brinell hardness). The part has four specified tolerance limits, consisting of one geometric and three-dimensional tolerance limits. In particular, the concentricity with the ∅0.05 mm tolerance is applied to the geometric tolerance. Based on the geometric entities obtained from the DXF, OOMFs are defined by the algorithms developed in Sect. 3. As shown in Fig. 22, 17 OOMFs are defined for this example, and 25 manufacturing faces are derived from the OOMFs.

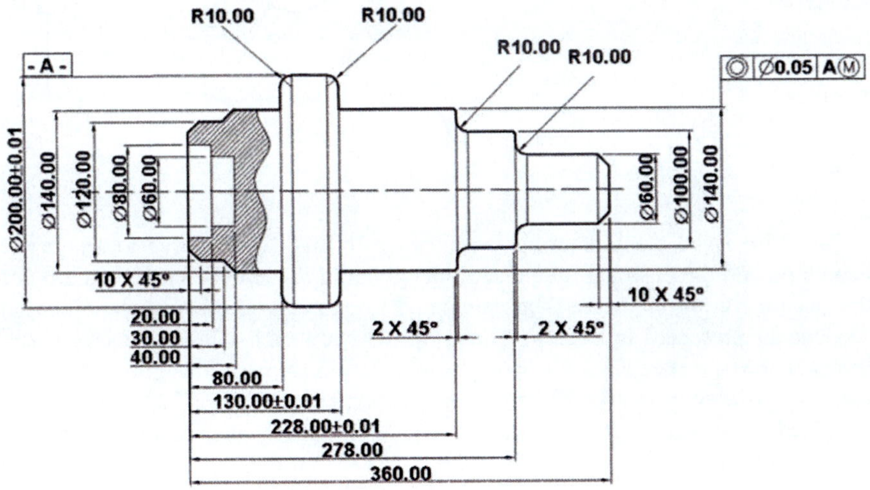

Fig. 21 Sample part

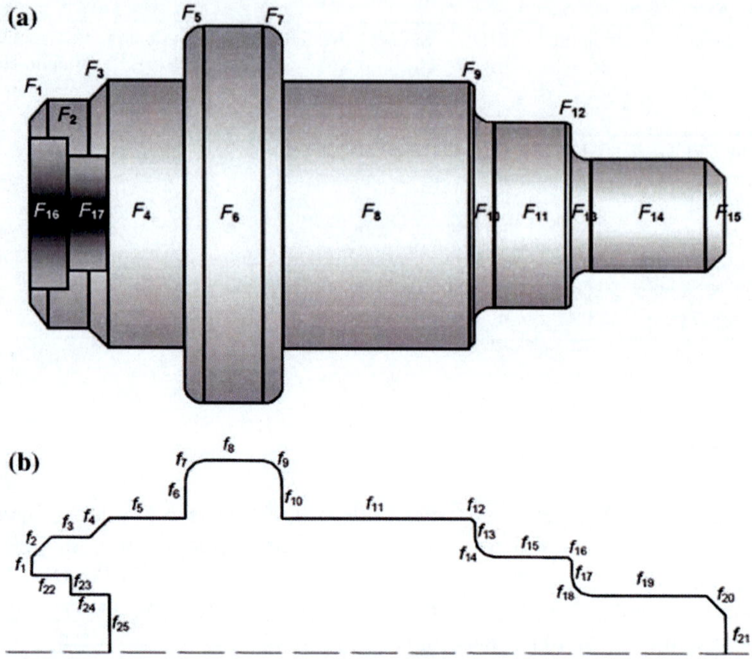

Fig. 22 Definition of OOMFs and manufacturing faces for the sample part: **a** 17 OOMFs generated from the DXF, and **b** 25 manufacturing faces derived from the OOMFs

Fig. 23 Tolerance factor graph based on tolerance limits that are assigned to the sample part

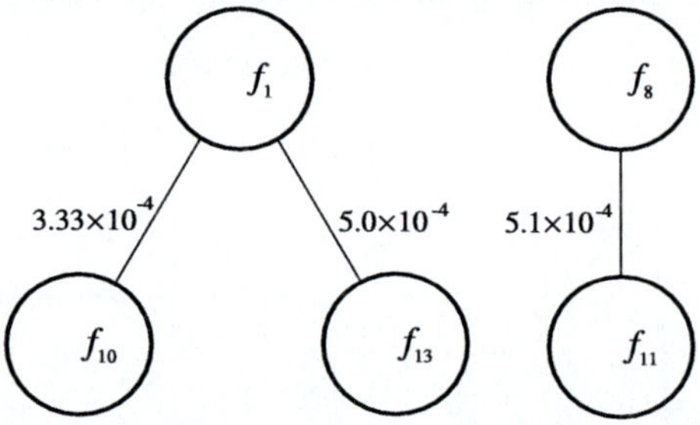

The 25 derived manufacturing faces shown in Fig. 22b are used to assign the geometric and dimensional tolerance limits given in Fig. 21. Tolerance factors for the manufacturing faces are then calculated to carry out the setup planning and sequencing presented in Sect. 4. Thus, Fig. 23 shows the tolerance factor graph based on the tolerance limits that are given to the sample part. This graph is used to make its adjacency matrix $\boldsymbol{T}$ (see Appendix 6). The tolerance factor graph is a weighted, undirected graph $\boldsymbol{G} = (\boldsymbol{V}, \boldsymbol{E})$, representing the relative tolerance information of a part. The manufacturing faces within the part are represented as the vertices $V = \{f_1, f_2, \ldots, f_n\}$. The relative tolerance information is represented using the weighted edges $E = \{e_1, e_2, \ldots, e_m\}$. If the value of a weighted edge

Table 2 Generated setup sequences for the sample part

Setup ID	Datum	Manufacturing faces
1	f_8, f_{21}	$f_1, f_2, f_3, f_4, f_5, f_6, f_7, f_{22}, f_{23}, f_{24}, f_{25}$
2	f_1, f_5	$f_{21}, f_{20}, f_{19}, f_{18}, f_{17}, f_{16}, f_{15}, f_{14}, f_{13}, f_{12}, f_{11}, f_{10}, f_9, f_8$

between two manufacturing faces, f_i and f_j, is small, it might be better if these two faces belong to the same set of sequences to be machined. Thus, the machining of the two manufacturing faces at one setup is likely to satisfy the given tight tolerance requirement.

As shown in Fig. 22, the manufacturing faces f_1 and f_{10} have small relative tolerance values associated by weighted edges. However, the two manufacturing faces cannot be machined at one setup owing to their different tool approach directions. As shown in Fig. 22b, the manufacturing face f_1 can be machined by a left-approaching tool, while the manufacturing face f_{10} requires a right-approaching tool. Similarly, the manufacturing faces f_1 and f_{13} also cannot belong to the same setup sequence. On the other hand, it is highly recommended that the manufacturing faces f_8 and f_{11} be put into the same setup sequence.

Based on the setup planning and sequencing algorithm used, two set-up sequences are generated with datum information. Table 2 describes the generated setup sequences for the sample part shown in Fig. 21.

In Table 2, the first setup sequence includes both external and internal turning processes, while the second setup sequence includes only an external turning process. As discussed before, the manufacturing faces f_8 and f_{11} in Table 2 belong to the second setup sequence of the setup planning and sequencing process.

In addition, manufacturing faces at each setup in Table 2 are used to form a manufacturing face direction to search tool alternatives via the DB search algorithm, which is discussed in Sect. 5.1. For instance, an external manufacturing face direction is initially generated with manufacturing faces $f_1, f_2, f_3, f_4, f_5, f_6$, and f_7, and its numeral values become *345453C* in the right-approaching tool direction, where *C* represents CCW. This implies that a tool cuts all the external faces at the first setup. The face direction is then inserted in a dynamic SQL query with other search criteria, including the required tool material type. This SQL query is sent to the DB server, and is used to search the matched turning tool alternatives. If there is no tool matched, the number of manufacturing faces is reduced, and a new SQL query is formed to search the matched tool alternatives for the reduced manufacturing faces.

Possible tool sets are created with the tool alternatives that are retrieved by the developed DB search algorithm. These tool sets are used for the ranking process (see Appendix 10), including the creation of MVs based on the setup sequences in Table 2. In particular, four user-defined parameters are used in this ranking process: (1) a tool setup time of 12 min, (2) a pay rate of $0.2 per minute, (3) a weighting factor for production cost of 0.5, and (4) a weighting factor for production time of 0.5. The calculation of production time for the ranking process uses a standard tool-indexing time that is pre-stored in the tool DB. This indexing

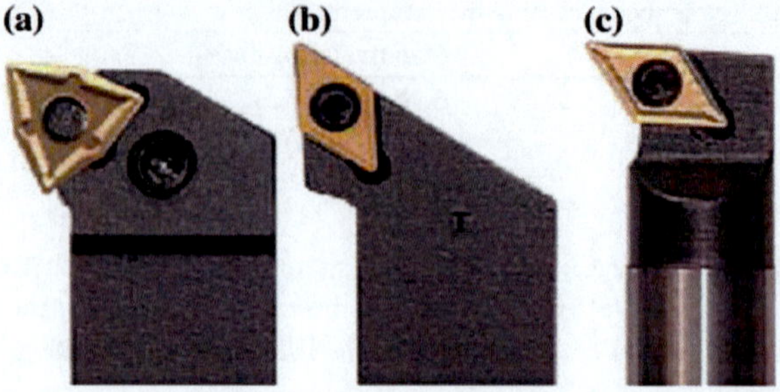

Fig. 24 Best tool set selected for the sample part: **a** *E3941*: PWLNR-1616H06-S, **b** *E2579*: SDJCR-1616H11-S, and **c** *I164*: SDUCR07-S

(a)

Set up Sequence ID: 1
Datum plane ID: f21
Clamping Face ID: f8
Process: Roughing

MVID	FaceType	depth(mm)	feed(mm/rev)	speed(m/min)	Xs	Xe	Zs	Ze	Tool
MV6	CYL	4.0	0.6	25.560001	100.0	90.0	0.0	80.0	E3941
MV5	CYL	4.0	0.6	25.560001	90.0	70.0	0.0	80.0	E3941
				⋮					
f4	TCYL	4.0	0.6	25.560001	70.0	60.0	30.0	40.0	E3941
f7	CIR	4.0	0.6	25.560001	100.0	90.0	80.0	90.0	E3941

Set up Sequence ID: 2
Datum plane ID: f1
Clamping Face ID: f5
Process: Roughing

MVID	FaceType	depth(mm)	feed(mm/rev)	speed(m/min)	Xs	Xe	Zs	Ze	Tool
MV20	CYL	4.0	0.6	21.9816	100.0	90.0	360.0	130.0	E2579
MV19	CYL	4.0	0.6	21.9816	90.0	70.0	360.0	130.0	E2579
				⋮					
f12	TCYL	4.0	0.6	21.9816	70.0	68.0	230.0	228.0	E2579
f9	CIR	4.0	0.6	21.9816	100.0	90.0	130.0	120.0	E2579

(b)

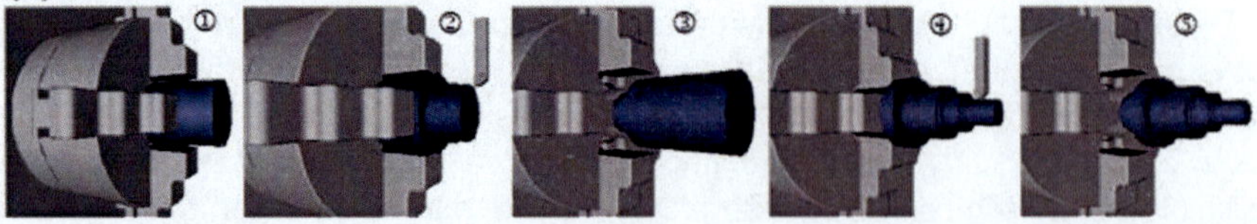

Fig. 25 Generated process plan (**a**) and its virtual simulation (**b**), including ① the first setup, ② the rough-turning process at the first setup, ③ the second setup, ④ the rough-turning process at the second setup, and ⑤ completion of the simulation

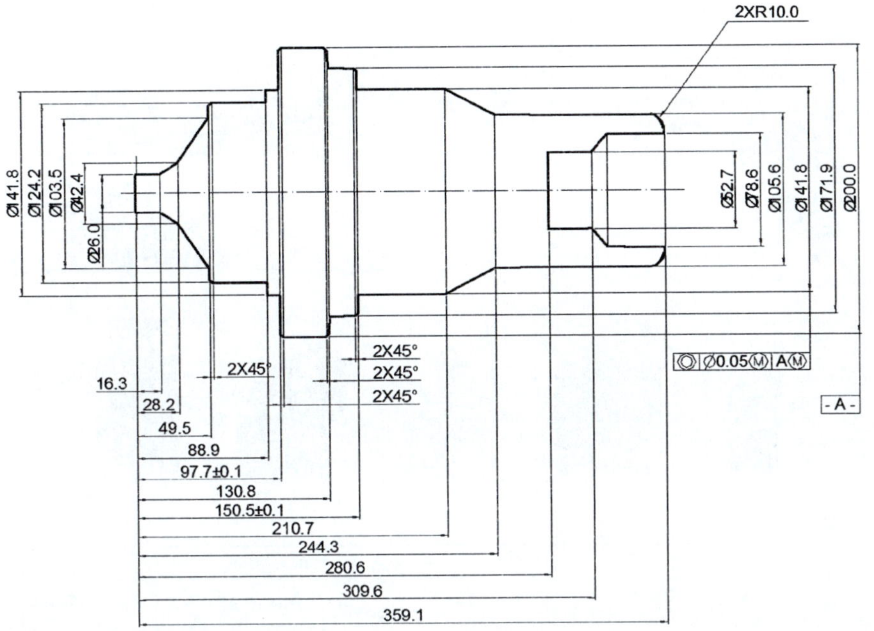

Fig. 26 Sample part

time depends on the type of tool insert and tool holder. Figure 24 shows the best tool set selected to machine the sample part according to the ranking process described here.

Figure 24 shows a tool set consisting of three turning tools. This tool set ranks first in possible tool sets. As an external turning tool, the tool *E3941* shown in Fig. 24a is selected based on manufacturing faces f_1, f_2, f_3, f_4, f_5, f_6, and f_7 in Fig. 22b. At the first setup in Table 2, this tool is combined with the internal turning tool *I164* shown in Fig. 24c. The tool *E2579* shown in Fig. 24b is selected based on all the manufacturing faces at the second setup in Table 2.

All the tools in Fig. 24 entail the use of the same insert grade of 836, which requires TiN multi-layers coated via the physical vapor deposition (PVD) process. These tool-clamping types require the use of large rigidity tools such as *P*- and *S*-types, which allow a stable machining process. The inserts used, of *W*- and *D*-types, are versatile enough to reduce production time and cost by minimizing tool setup time.

In the machine selection with the tool set shown in Fig. 24, the 10HC DAEWOO horizontal lathe is suggested as the most economical machine, despite the fact that its machine capital cost and maintenance cost are the most expensive among the machines in the machine DB, owing to the 10HC DAEWOO horizontal lathe having the highest MTBF and the lowest MTTR among the machines. However, the appropriateness of assumed data in machine selection can be also verified by applying traditional analysis procedures such as failure mode analysis,

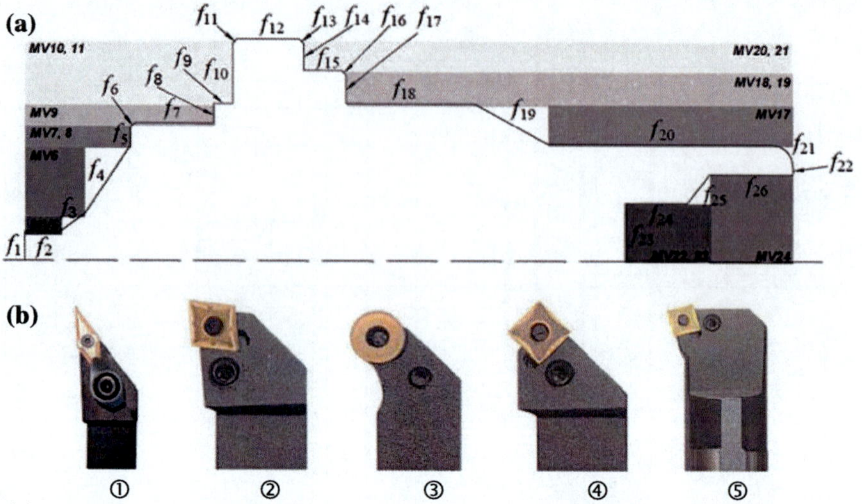

(c)

Setup ID	Clamp	Datum	Sequence of operations					
			Seq. ID	Machining Volume	Tool ID	Speed (m/min)	Feed Rate (mm)	Cut-depth (mm)
1	f_{12}	f_{22}	1-1	MV 10, 11	E760	138	0.6	4.0
			1-2	MV 9				
			1-3	MV 7, 8				
			1-4	MV 6				
			1-5	MV 5				
			1-6	f_3, f_4				
			1-7	f_6, f_{11}	E2382	138	0.6	4.0
2	f_9	F_1	2-1	MV 20, 21	E1450	138	0.6	4.0
			2-2	MV 18, 19				
			2-3	MV 17				
			2-4	f_{19}				
			2-5	f_{16}, f_{13}	E2382	138	0.6	4.0
			2-6	f_{21}	E2148	109	0.6	4.0
			2-7	MV 24	I59	138	0.6	4.0
			2-8	MV 22, 23				
			2-9	f_{25}				

Fig. 27 Summary of the process planning for the part illustrated in Fig. 26: **a** generated MVs, **b** selected tool set (① *E760*: MVJNR-2525M16-S, ② *E1450*: PCLNR-2020K12-S, ③ *E2148*: PRSCR-3225P16-S, ④ *E2382*: PSSNR-2020K12-S, and ⑤ *I59*: A32S-PSKNR12-S), and **c** detailed process plan

quality effects and criticality analysis, reliability and maintainability analysis, and maintenance task analysis. Figure 25 shows a generated process plan for this sample part, and its simulation process.

7.2 *Example 2*

As shown in Fig. 26, another part is tested to evaluate the approach developed. The part is made of non-alloyed carbon steel that is classified in the P1 group (see Appendix 8) based on ISO 513, and has a hardness of 300 HB.

This part is converted to 18 OOMFs, from which 26 manufacturing faces are derived. These faces are used to assign the geometric and dimensional tolerance limits given in Fig. 27. After the setup planning, sequencing processing, and tool and machine selection, a process plan is generated with a recommended tool set and machine. The same resources and parameters as those used in the first example are applied here. Figure 27 shows a summary of the process planning for the part illustrated in Fig. 26. As shown in Fig. 27b, five tools with the first rank in the possible tool sets are selected; their inserts are *V*-, *C*-, and *S*-types that are versatile and economical enough to reduce production time and cost. The used insert grades of these tools can be 320P, 525P, or TiC/TiCN/TiN multi-layer coated grade which is produced through the chemical vapor deposition (CVD) process. In the machine selection, the 10HC DAEWOO horizontal lathe is also suggested as the most economical machine, owing to its minimal maintenance cost.

8 Summary

This chapter presented an approach to process planning that proposes to generate a process plan with tool and machine alternatives based on a dynamic manufacturing environment.

For the tool and machine selection discussed in this chapter, data search algorithms and knowledge have been developed and implemented via relational database (RDB) technologies. The RDB technologies help to efficiently define and transact a large amount of tool- and machine-related information in the database. Various OODBs based on the OOP technique are used to efficiently manipulate the retrieved tool- and machine-related information. The OODBs are subsequently applied in tool and machine selection based on weighting factors of production cost and time. Once tools and machines are selected for the process plan of a part, they become unavailable until the part is completely manufactured on the shop floor. An UPDATE statement based on the DML of the SQL is used to change the availability state of the selected tools and machines in the DBs—it simply sets DB fields related to tool quantity and date for machine usage.

Based on the approach presented in this chapter, a process plan can be generated with feasible tool and machine alternatives. This process plan can be used to quickly exploit manufacturing opportunities for a specific product at the inter-enterprise level, and to realize fabrication at the shop floor level. In addition, the process plan can provide manufacturing systems with a reduced number of feasible "what-if" scenarios for fabrication, and can execute a practical assessment of manufacturability and a reliable estimation of manufacturing time and cost. The reliable estimation is efficiently reached by using the selected tools and machines, which are used to calculate machining cycle data such as the depth of cut, speed, feed rate, and cutting time. Although the rotational components are only used in examples for the proposed methods, the described concepts can be extended to process planning for other shapes of components such as non-rotational parts.

Appendix 1 Database Models for Process Planning

A.1.1 Data Modeling for External Insert Holder

Figure 28

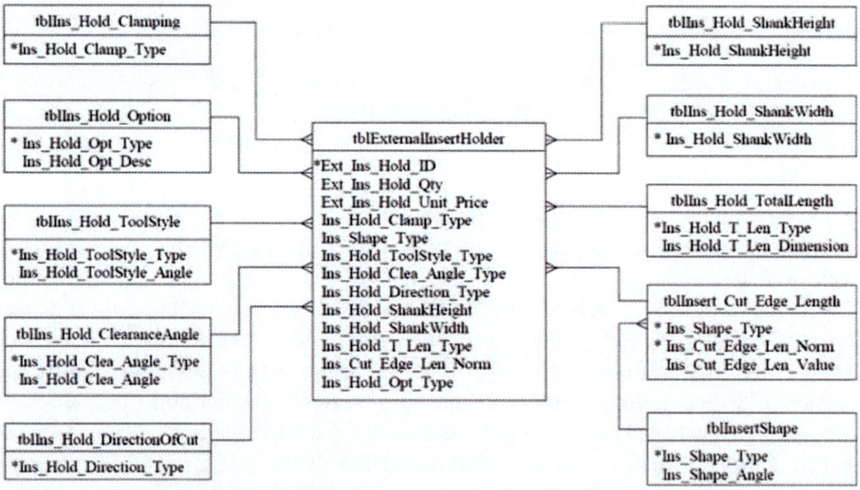

Fig. 28 Data model external insert holder

A.1.2 Data Modeling of Insert for Turning

Figure 29

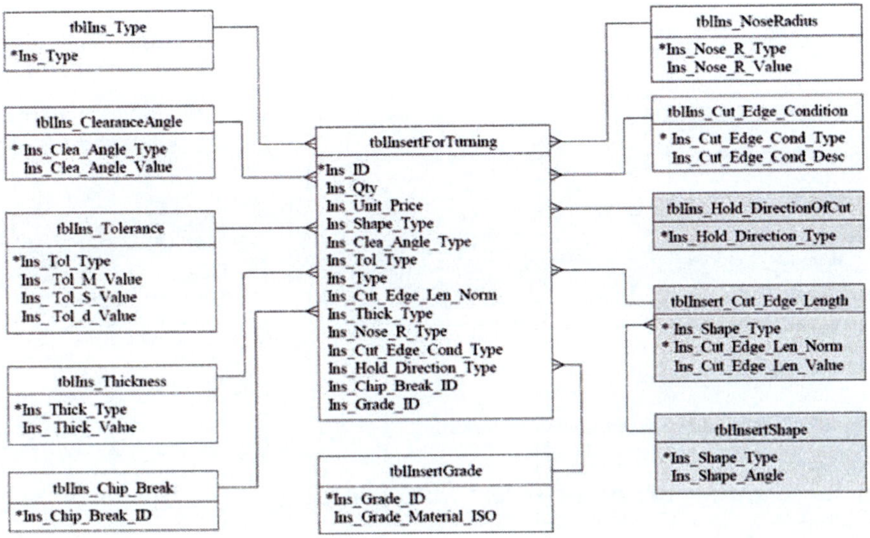

Fig. 29 Data model of insert for turning

A.1.3 Data Modeling for Internal Insert Holder

Figure 30

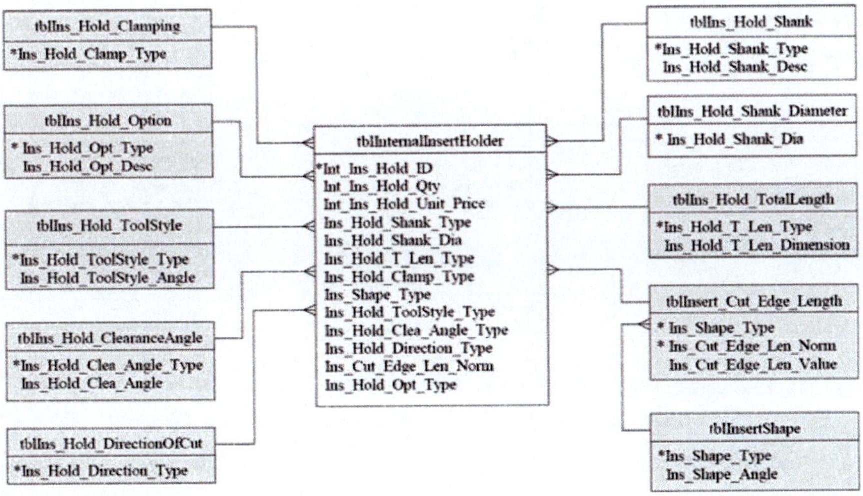

Fig. 30 Data model of internal insert holder

A.1.4 Data Modeling of Insert Holder for Threading

Figure 31

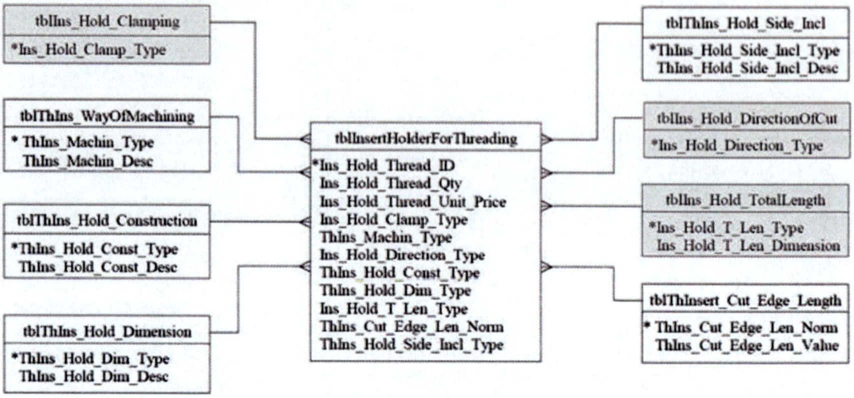

Fig. 31 Data model of insert holder for threading

A.1.5 Data Modeling of Insert for Threading

Figure 32

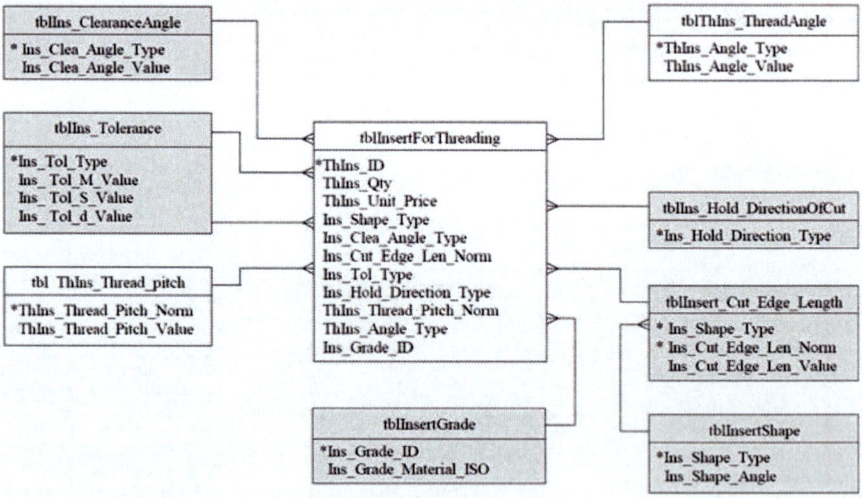

Fig. 32 Data model of insert for threading

A.1.6 Data Modeling of Turning Tool

Figure 33

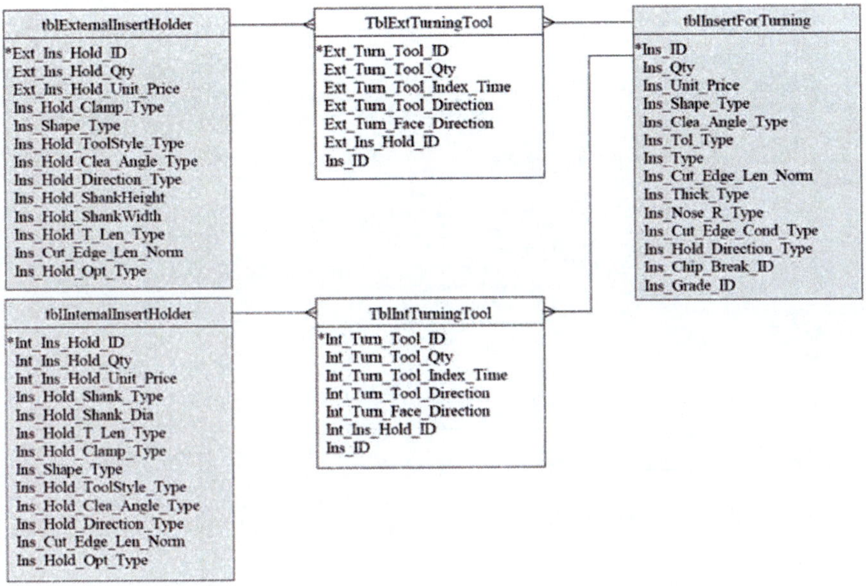

Fig. 33 Data model of turning tool

A.1.7 Data Modeling of Threading Tool

Figure 34

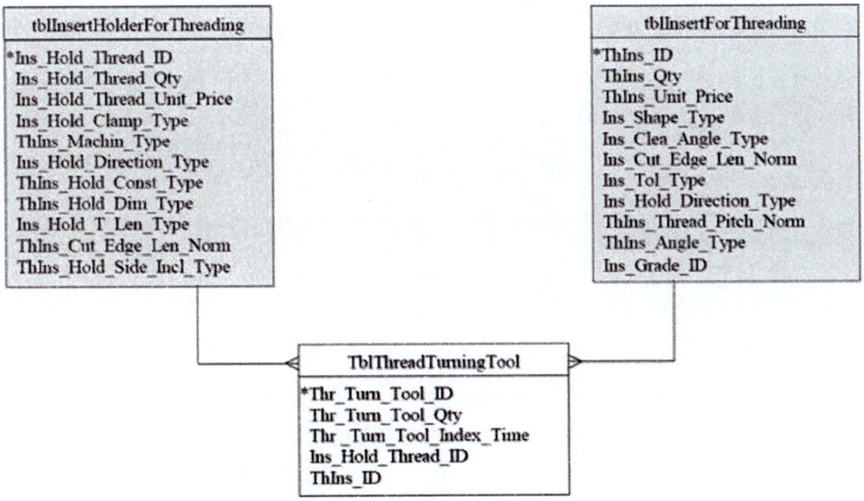

Fig. 34 Data model of threading tool

A.1.8 Data Modeling of Numerical Control Lathe Machine

Figure 35

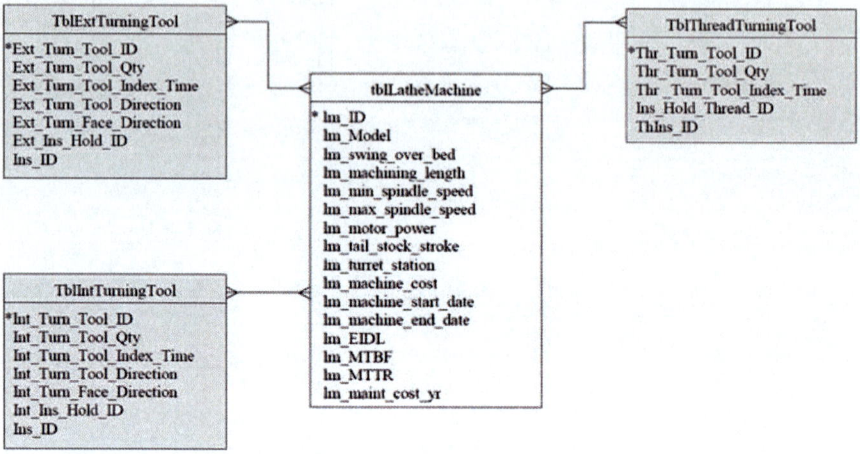

Fig. 35 Data model of NC lathe machine

Appendix 2 Flowchart of Obtaining Geometry Entities from DXF Representations

Figure 36

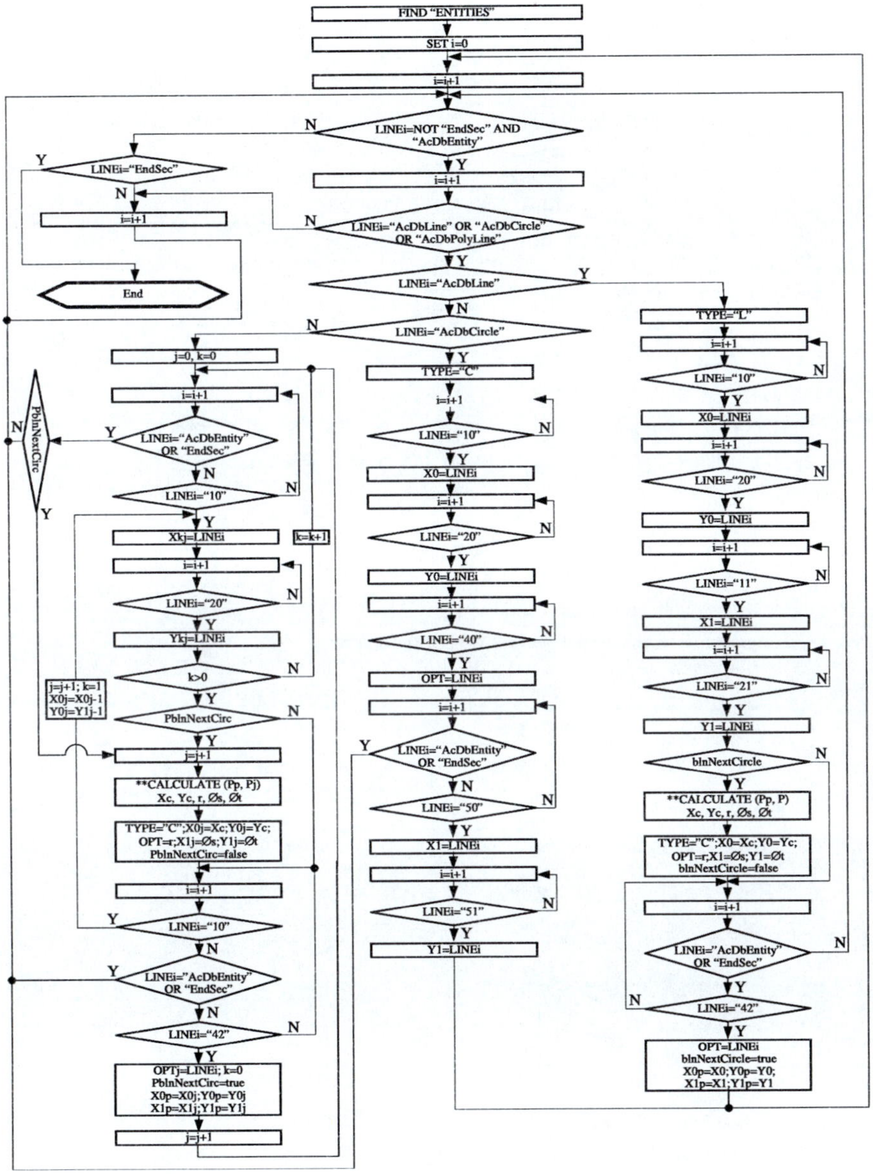

Fig. 36 Flowchart of obtaining geometry entities from DXF representations

Appendix 3 Decision-Making Diagrams for Feature Definition Based on Obtained Geometric Entities

A.3.1 Tangent Line Entity-Based Decision-Making Diagram for Feature Definition

Figure 37

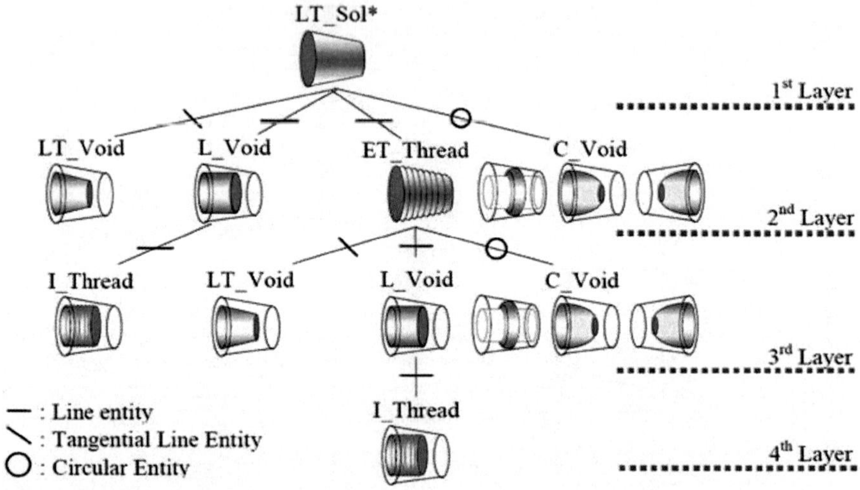

Fig. 37 Tangent line entity-based feature definition diagram

A.3.2 Circular Entity-Based Decision-Making Diagram for Feature Definition

Figure 38

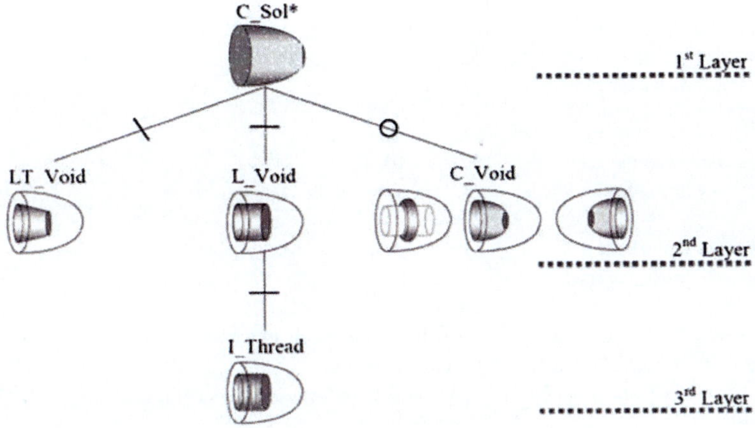

Fig. 38 Circular entity-based feature definition diagram

Appendix 4 Structure of ManufFeature Java Class

```
Public class ManufFeature implements FeatureManagement {

    /* General information-related properties */
    private int projectID;             //project ID
    private String projectName;        //project Name
    private String dxfID;              //DXF file (drawing) number
    private String partNumber;         //part number
    private String partName;           //part name

    /* Feature information-related properties */
    private int featureID;             //feature ID
    private String featureShape;       //feature shape in the same inheritance
                                       //level
    private float xStart;              //feature position in x-coordinate
    private float[] parameter;         //parameter for a feature defined in this
                                       //dissertation
    private float[][] transf2D;        //transformation matrix of a feature
    private String fillingType;        //'void' or 'solid'
    private MafFace[] faceList;        //generated manufacturing faces of this
                                       //feature
    private int prevFeatureID;         //an adjacent feature at the left side
    private String ingredientList;     //embedded features in this feature

    /* methods */
    public void setProjectID (int pID) {...}
    public void setProjectName (String pName) {...}
    public void setDxfID (String dxfID) {...}

                  ⋮

    public MafFace[] getMafFace () {...}
    public int getPrevFeatureID () {...}
    public int[] getIngredientFeature () {...}
    public MafFace[] getManufacturingFace () {...}
}
```

Appendix 5 Several Examples of Tolerance Factors Defined in this Chapter

Figure 39

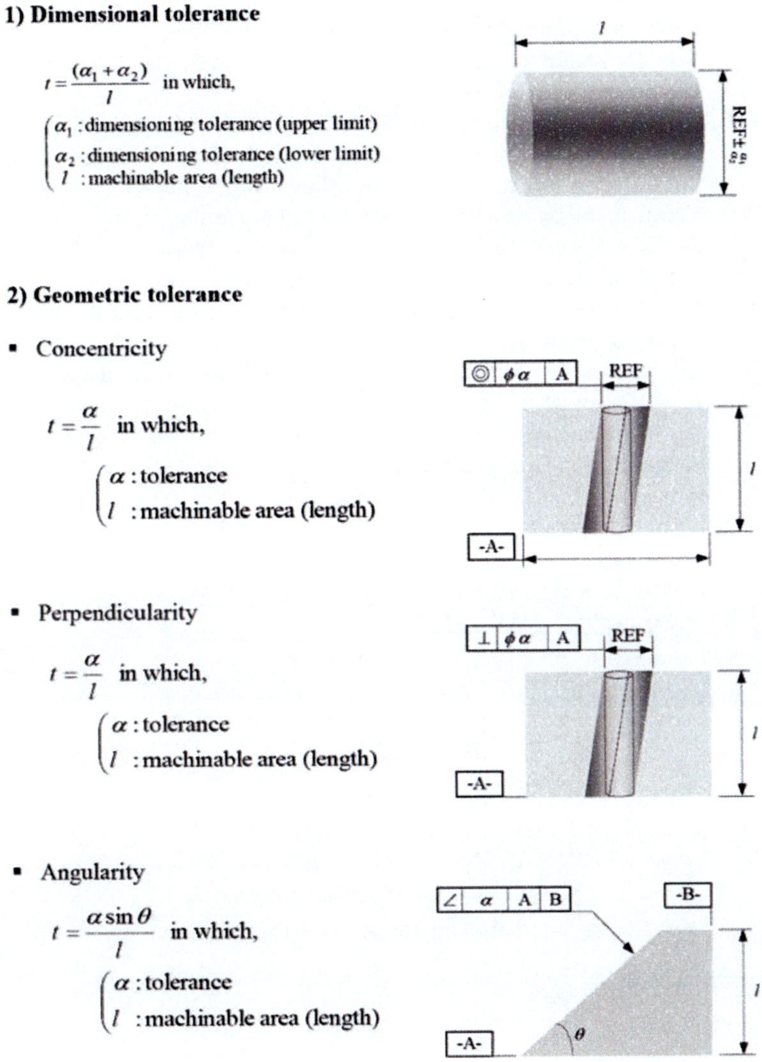

Fig. 39 Example tolerance factors

Appendix 6 Setup Planning and Sequencing Algorithms

A.6.1 Nomenclatures

n	Number of manufacturing faces within a rotational component
f_i	Manufacturing faces, $i = 1, 2, \cdots, n$
$\mathbf{K_0}$	Set of manufacturing faces that exist on the stock
$\mathbf{K}_i$	Set of manufacturing faces that exist after the work-piece was machined in the ith setup
$\mathbf{X}$	Set of manufacturing faces that are suitable for location or clamping
$\mathbf{X}(\mathbf{K}_i)$	Set of manufacturing faces that are suitable for location or clamping after the work-piece has been machined in the ith setup
$\mathbf{C}$	Set of manufacturing faces that are cylindrical surfaces
$\mathbf{P}$	Set of manufacturing faces that are plane surfaces
$\mathbf{O}$	Set of manufacturing faces that are cone surfaces
$\mathbf{A}_1$	Set of manufacturing faces that can be machined from the left side of the component
$\mathbf{A}_2$	Set of manufacturing faces that can be machined from the right side of the component
$\mathbf{A}_3$	Set of manufacturing faces that can be machined from either the right side or left side of the component
$\mathbf{T} = [t_{ij}]$	Adjacency matrix of the tolerance factor graph, $i, j = 1, 2, \ldots, n$

A.6.2 Mathematical Formulation

Set of manufacturing faces:

$$\mathbf{F} = \{f_1, f_2, \ldots, f_n\}$$

Stock geometry vector:

$$\mathbf{K} = \{k_1, k_2, \ldots, k_n\}$$

$$k_i = \begin{cases} 1 & \text{if face } f_i \text{ exists on the stock} \\ 0 & \text{otherwise} \quad i = 1, 2, \ldots, n \end{cases}$$

Fixturing vector:

$$\mathbf{X} = \{x_1, x_2, \ldots, x_n\}$$

$$x_i = \begin{cases} 1 & \text{if face } f_i \text{ is suitable for locating or clamping} \\ 0 & \text{otherwise} \end{cases}$$

$$(i = 1, 2, \ldots, n)$$

Cylindrical vector:

$$\mathbf{C} = \{c_1, c_2, \ldots, c_n\}$$
$$c_i = \begin{cases} 1 & \text{if face } f_i \text{ is a cylindrical surface} \\ 0 & \text{otherwise} \end{cases}$$
$$(i = 1, 2, \ldots, n)$$

Plane vector:

$$\mathbf{P} = \{p_1, p_2, \ldots, p_n\}$$
$$p_i = \begin{cases} 1 & \text{if face } f_i \text{ is a plane surface} \\ 0 & \text{otherwise} \end{cases}$$
$$(i = 1, 2, \ldots, n)$$

Cone vector:

$$\mathbf{O} = \{o_1, o_2, \ldots, o_n\}$$
$$o_i = \begin{cases} 1 & \text{if face } f_i \text{ is a cone surface} \\ 0 & \text{otherwise} \end{cases}$$
$$(i = 1, 2, \ldots, n)$$

Tool approach vector:

$$\mathbf{A} = \{a_1, a_2, \ldots, a_n\}^{\mathrm{T}}$$
$$a_i = \begin{cases} [1, 0] & \text{if } f_i \text{ can be machined only from the left side} \\ [0, 1] & \text{if } f_i \text{ can be machined only from the right side} \\ [1, 1] & \text{if } f_i \text{ can be machined either from the left side} \\ & \text{or from the right side} \end{cases}$$

Tolerance factor (Relative tolerance):

$$\mathbf{T} = [t_{ij}], \quad t = \frac{1}{\sum_{i=1}^{m} \frac{1}{t_i}}$$

A.6.3 Setup Planning and Sequencing Algorithm

Algorithm A.6.1 Setup Formation

Let $S' = A_1 - A_3$, $S'' = A_2 - A_3$, $S^* = A_3$
If $S^* = \phi$ Then
  Exit
Else
  Find $t_{xy} = \min[t_{ij}]$, in which $f_i \in A_3, f_j \notin A_3, t_{ij} \neq 0, i, j = 1,2,\cdots,n$
  If such a t_{xy} is found then
    If $f_y \in A_1$ Then
      Let $S' = S' \cup \{f_x\}$
    Else
      Let $S'' = S'' \cup \{f_x\}$
    Let $S^* = S^* - \{f_x\}$
    Goto statement "If $S^* = \phi$ Then..."
  Else
    Let a and b denote the number of elements in S' and S'', respectively
    If $a \geq b$ Then
      Let $S' = S' \cup S^*, S^* = \phi$, Exit
    Else
      Let $S'' = S'' \cup S^*, S^* = \phi$, Exit

Algorithm A.6.2 Datum selection-clamping position (c')

Let $H = C \cap X \cap S''$
If $H = \phi$ Then
  Let $H = O \cap X \cap S''$
  If $H \neq \phi$ Then
    Goto statement "Find $t_{xy} = \min[t_{ij}]$..."
  Else
    Find f_x such that $\sum_{j=1}^{n} t_{xy} = 0$, in which $f_x \in S' \cap A_3 \cap C \cap X, f_j \in S'$
    If such a f_x is found then
      Let $S' = S' - \{f_x\}, S'' = S'' + \{f_x\}$
      Let $c' = f_x$, Exit
    Else
      Let $q(f_i) = \min[t_{ij}], \forall f_i \in S' \cap A_3 \cap C \cap X$, in which $f_j \in S', t_{ij} \neq 0$
      Find $q(f_x) = \max[q(f_i)]$
      Let $S' = S' - \{f_x\}, S'' = S'' + \{f_x\}$
      Let $c' = f_x$, Exit
Else
  Find $t_{xy} = \min[t_{ij}]$, in which $f_i \in H, f_j \in S', t_{ij} \neq 0; i, j = 1,2,\cdots,n$
  If such a t_{xy} is found then
    Let $c' = f_x$, Exit
  Else
    Let $c' = f_u$, in which $f_i \in H$ and its diameter is the largest, Exit

Algorithm A.6.3 Datum selection-stopping position (p′)

Let $\boldsymbol{H} = \boldsymbol{P} \cap \boldsymbol{X} \cap \boldsymbol{S}''$

If $\boldsymbol{H} \neq \phi$ Then

Find $t_{xy} = \min[t_{ij}]$, in which $f_i \in \boldsymbol{H},\ f_j \in \boldsymbol{S}',\ t_{ij} \neq 0; i, j = 1,2,\cdots,n$

If such a t_{xy} is found then

Let $p' = f_x$, Exit

Algorithm A.6.4 Setup sequencing

Let $d(f_i) = \sum_{j=1}^{n} w_{ij}, \quad \forall f_i \in \{c', p', c'', p''\}$, in which

$$w_{ij} = \begin{cases} 1 & if\ t_{ij} \neq 0, \{f_i, f_j\} \not\subset \boldsymbol{S}', \{f_i, f_j\} \not\subset \boldsymbol{S}'' \\ 0 & otherwise \end{cases}$$

Find $d(f_x) = \max[d(f_i)]$, in which $f_i \in \{c', p', c'', p''\}$

If $f_x \in \{c'', p''\}$ then

Let $\boldsymbol{S}_1 = \boldsymbol{S}', c_1 = c', p_1 = p'$

Let $\boldsymbol{S}_2 = \boldsymbol{S}'', c_2 = c'', p_2 = p''$

Else

Let $\boldsymbol{S}_1 = \boldsymbol{S}'', c_1 = c'', p_1 = p''$

Let $\boldsymbol{S}_2 = \boldsymbol{S}', c_2 = c', p_2 = p'$

If $c_2 \notin \boldsymbol{X}(\boldsymbol{K}_1)$ or $p_2 \notin \boldsymbol{X}(\boldsymbol{K}_1)$ then

Swap $\boldsymbol{S}_1$ with $\boldsymbol{S}_2$, c_1 with c_2, and c_1 with c_2

If $c_1 \notin \boldsymbol{K}_0 \cap \boldsymbol{X}$ then

Let $\boldsymbol{H}_c = \boldsymbol{C} \cap \boldsymbol{X} \cap \boldsymbol{S}_2 \cap \boldsymbol{K}_0$

If $\boldsymbol{H}_c \neq \phi$ then

Let $c_1 = f_x$, in which $f_x \in \boldsymbol{H}_c$ and its diameter is the largest

Else

Let $\boldsymbol{S}_3 = \boldsymbol{S}_2, c_3 = c_2, p_3 = p_2$

Let $\boldsymbol{S}_2 = \boldsymbol{S}_1, c_2 = c_1, p_2 = p_1$

Let $\boldsymbol{S}_1 = \{c_2\}$

Let $c_1 = f_x$, in which $f_x \in \boldsymbol{C} \cap \boldsymbol{X} \cap \boldsymbol{K}_0$ and it is practical for machining $\boldsymbol{S}_1$

Let $p_1 = f_y$, in which $f_y \in \boldsymbol{P} \cap \boldsymbol{X} \cap \boldsymbol{K}_0, \{f_y, p_2\} \not\subset \boldsymbol{A}_1, \{f_y, p_2\} \not\subset \boldsymbol{A}_2$

If $p_1 \notin \boldsymbol{K}_0 \cap \boldsymbol{X}$ then

Let $p_1 = f_u$, in which $f_u \in \boldsymbol{P} \cap \boldsymbol{X} \cap \boldsymbol{S}_2 \cap \boldsymbol{K}_0$ and it is an end face

Appendix 7 Tolerance-Grade Conversion Table with Corresponding Machining Processes

A.7.1 IT-Numbers Based on Tolerance Coefficients

Grade	Tolerance coefficient	Grade	Tolerance coefficient
IT5	7	IT11	100
IT6	10	IT12	160
IT7	16	IT13	250
IT8	25	IT14	400
IT9	40	IT15	640
IT10	64	IT16	1,000

A.7.2 Machining Processes Corresponding to IT-Numbers

Machining process	Tolerance grade (IT-number)								
	IT4	IT5	IT6	IT7	IT8	IT9	IT10	IT11	…
Lapping and honing	⊕	⊕							
Cylindrical grinding		⊕	⊕	⊕					
Surface grinding		⊕	⊕	⊕	⊕				
Diamond turning and boring		⊕	⊕	⊕					
Broaching		⊕	⊕	⊕	⊕				
Reaming		⊕	⊕	⊕	⊕	⊕	⊕		
Fine turning				⊕	⊕				
Finishing turning					⊕	⊕			
Semi-rough turning						⊕	⊕		
Rough turning							⊕	⊕	
Milling							⊕	⊕	
⋮									

Appendix 8 Summary of Knowledge Tables for Calculating Machining Parameters

A.8.1 Material Classifications According to ISO 513

Material	Description
P_I	Carbon steels non-alloyed
	Carbon cast steels
	Carbon tool steels
	Low alloyed steels
P_{II}	Alloyed and medium alloyed steels
	Low and medium alloyed steels
	Alloyed tool steels
	Ferritic and martensitic corrosion-resistant steels
M_I	Austenitic and Ferritic-Austenitic corrosion-resistant, heat-resistant, and creep-resistant steels
	Nonmagnetic and abrasive resistant steels
M_{II}	Special creep-resistant Ni, Co, Fe, and Ti-based alloys
M_{III}	Heat-treated steels with hardness 48–60 HRC
	Hardened ingot-mould iron with hardness 55–85 HSH
K_I	Grey cast iron alloyed and non-alloyed
	Nodular cast iron
	Malleable cast iron
K_{II}	Non-ferrous metals
	Al alloys
	Cu alloys

A.8.2 Insert Grades According to ISO 513

Chemical vapour deposition (CVD)	Physical vapour deposition (PVD)		Uncoated
320P 210K 525P 530P 535P	816	836	S10 S20 HF7 HF10
TiC/TiCN/TiN Multi-layer coated	TiN coated		

A.8.3 Basic Cutting Speed in Turning

M	Turning	Tool Materials/cutting speed, v_{c15} (m-min^{-1})											
		Type	Grade										
			320P	210 K	525P	530P	535P	816	S10	S20	836	HT7	HF10
P_I	Fine turning	**S**	340	290	280	–	–	220	210	180	–	–	–
		C,W	340	290	280	–	–	220	210	180	–	–	–
		T	340	290	260	–	–	220	190	170	–	–	–
		D	330	270	260	–	–	210	190	170	–	–	–
		V	330	270	260	–	–	200	190	160	–	–	–
		R	340	290	280	–	–	220	210	180	–	–	–
	Finishing	**S**	300	250	245	240	–	–	–	–	–	–	–
		C,W	300	250	245	240	–	–	–	–	–	–	
		T	290	240	230	225	–	–	–	–	–	–	–
		D	280	240	230	225	–	–	–	–	–	–	–
		V	290	230	220	210	–	–	–	–	–	–	–
		R	300	250	245	240	–	–	–	–	–	–	–
	Semi-roughing	**S**	235	195	180	175	–	–	–	–	–	–	–
		C,W	235	195	180	175	–	–	–	–	–	–	–
		T	225	185	170	165	–	–	–	–	–	–	–
		D	225	185	170	165	–	–	–	–	–	–	–
		V	215	175	160	155	–	–	–	–	–	–	–
		R	235	195	180	175	–	–	–	–	–	–	–
	Roughing	**S**	165	–	135	130	120	–	–	–	–	–	–
		C,W	165	–	135	130	120	–	–	–	–	–	–
		T	155	–	125	125	110	–	–	–	–	–	–
		D	155	–	125	125	110	–	–	–	–	–	–
		V	155	–	125	125	110	–	–	–	–	–	–
		R	165	–	135	130	120	–	–	–	–	–	–
P_{II}	Fine turning	**S**	260	225	210	–	–	165	155	135	–	–	–
		C,W	260	225	210	–	–	165	155	135	–	–	–
		T	260	225	195	–	–	165	140	125	–	–	–
		D	240	210	195	–	–	155	140	125	–	–	–
		V	240	210	195	–	–	150	130	115	–	–	–
		R	260	225	210	–	–	165	155	135	–	–	–
	Finishing	**S**	230	190	180	175	–	–	–	–	–	–	–
		C, W	230	190	180	175	–	–	–	–	–	–	–
		T	220	180	170	165	–	–	–	–	–	–	–
		D	220	180	170	165	–	–	–	–	–	–	–
		V	220	180	165	160	–	–	–	–	–		–
		R	230	190	180	175	–	–	–	–	–	–	–

M	Turning	Tool Materials/cutting speed, v_{c15} (m-min^{-1})											
		Type	Grade										
			320P	210 K	525P	530P	535P	816	S10	S20	836	HF7	HF10
P_{II}	Semi-roughing	**S**	175	145	135	130	–	–	–	–	–	–	–
		C,W	175	145	135	130	–	–	–	–	–	–	–
		T	170	140	130	125	–	–	–	–	–	–	–
		D	170	140	130	125	–	–	–	–	–	–	–
		V	160	130	120	115	–	–	–	–	–	–	–
		R	175	145	135	130	–	–	–	–	–	–	–
	Roughing	**S**	125	–	100	95	90	–	–	–	–	–	–
		C,W	125	–	100	95	90	–	–	–	–	–	–
		T	115	–	95	90	85	–	–	–	–	–	–
		D	115	–	95	90	85	–	–	–	–	–	–
		V	115	–	95	90	85	–	–	–	–	–	–
		R	125	–	100	95	90	–	–	–	–	–	–
M_I	Fine turning	**S**	230	–	–	180	–	100	–	–	80	60	45
		C,W	230	–	–	180	–	100	–	–	80	60	45
		T	210	–	–	170	–	85	–	–	80	60	45
		D	210	–	–	170	–	85	–	–	70	55	40
		V	200	–	–	160	–	80	–	–	70	55	40
		R	230	–	–	180	–	100	–	–	80	60	45
	Finishing	**S**	–	160	–	140	–	100	–	–	80	–	–
		C,W	–	160	–	140	–	100	–	–	80	–	–
		T	–	155	–	135	–	90	–	–	75	–	–
		D	–	150	–	135	–	90	–	–	75	–	–
		V	–	150	–	130	–	90	–	–	70	–	–
		R	–	160	–	140	–	100	–	–	80	–	–
	Semi-roughing	**S**	–	120	110	105	–	85	–	–	55	–	–
		C,W	–	120	110	105	–	85	–	–	55	–	–
		T	–	115	100	95	–	80	–	–	50	–	–
		D	–	115	100	95	–	80	–	–	50	–	–
		V	–	105	95	90	–	75	–	–	45	–	–
		R	–	120	110	105	–	85	–	–	55	–	–
	Roughing	**S**	–	–	–	80	–	50	–	–	43	–	–
		C, W	–	–	–	80	–	50	–	–	43	–	–
		T	–	–	–	75	–	45	–	–	38	–	–
		D	–	–	–	75	–	45	–	–	38	–	–
		V	–	–	–	75	–	45	–	–	38	–	–
		R	–	–	–	80	–	50	–	–	43	–	–

M	Turning	Tool Materials/cutting speed, v_{c15} (m-min^{-1})											
		Type	Grade										
			320P	210 K	525P	530P	535P	816	S10	S20	836	HF7	HF10
M_{II}	Finishing	**S**	–	–	–	–	–	50	–	–	45	40	35
		C,W	–	–	–	–	–	50	–	–	45	40	35
		T	–	–	–	–	–	45	–	–	40	35	30
		D	–	–	–	–	–	45	–	–	40	35	30
		V	–	–	–	–	–	40	–	–	35	30	25
		R	–	–	–	–	–	50	–	–	45	40	35
	Semi-roughing	**S**	–	–	–	–	–	35	–	–	30	27	20
		c,w	–	–	–	–	–	35	–	–	30	27	20
		T	–	–	–	–	–	30	–	–	25	20	18
		D	–	–	–	–	–	30	–	–	25	20	18
		V	–	–	–	–	–	25	–	–	20	15	12
		R	–	–	–	–	–	35	–	–	30	27	20
	Roughing	**S**	–	–	–	–	–	30	–	–	25	20	18
		c,w	–	–	–	–	–	30	–	–	25	20	18
		T	–	–	–	–	–	25	–	–	20	18	15
		D	–	–	–	–	–	25	–	–	20	18	15
		V	–	–	–	–	–	25	–	–	20	18	15
		R	–	–	–	–	–	30	–	–	25	20	18
M_{III}	Finishing	**S**	–	50	–	–	–	50	–	–	35	45	30
		c,w	–	50	–	–	–	50	–	–	35	45	30
		T	–	45	–	–	–	45	–	–	35	40	20
		D	–	45	–	–	–	45	–	–	35	40	20
		V	–	40	–	–	–	40	–	–	30	35	20
		R	–	50	–	–	–	50	–	–	35	45	30
	Semi-roughing	**S**	–	35	–	–	–	35	–	–	22	30	18
		c,w	–	35	–	–	–	35	–	–	22	30	18
		T	–	30	–	–	–	30	–	–	18	25	15
		D	–	30	–	–	–	30	–	–	18	25	15
		V	–	25	–	–	–	25	–	–	15	20	10
		R	–	35	–	–	–	35	–	–	22	30	18
K_{I}	Fine turning	**S**	–	250	–	210	–	170	–	–	145	140	125
		c,w	–	250	–	210	–	170	–	–	145	140	125
		T	–	230	–	200	–	160	–	–	135	135	120
		D	–	230	–	200	–	160	–	–	135	135	120
		V	–	225	–	195	–	155	–	–	130	130	115
		R	–	250	–	210	–	170	–	–	145	140	125

M	Turning	Tool Materials/cutting speed, v_{c15} (m-min^{-1})											
		Type	Grade										
			320P	210 K	525P	530P	535P	816	S10	S20	836	HF7	HF10
K_I	Finishing	**S**	–	220	–	185	–	145	–	–	125	–	–
		C, W	–	220	–	185	–	145	–	–	125	–	–
		T	–	210	–	175	–	135	–	–	115	–	–
		D	–	210	–	175	–	135	–	–	115	–	–
		V	–	200	–	160	–	125	–	–	105	–	–
		R	–	220	–	185	–	145	–	–	125	–	–
	Semi-roughing	**S**	–	175	–	150	–	120	–	–	105	–	–
		C,W	–	175	–	150	–	120	–	–	105	–	–
		T	–	165	–	140	–	110	–	–	95	–	–
		D	–	165	–	140	–	110	–	–	95	–	–
		V	–	155	–	130	–	100	–	–	85	–	–
		R	–	175	–	150	–	120	–	–	105	–	–
	Roughing	**S**	–	130	–	115	–	90	–	–	80	–	–
		C,W	–	130	–	115	–	90	–	–	80	–	–
		T	–	120	–	105	–	80	–	–	70	–	–
		D	–	120	–	105	–	80	–	–	70	–	–
		R	–	130	–	115	–	90	–	–	80	–	–
K_{II} (Al) HB 100	Finishing	**S**	–	–	–	–	–	800	–	–	–	680	–
		C,W	–	–	–	–	–	800	–	–	–	680	–
		T	–	–	–	–	–	800	–	–	–	680	–
		D	–	–	–	–	–	750	–	–	–	600	–
		V	–	–	–	–	–	700	–	–	–	550	–
		R	–	–	–	–	–	800	–	–	–	680	–
	Semi-roughing	**S**	–	–	–	–	–	600	–	–	–	480	–
		C,W	–	–	–	–	–	600	–	–	–	480	–
		T	–	–	–	–	–	600	–	–	–	480	–
		D	–	–	–	–	–	550	–	–	–	450	–
		V	–	–	–	–	–	500	–	–	–	400	–
		R	–	–	–	–	–	600	–	–	–	480	–
	Roughing	**S**	–	–	–	–	–	400	–	–	–	350	–
		C,W	–	–	–	–	–	400	–	–	–	350	–
		T	–	–	–	–	–	400	–	–	–	350	–
		D	–	–	–	–	–	350	–	–	–	320	–
		V	–	–	–	–	–	300	–	–	–	280	–
		R	–	–	–	–	–	400	–	–	–	350	–

M	Turning	Tool Materials/cutting speed, v_{c15} (m-min^{-1})											
		Type	Grade										
			320P	210 K	525P	530P	535P	816	S10	S20	836	HF7	HF10
K_I	Finishing	**S**	–	220	–	185	–	145	–	–	125	–	–
		C, W	–	220	–	185	–	145	–	–	125	–	–
		T	–	210	–	175	–	135	–	–	115	–	–
		D	–	210	–	175	–	135	–	–	115	–	–
		V	–	200	–	160	–	125	–	–	105	–	–
		R	–	220	–	185	–	145	–	–	125	–	–
	Semi-roughing	**S**	–	175	–	150	–	120	–	–	105	–	–
		C,W	–	175	–	150	–	120	–	–	105	–	–
		T	–	165	–	140	–	110	–	–	95	–	–
		D	–	165	–	140	–	110	–	–	95	–	–
		V	–	155	–	130	–	100	–	–	85	–	–
		R	–	175	–	150	–	120	–	–	105	–	–
	Roughing	**S**	–	130	–	115	–	90	–	–	80	–	–
		C,W	–	130	–	115	–	90	–	–	80	–	–
		T	–	120	–	105	–	80	–	–	70	–	–
		D	–	120	–	105	–	80	–	–	70	–	–
		R	–	130	–	115	–	90	–	–	80	–	–
K_{II} (Al) HB 100	Finishing	**S**	–	–	–	–	–	800	–	–	–	680	–
		C,W	–	–	–	–	–	800	–	–	–	680	–
		T	–	–	–	–	–	800	–	–	–	680	–
		D	–	–	–	–	–	750	–	–	–	600	–
		V	–	–	–	–	–	700	–	–	–	550	–
		R	–	–	–	–	–	800	–	–	–	680	–
	Semi-roughing	**S**	–	–	–	–	–	600	–	–	–	480	–
		C,W	–	–	–	–	–	600	–	–	–	480	–
		T	–	–	–	–	–	600	–	–	–	480	–
		D	–	–	–	–	–	550	–	–	–	450	–
		V	–	–	–	–	–	500	–	–	–	400	–
		R	–	–	–	–	–	600	–	–	–	480	–
	Roughing	**S**	–	–	–	–	–	400	–	–	–	350	–
		C,W	–	–	–	–	–	400	–	–	–	350	–
		T	–	–	–	–	–	400	–	–	–	350	–
		D	–	–	–	–	–	350	–	–	–	320	–
		V	–	–	–	–	–	300	–	–	–	280	–
		R	–	–	–	–	–	400	–	–	–	350	–

A.8.4 Feed Rate and Cutting Depth in Turning

Material	Feed rate, f_r (mm-rev^{-1})				Cutting depth, a_p (mm)			
	Fine turning	Finishing	Semi-roughing	Roughing	Fine turning	Finishing	Semi-roughing	Roughing
$P_{I,II}$	0.05–0.1	0.1–0.2	0.2–0.4	0.4–0.8	0.2–1.0	0.8–2.0	1.5–4.0	4.0–10.0
M_I	0.05–0.1	0.1–0.2	0.2–0.4	0.4–0.8	0.2–1.0	0.8–2.0	1.5–4.0	4.0–10.0
M_{II}	–	0.1–0.2	0.2–0.3	0.3–0.4	–	0.05–1.5	1.5–2.5	2.5–3.5
M_{II}	–	0.08–0.2	0.2–0.3	–	–	0.8–1.5′	1.5–2.5	–
K_I	0.05–0.1	0.1–0.2	0.2–0.4	0.4–0.8	0.2–1.0	0.8–2.0	1.5–4.0	4.0–10.0
K_{II}	–	0.1–0.2	0.2–0.4	0.4–0.8	–	0.8–2.0	1.5–4.0	4.0–10.0

A.8.5 Durability Correction Factor (P, M, K Types)

T_{min}	K_{VT}	T_{min}	K_{VT}
10	1.10	30	0.84
15	1.00	45	0.76
20	0.93	60	0.71

A.8.6 Work-Piece Hardness Correction Factor

A.8.6.1 *P* Type

HB	K_{VHB}	*HB*	K_{VHB}
120	1.18	220	0.90
140	1.12	240	0.86
160	1.05	260	0.82
180	1.00	280	0.80
200	0.95	300	0.77

A.8.6.2 *M* Type

HB	K_{VHB}	*HB*	K_{VHB}
<150	1.40	270–300	0.72
150–180	1.18	300–330	0.68
180–210	1.00	330–360	0.66
210–240	0.87	360–390	0.62
240–270	0.79		

A.8.6.3 *K* Type

HB	K_{VHB}	*HB*	K_{VHB}
Grey/nodular/malleable cast iron		Heat resistant/special cast iron	
160–200	1.26	200–300	0.50
200–240	1.00	300–360	0.40
240–280	0.80	360–450	0.30
280–330	0.60		

A.8.7 Optimized Cutting Speed

$$v_C = v_{C15} \times k_{VT} \times k_{VHB} \tag{6}$$

where

v_C Optimized cutting speed
v_{C15} Basic cutting speed
k_{VHB} Hardness correction factor
k_{VT} Durability correction factor

Appendix 9 Structure of Turning ToolOODB Java Class

```
public class TurningToolOODB extends ToolHolderDB {

        /* properties */
        private String tool_ID;
        private int tool_Qty;
        private int setUpID;
        private String tool_material;
        private float tool_Index_Time;
        private float tool_Cost;
        private String tool_Type;
        private String tool_Direction;
        private String face_Direction;
        private String hold_direction_type;
        private String faceSet;
        private float max_production_time;
        private float used_time;
        private int insert_edge_len;
        private float speed;
        private float feedrate;

        /* methods */
        public void setTool_ID(String tID) {...}
        public void setTool_Qty(int n) {...}
        public void setTool_material(String m) {...}
        public void setTool_Index_Time(float t) {...}
        public void setTool_Cost(float c) {...}

                        ⋮

        public String getFace_Direction() {...}
        public String getHold_direction_type() {...}
        public String getFaceSet() {...}
        public float getMax_production_time() {...}
        public float getUsed_time() {...}
}
```

Appendix 10 Calculation of Production Time and Cost

A.10.1 Production Time Factor, T_p

$$T_p = T_m + T_i \times N_t + T_{su} \times N_{su} \tag{7}$$

where

T_m Machining time $(L/(f_r \times N))$
L Feature length
N Cutting speed in rpm
f_r Maximum permissible feed in mm/rev
T_i Time to perform an index
N_t Number of tools to machine a part
T_{su} Time to perform a set-up
N_{su} Number of tools that must be added to the machine's magazine

A.10.2 Production Cost Factor, C_t

$$C_t = C_i \times N_t + C_{su} \times N_{su} + \sum_{i=1}^{n} C_t^{(i)} \tag{8}$$

where

C_i Cost of performing an index for a tool, $(P \times T_i)$
C_{su} Cost of performing a set-up for a tool, $(P \times T_{su})$
N_{su} Number of tools that must be added to the machine's magazine
$C_t^{(i)}$ Cost of a tool i
n Number of tools required to perform operations
P Pay rate

A.10.3 Rank of Tool Alternatives, Score

$$\text{Score} = \alpha \times C_t + \beta \times T_p \tag{9}$$

where

Weighting condition	α	β
Normal condition	0.5	0.5
Cost weighted condition	1.0	0.0
Production time weighted condition	0.0	1.0

Appendix 11 Knowledge Tables and Equation for Calculating the Maximum Power

A.11.1 Cutting Resistant and Feed Influence Exponent

Material group	P_{S1} [MPa]		$1 - Z$
P_I	1,760		0.76
P_{II}	1,770		0.75
M_I	2,530		0.75
M_{II}	Ni, Co alloys	2,895	0.76
	Ti alloys	1,860	0.79
M_{III}	2,060		0.86
K_I	Grey cast iron	1,120	0.78
	Malleable cast iron	1,190	0.77
	Nodular cast iron	1,430	0.76
K_{II}	Cu alloys	710	0.76
	Al alloys	508	0.78
	Mg alloy	250	0.78

A.11.2 Coefficient, k_{kr}

Approach angle	k_{kr}
90	1.00
80	1.015
70	1.02
60	1.04
55	1.06
50	1.08
45	1.10

A.11.3 Required Motor Power, P

$$P = \frac{a_p \cdot f_r^{1-Z} \cdot P_{S1} \cdot k_{\mathrm{kr}} \cdot v_C}{42{,}000} \,[\mathrm{kW}] \tag{10}$$

where

v_C	Optimized cutting speed [m/min] [Refer to Eq. (6)]
f_r	Feed rate [mm/rev]
a_p	Cutting depth [mm]
$1 - Z$	Feed influence exponent for different materials machined
P_{S1}	Specific cutting resistance (cutting force for feed $f = 1$ mm/rev at $kr = 90$)
k_{kr}	Coefficient representing the approach angle kr influence

Appendix 12 Structure of MachineOODB Java Class

```
public class MachineOODB {

        /* properties */
        private int lm_ID;
        private String lm_Model;
        private int lm_swing_over_bed;
        private int lm_machining_length;
        private int lm_min_spindle_speed;
        private int lm_max_spindle_speed;
        private float lm_motor_power;
        private int lm_tail_stock_stroke;
        private int lm_turret_station;
        private float lm_machine_cost;
        private int lm_EIDL;
        private int lm_MTBF;
        private float lm_MTTR;
        private float lm_maint_cost_yr;
        private float lm_pay_rate;

        /* methods */
        public void setLm_swing_over_bed(int d) {...}
        public void setLm_machining_length(int d) {...}
        public void setLm_min_spindle_speed(int s) {...}
        public void setLm_max_spindle_speed(int s) {...}
        public void setLm_motor_power(float p) {...}

                        ⋮

        public float getLm_machine_cost() {...}
        public int getLm_EIDL() {...}
        public int getLm_MTBF() {...}
        public float getLm_MTTR() {...}
        public float getLm_maint_cost_yr() {...}
        public float getLm_pay_rate() {...}
}
```

References

1. Ahmad N, Haque A, Hasin A (2001) Current trend in computer aided process planning. In: Proceedings of the 7th annual paper meeting and 2nd international conference, pp 81–92
2. Cao Q, Dowlatshahi S (2005) The impact of alignment between virtual enterprise and information technology on business performance in an agile manufacturing environment. J Oper Manag 23:531–550
3. Chang T, Wysk RA, Wang H (1997) Computer-aided manufacturing, 2nd edn. Prentice Hall, Englewood Cliffs

4. Chung C, Peng Q (2004) The selection of tools and machines on web-based manufacturing environments. Int J Mach Tools Manuf 44:315–324
5. Huang SH (1998) Automated set-up planning for lathe machining. Int J Manuf Syst 17:196–208
6. Huang SH, Zhang H (1996) Tolerance analysis in set-up planning for rotational parts. Int J Manuf Syst 15:340–350
7. Fernandes JK, Raja HV (2000) Incorporated tool selection system using object technology. Int J Mach Tools Manuf 40:1547–1555
8. Kumar M, Rajotia S (2003) Integration of scheduling with computer aided process planning. J Mat Proc Technol 138:297–300
9. Moon C, Seo Y (2005) Evolutionary algorithm for advanced process planning and scheduling in a multi-plant. Comput Ind Eng 48:311–325
10. Moon C, Lee M, Seo Y, Lee YH (2002) Integrated machine tool selection and operation sequencing with capacity and precedence constraints using genetic algorithm. Comput Ind Eng 43:605–621
11. Peng Q, Hall RF, Lister MP (2000) Application and evaluation of VR-based CAPP system. J Mat Proc Technol 107:153–159
12. Pramet (2001) Lathe turning tool handbook, Czech Republic
13. Pramet (2001) Lathe turning tool catalogue, Czech Republic
14. Singh N (1996) System approach to computer-integrated design and manufacturing. Wiley, New York
15. Soromaz D, Khosknevis B (1997) Machine and tool constraint specification for integrated process planning system. In: Proceedings of the 7th industrial engineering research conference, pp 901–906
16. Usher MJ, Fernandes JK (1999) An object-oriented application of tool selection in dynamic process planning. Int J Prod Res 37:2879–2894
17. Watson TR (1999) Data management, database and organization. Wiley, New York
18. Texas A&M University, (2012) Tolerances. Lecture notes (ENG105). http://edg.tamu.edu/PEOPLE/schreuders/ENDG105/Lecture%206.2%20Tolerances.pdf

Computation of Offset Curves Using a Distance Function: Addressing a Key Challenge in Cutting Tool Path Generation

C. K. Au and Y.-S. Ma

1 Introduction

Tool path generation plays an important role in manufacturing. Tool paths should provide the cutter machine with the required geometry safely, efficiently, and economically, and ensure that the specified surface finish is achieved. The techniques used to generate tool paths are essential for the automation and optimization of machining processes. With respect to computer-aided manufacturing (CAM) software design and implementation, tool path generation is a major function to support the efficient application of modern computer numerical control (CNC) machine tools. Improvements in tool path generation methods may lead to substantial gains in reduced machining time and cost [11]. Yet in the tool path generation process, there exists a significant and nontrivial mathematical challenge: offset curve computation [4].

Curve representation and self-intersection are two major issues of offset curve computation. These two issues are processed separately. An offset curve is a set of points that lie a distance r perpendicular from a progenitor curve in $\mathbf{R}^2$. If a curve is expressed parametrically as

$$\mathbf{C}(t) = (x(t), y(t)) \tag{1}$$

then its offset curve $\mathbf{C}_r(t)$ is given by

$$\mathbf{C}_r(t) = C(t) + r \cdot n \tag{2}$$

C. K. Au (✉)
Faculty of Engineering, University of Waikato, Hamilton, New Zealand
e-mail: ckau@waikato.ac.nz

Y.-S. Ma
Department of Mechanical Engineering, University of Alberta, Edmonton, Canada
e-mail: yongsheng.ma@ualberta.ca

Y.-S. Ma (ed.), *Semantic Modeling and Interoperability in Product and Process Engineering*, Springer Series in Advanced Manufacturing,
DOI: 10.1007/978-1-4471-5073-2_8,

where

$$\mathbf{n} = \frac{(\dot{y}(t), -\dot{x}(t))}{\sqrt{(\dot{x}(t))^2 + (\dot{y}(t))^2}} \tag{3}$$

n is the unit normal of the curve $\mathbf{C}(t)$.

Note that in this definition, the offset distance r can be positive or negative, yielding an offset curve on the either side of the progenitor curve. The parametric curve $\mathbf{C}(t)$ is usually a polynomial or rational in most geometric modeling systems. Obviously, an offset curve cannot be represented by a polynomial or rational curve due to the square root term in the denominator, unless the term $(\dot{x}(t))^2 + (\dot{y}(t))^2$ is a perfect square of a polynomial. This special type of curve is identified by Farouki [3] and is termed a Pythagorean Hodograph curve. Usually, the offset curves are approximated by polynomial or rational functions while the self-intersections are searched and trimmed afterward.

2 Research Background

Geometric modeling in engineering design and manufacturing applications frequently requires curve-offsetting algorithms, such as tool path generation in CAM software tools [4, 10, 11]. Various techniques have therefore been developed to approximate an offset curve to a polynomial or rational curve so that it can be represented in the geometric modeling system. Tiller and Hanson [19] and Coquillart [2] have proposed various approaches to translate the control polygon of a B-spline curve to obtain an offset curve. Klass [5] used a cubic Hermite curve to approximate an offset curve, while Hoschek suggests using a least square method to fit an offset curve. Piegl and Tiller [15] interpolated the offset sample points with a B-spline curve. The excess knots are eliminated so that the offset curve has an appropriate number of control points. Lee et al. [7–9] addressed the problem by sweeping a circle along the progenitor curve, which yielded a convolution as the offset curve. Zhao and Wang [21] generated the offset approximation using the same approach with different parameterization. The accuracy, degree, and number of control points of the offset curve are major concerns in curve representation.

Self-intersection of the offset curve occurs locally, with the offset distance larger than the radius of curvature of the progenitor curve, and the phenomenon can also be observed globally if the distance between two parts of the progenitor curve is short. These self-intersections are referred to as topological changes [10] between the progenitor curve and the offset curve, and are usually trimmed away to yield a proper offset curve in most applications.

Lee et al. [7–9] approximated the progenitor curve with a set of line segments. A plane sweep algorithm was employed to detect all self-intersections. The offset curve was obtained with a trimming operation. Park and Choi [13] used expanded pairwise interference detection tests to detect and remove the local self-intersections. A raw offset curve was constructed, and the global self-intersections were checked and eliminated by inclusion relationship. Seong et al. [17, 18] defined a function among the offset distance, parameters of the progenitor curve, and the offset curve. The critical points of the zero set of this function in the parametric space were used to determine the self-intersections. A trimmed offset curve was then created. Li and Martin [10] used a method based on medial axis transformation to determine the self-intersections. Lai et al. [6] proposed a two-way detecting algorithm to find the intersections so as to shorten the searching time. The literature reflects a common practice of handling the topological changes by detecting and trimming away the self-intersections. However, the task of detecting these self-intersections is nontrivial [10]. The real application of tool path generation requires advanced extension of curve offsetting algorithms. Some of the functions required in pocket machining include covering and window-crossing techniques, especially when the step-over is greater than the cutter radius. Generation of cutter traversing routes has to consider cutter tooth entry/exit conditions, the actual radial width of cut, and further cutting strategies for machining other complex features [11].

This chapter presents an approach for computing the offset curves by using a distance function. A distance function of a planar curve is defined to compute the distances from any point on a plane to that curve. A progenitor curve is approximated by a set of arcs. The distance function of the progenitor curve, which is expressed as a set, is plotted based on the distance function of a point. Using the set theory and this distance function, the set of points with the constant offset distance from the approximated progenitor curve can be obtained. Hence, an offset curve is generated by the Boolean set operations. The approach is implemented by using solid modeling, which is an application of set theory. The advantage of this approach is the simplicity of the computation algorithm, as no searching and trimming routines are required to generate the offset curve and the offset curve can be represented by a nonuniform rational B-spline (NURBS) curve.

3 Distance Function

The distance function of a geometric entity yields the Euclidean distance from any point to that geometric entity. The complexity of the distance function is largely dependent upon the geometric entity. This is a simple function for a point, but numerical computation and approximation are usually required for a more complex geometry such as a free-form NURBS curve or surface.

3.1 Distance Function of a Point

The distance function dist : $\mathbf{R}^2 \to \mathbf{R}$ of a point $\mathbf{P}$ defines the Euclidean distance from a point $\mathbf{Q}(\in \mathbf{R}^2)$ to point $\mathbf{P}(\in \mathbf{R}^2)$. Hence, $\forall \mathbf{Q} \in \mathbf{R}^2$, $\exists d = |\mathbf{PQ}| \in \mathbf{R}$ such that

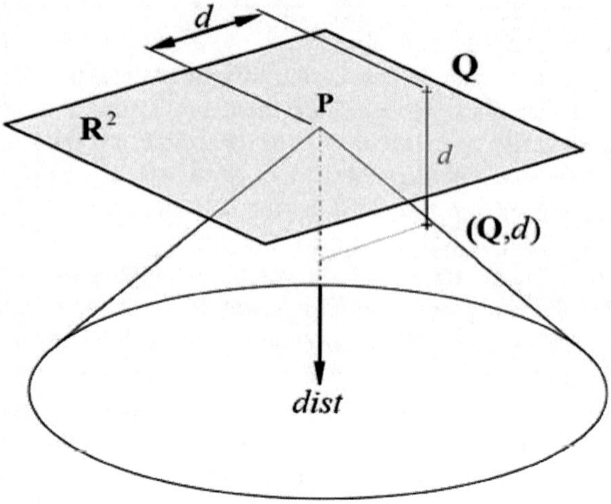

Fig. 1 The distance function of a point **P**

$$d = \text{dist}(\mathbf{P}, \mathbf{Q}) \tag{4}$$

Plotting the distance function of a point **P** over a two-dimensional plane $\mathbf{R}^2$ yields a cone with a half angle of $\frac{\pi}{4}$ and with the apex at the point **P**, as shown in Fig. 1. For any point **Q** ($\in \mathbf{R}^2$), its distance d from point **P** is given by the distance function dist.

The conical face of the distance function dist(**P**, **Q**) shown in Fig. 1 divides the three-dimensional space $\mathbf{R}^2 \times \mathbf{R}$ into two sub-spaces, K and its complement K':

$$K = \{(\mathbf{Q}, d) : d \geq \text{dist}(\mathbf{P}, \mathbf{Q}), \mathbf{P} \in \mathbf{R}^2, \forall \mathbf{Q} \in \mathbf{R}^2\} \tag{5}$$

and

$$K' = \{(\mathbf{Q}, d) : d < \text{dist}(\mathbf{P}, \mathbf{Q}), \mathbf{P} \in \mathbf{R}^2, \forall \mathbf{Q} \in \mathbf{R}^2\} \tag{6}$$

Hence, the distance function of a pointis written as

$$\partial K = \{(\mathbf{Q}, d) : d = \text{dist}(\mathbf{P}, \mathbf{Q}), \mathbf{P} \in \mathbf{R}^2, \forall \mathbf{Q} \in \mathbf{R}^2\} \tag{7}$$

3.2 Distance Function of Two Points

A distance function of two points, $\mathbf{P}_1$ and $\mathbf{P}_2$, yields the minimum Euclidean distance from a point $\mathbf{Q}(\in \mathbf{R}^2)$ to either point $\mathbf{P}_1$ or $\mathbf{P}_2$. Hence the distance function dist : $\mathbf{R}^2 \to \mathbf{R}$ of two points $\mathbf{P}_1$ and $\mathbf{P}_2$ is defined as

$$\text{dist}(\mathbf{P}_1|\mathbf{P}_2, \mathbf{Q}) = \min_{\mathbf{P} \in \{\mathbf{P}_1, \mathbf{P}_2\}} \text{dist}(\mathbf{P}, \mathbf{Q}), \forall \mathbf{Q} \in \mathbf{R}^2 \tag{8}$$

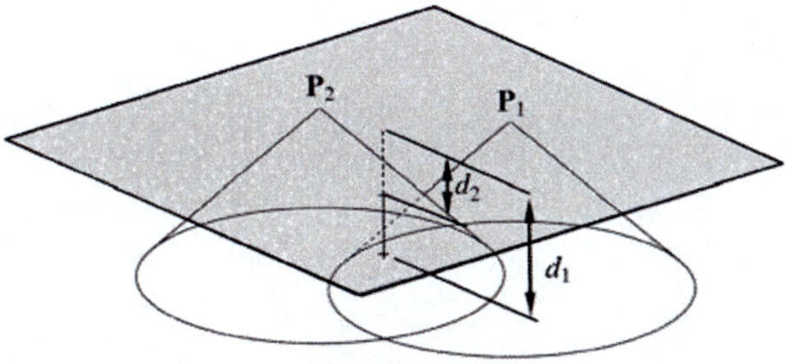

Fig. 2 The distance function of two points

Figure 2 shows the distance functions of points $\mathbf{P}_1$ and $\mathbf{P}_2$. The distances of point $\mathbf{Q}$ from point $\mathbf{P}_1$ and $\mathbf{P}_2$ is given as $\mathrm{dist}(\mathbf{P}_1, \mathbf{Q})$ and $\mathrm{dist}(\mathbf{P}_2, \mathbf{Q})$. As a result,

$$K_1 = \{(\mathbf{Q}, d_1) : d_1 \geq \mathrm{dist}(\mathbf{P}_1, \mathbf{Q}), \mathbf{P}_1 \in \mathbf{R}^2, \forall \mathbf{Q} \in \mathbf{R}^2\} \tag{9}$$

$$K_2 = \{(\mathbf{Q}, d_2) : d_2 \geq \mathrm{dist}(\mathbf{P}_2, \mathbf{Q}), \mathbf{P}_2 \in \mathbf{R}^2, \forall \mathbf{Q} \in \mathbf{R}^2\} \tag{10}$$

and

$$K_1 \cup K_2 = \left\{(\mathbf{Q}, d) : d \geq \min_{\mathbf{P} \in \{\mathbf{P}_1, \mathbf{P}_2\}} \mathrm{dist}(\mathbf{P}, \mathbf{Q}), \forall \mathbf{Q} \in \mathbf{R}^2\right\} \tag{11}$$

The closure of the set $K_1 \cup K_2$ is expressed as

$$\partial(K_1 \cup K_2) = \left\{(\mathbf{Q}, d) : d = \min_{\mathbf{P} \in \{\mathbf{P}_1, \mathbf{P}_2\}} \mathrm{dist}(\mathbf{P}, \mathbf{Q}), \forall \mathbf{Q} \in \mathbf{R}^2\right\} \tag{12}$$

which is the distance function of the points $\mathbf{P}_1$ and $\mathbf{P}_2$ as defined in Eq. (8).

3.3 Distance Function of a Planar Curve

A planar curve $\mathbf{C}(t)$ can be expressed as a set $\mathbf{E}$ of points:

$$\mathbf{E} = \{\mathbf{P}_i : \mathbf{P}_i \in \mathbf{C}(t)(\subset \mathbf{R}^2), \forall t\} \tag{13}$$

Hence, the distance function $\mathrm{dist} : \mathbf{R}^2 \to \mathbf{R}$ of a curve $\mathbf{C}(t)$ is defined as

$$\mathrm{dist}(\mathbf{C}(t), \mathbf{Q}) = \min_{\mathbf{P}_i \in \mathbf{E}} \mathrm{dist}(\mathbf{P}_i, \mathbf{Q}), \quad \forall \mathbf{Q} \in \mathbf{R}^2 \tag{14}$$

If the geometry of the distance function of a point $\mathbf{P}_i (\in \mathbf{E})$ is expressed as a ∂K_i such that

$$K_i = \{(\mathbf{Q}, d_i) : d_i \geq \mathrm{dist}(\mathbf{P}_i, \mathbf{Q}), \mathbf{P}_i \in \mathbf{E}, \quad \forall \mathbf{Q} \in \mathbf{R}^2\} \tag{15}$$

then the geometry of the distance function of a curve $\mathbf{C}(t)$ is

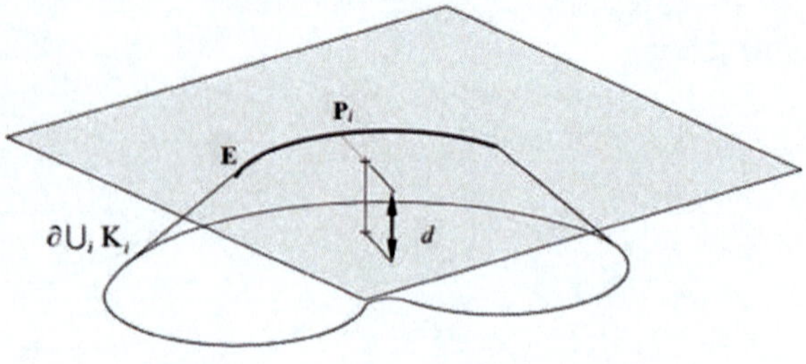

Fig. 3 The distance function of a curve

$$\partial(\cup_i K_i) = \left\{(\mathbf{Q}, d) : d = \min_{\mathbf{P}_i \in \mathbf{E}} \operatorname{dist}(\mathbf{P}_i, \mathbf{Q}), \forall \mathbf{P}_i \in \mathbf{E}, \quad \forall \mathbf{Q} \in \mathbf{R}^2\right\} \tag{16}$$

The distance function $\partial(\cup K_i)$ of the curve $\mathbf{C}(t)$ (expressed as the set $\boldsymbol{E}$) is shown in Fig. 3.

4 Curve Offset

A plane parallel to the planar curve $\mathbf{C}(t)$ at a distance r is expressed as a set $\pi_r(\in \mathbf{R}^2 \times \mathbf{R})$

$$\pi_r = \{(\mathbf{Q}_i, d) : d = r, \forall r \in \mathbf{R}, \forall \mathbf{Q}_i \in \mathbf{R}^2\} \tag{17}$$

Intersecting the sets $\partial(\cup_i K_i)$ and π_r yields

$$(\partial(\cup_i K_i)) \cap \pi_r = \left\{(\mathbf{Q}_i, r) \in \pi_r{:}r = \min_{P_i \in \mathbf{E}} \operatorname{dist}(\mathbf{P}_i, \mathbf{Q}_i), \mathbf{P}_i \in \mathbf{E}, \forall \mathbf{Q}_i \in \mathbf{R}^2\right\} \tag{18}$$

where $\cap$ is the Boolean intersection operator.

The minimum of $\operatorname{dist}(\mathbf{P}_i, \mathbf{Q}_i)$ for all $\mathbf{P}_i \in \mathbf{E}$ and $\forall \mathbf{Q}_i \in \mathbf{R}^2$ implies that the line $\mathbf{P}_i\mathbf{Q}_i$ is perpendicular to the curve $\mathbf{C}(t)$, and

$$(\mathbf{Q}_i, r) = (\mathbf{P}_i + r \cdot \mathbf{n}_i, r) \tag{19}$$

where $\mathbf{n}_i$ is the unit normal of curve $\mathbf{C}(t)$ at point $\mathbf{P}_{i\circ}$

Equation (17) can be rewritten as

$$\mathbf{Q}_i = \mathbf{P}_i + r \cdot \mathbf{n}_i \tag{20}$$

which is equivalent to projecting the set $(\partial(\cup_i K_i)) \cap \pi_r$ onto $\mathbf{R}^2$, which yields

$$\mathbf{C}_r(t) = \mathbf{C}(t) + r \cdot \mathbf{n}(t) \tag{21}$$

where $\mathbf{C}_r(t) = \{\mathbf{Q}_i : \forall(\mathbf{Q}_i, r) \in \pi_r, \forall i\}$

Equation (19) is an offset curve with offset distance r from the progenitor curve $\mathbf{C}_r(t)$.

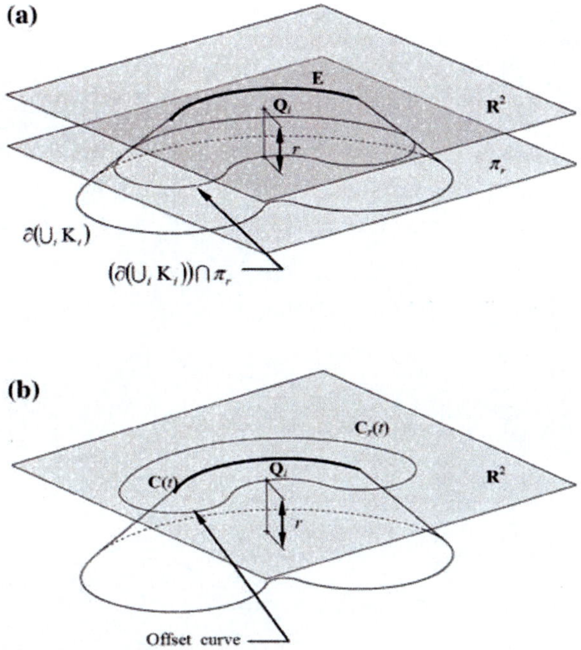

Fig. 4 Obtaining an offset curve. (**a**) Interaction operation to obtained $(\partial(\cup_i K_i)) \cap \pi_r$. (**b**) Offset curve

In Fig. 4a, the set $(\partial(\cup_i K_i)) \cap \pi_r$ is obtained by intersecting the sets $\partial(\cup K_i)$ and π_r. The resultant set is then projected onto the space $\mathbf{R}^2$, as shown in Fig. 4b.

Figure 5 shows an example of obtaining the offset curves based on set theory. A distance function of a closed curve (of a horse shape) is plotted in Fig. 5a, which shows the distance of the points inside the closed curve to the curve. A family of inward offset curves is shown in Fig. 5b.

Algorithm I

```
Input:   E                        /* curve C(t) on plane π */
         r                        /* offset distance */
S = ∅
for each point P ∈ E
 create set K at point P          /* create distance function of point P */
 S = S ∪ K                        /* create distance function of curve C(t) */
end for
create set π_r                    /* create a plane parallel to π at an offset distance d */
Q_d = S ∩ π_r                     /* generate the point set (Q, d) */
Output: Q_0                       /* output the offset curve on plan π */
```

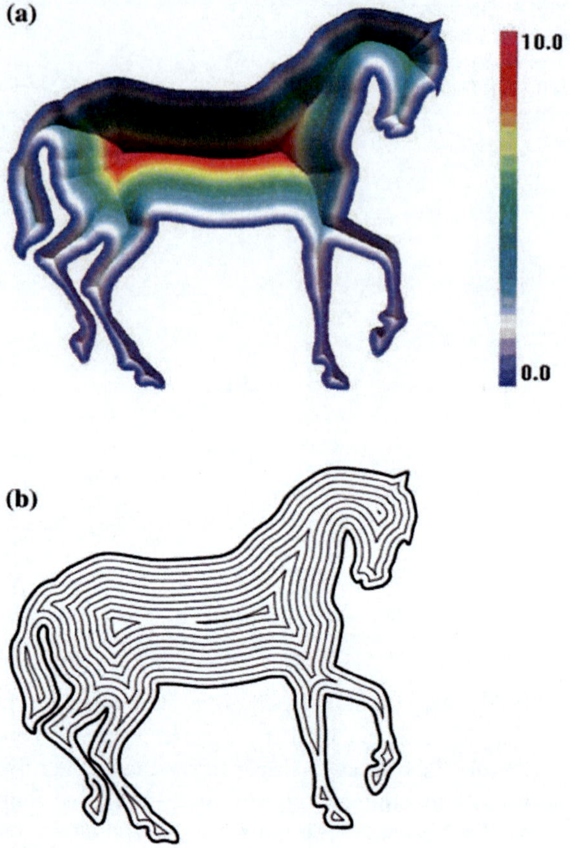

Fig. 5 Offset curve example. **(a)** Distance function inside a *closed curve*. **(b)** A family of inward *offset curve*

5 Algorithm and Implementation

A simple algorithm is proposed to generate the offset curves of a planar curve:

The algorithm is not easy to implement, as the sets $\boldsymbol{E}$, $\boldsymbol{K}$, $\mathbf{S}$, and π_r consist of an infinite number of elements. Discretization of the sets causes accuracy issues.

A geometric approach is employed to implement the algorithm. The geometric meanings of these sets are modelled using solid modeling representation. The solid modeling representation of an object is based on the idea that a physical object divides a three-dimensional space, $\mathbf{R}^3$, into two regions: interior and exterior, separated by the object boundaries. Hence, the sets K and S are represented by two solid models in the three-dimensional space. Several solid modeling operations can be employed to manipulate the solids (hence the sets) to obtain the offset curves.

The code fragment

$S = \varnothing$
for each point $\mathbf{P} \in \mathbf{E}$
create set K *at point* ***P***
$\mathbf{S} = \mathbf{S} \cup \mathrm{K}$
end for

is an infinite union operation. This is equivalent to sweeping a solid cone, K, along the curve $\mathbf{C}(t)$ as the trajectory.

Sweeping a tool solid along a trajectory to produce another solid is a complicated operation. The accuracy of the resultant solid from a sweeping operation largely depends upon the geometry of the tool solid and on the trajectory. Topologically, the resultant solid cannot have any self-intersection, since this violates the 2-manifold data structure of a solid.

In this implementation, the tool solid is a cone that has standard analytical geometry. As a result, the geometry of the trajectory, which is the progenitor curve of the offset, is the key factor for the accuracy and validity of the resultant solid. Basically, the varying radius of curvature of the progenitor affects both the accuracy and the topology of the resultant solid. The cone must be kept in an appropriate orientation to yield an accurate surface representation. The progenitor curvature also determines if there is a self-intersection during the sweep operation.

In order to increase the modeling efficiency, bi-arc approximation [12, 16] is employed to convert the progenitor curve into a set of arcs to facilitate the implementation. The arcs are paired up so that each bi-arc matches two end points and two end tangents of a curve segment. This method is appropriate in discretizing the curvature of a free-form curve.

Assume that $\mathbf{C}(t)$ is a NURBS curve which is approximated by a set of arcs $\mathbf{A}_i(\forall i)$. Hence, $\mathbf{C}(t) = \cup_i \mathbf{A}_i(s)$ for $0 \leq t \leq 1$ and $0 \leq s \leq 1$. A sweep operation is defined as

$$\mathbf{S}_i = \mathrm{sweep}(\mathbf{A}_i) \tag{22}$$

which yields a solid representation, $\mathbf{S}_i$, of the distance function of a curve $\mathbf{A}_i$.

Representing the curve $\mathbf{C}(t)$ by a set of arcs $\mathbf{A}_i(\forall i)$ simplifies the sweep operation, since both tool solid (the cone) and trajectory (the arc) are analytical geometries. Furthermore, this application of bi-arc approbation of a curve generates a standard solid representation of a distance function, since sweeping a cone along an arc yields the same solid geometry as revolving a cone about the axis through the center of the arc.

Figure 6a shows the sweeping of a cone along an arc, $\mathbf{A}_i$, which is equivalent to rotating a cone about the axis through the center of the arc. The height of the cone is chosen to be larger than the offset distance. One of the choices of the height can be the maximum of the largest radii of all the arcs and the offset distance. Figure 6b shows the resultant solid geometry $\mathbf{S}_i$ obtained by sweeping a cone along one of the arcs, $\mathbf{A}_i$. The two planar faces of the solid $\mathbf{S}_i$ will join the faces of the solids $\mathbf{S}_{i-1}$ (which is generated by sweeping a cone along arc $\mathbf{A}_{i-1}$ with radius r_{i-1}) and $\mathbf{S}_{i+1}$ (which is generated by sweeping a cone along arc $\mathbf{A}_{i+1}$ with radius r_{i+1}).

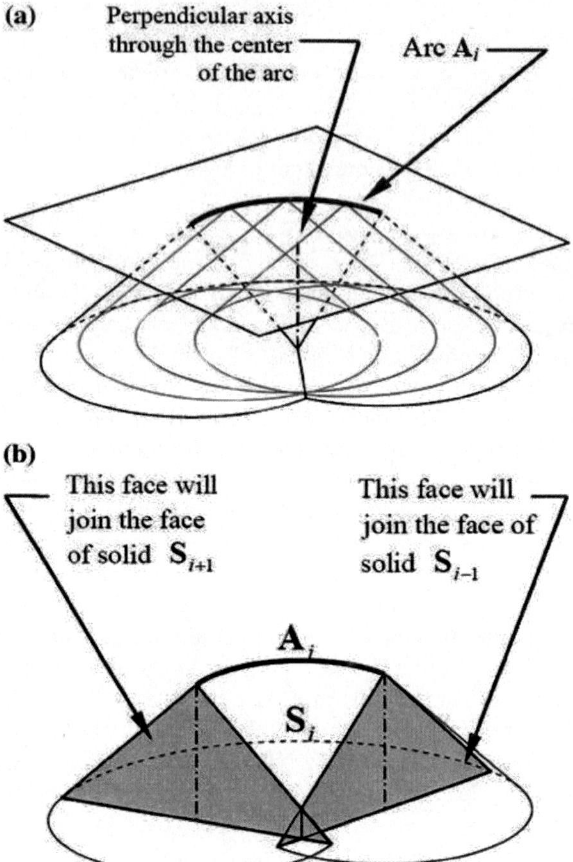

Fig. 6 Sweeping a *cone* along an *arc*. (**a**) Rotation as sweeping. (**b**) Resultant solid geometry

The other code fragment

create set π_r

$$\mathbf{Q}_r = \mathbf{S} \cap \pi_r$$

is to intersect the solid **S** with a plane, π_r, which is at a parallel distance r from the planar curve $\mathbf{C}(t)$. Since the solid **S** and the plane π_r are of different dimensions, this operation is replaced by sectioning the solid S by the plane π_r. The edges of the sectioned solid are extracted and projected onto the plane of the curve $\mathbf{C}(t)$. A section operation and project operation are defined as

$$\mathbf{Q}_r = \text{section}(\mathbf{S}, r) \tag{23}$$

which splits a solid **S** with a plane parallel at distance r to the plane of the curve $\mathbf{C}(t)$ and

$$\mathbf{Q}_o = \text{project}(\mathbf{Q}_r) \tag{24}$$

which projects the edge of the solid geometry $\mathbf{Q}_r$ onto the plane of curve $\mathbf{C}(t)$.

The algorithm is rewritten as

Algorithm II

Input: $\mathbf{C}(t)$ /* curve $\mathbf{C}(t)$ on plane π */
r /* offset distance */

Approximate $\mathbf{C}(t)$ by a set of $2n$ arcs $\mathbf{A}_i\ (i = 1, 2, \ldots, n)$
$\mathbf{S}=\varnothing$
For i =1, $2n$
$\mathbf{S} = \mathbf{S} \cup \text{sweep}(\mathbf{A}_i)$ /* create the solid geometry of the distance function*/
$\mathbf{Q}_o = \text{project}(\text{section}(\mathbf{S}, r))$
/* section the solid geometry and output the offset curve */

This algorithm is implemented with ParaSolid kernel, which provides the geometric modeling operation routines of sweep, section, and project.

Figure 7 shows the processes of obtaining the offset curves. A curve is approximated by two arcs, $\mathbf{A}_1$ and $\mathbf{A}_2$. The solid geometry $\mathbf{S}_1$ and $\mathbf{S}_2$, representing the distance functions for arc $\mathbf{A}_1$ and $\mathbf{A}_2$, are shown in Fig. 7a, b. Figure 7c, d depict their positions and the solid geometry of the resultant distance function $\mathbf{S}_1 \cup \mathbf{S}_2$, respectively. Sectioning the solid geometry of the resultant distance function yields the offset curves illustrated in Fig. 7e. The family of offset curves is projected onto the plane of the progenitor curve as shown in Fig. 7f.

Figures 8 and 9 show two examples of curve offsetting. Figure 8a is an open curve and its offset curves are shown in Fig. 8b. Figure 8c, d depict the distance function and its solid geometry with sections. In comparison, Fig. 9a, b show a closed curve with an island inside, and its inward offset curves. The distance function and its solid geometry with sections are shown in Fig. 9c, d, respectively. For visualization purposes, only the distance function inside the loop of the closed curve is shown.

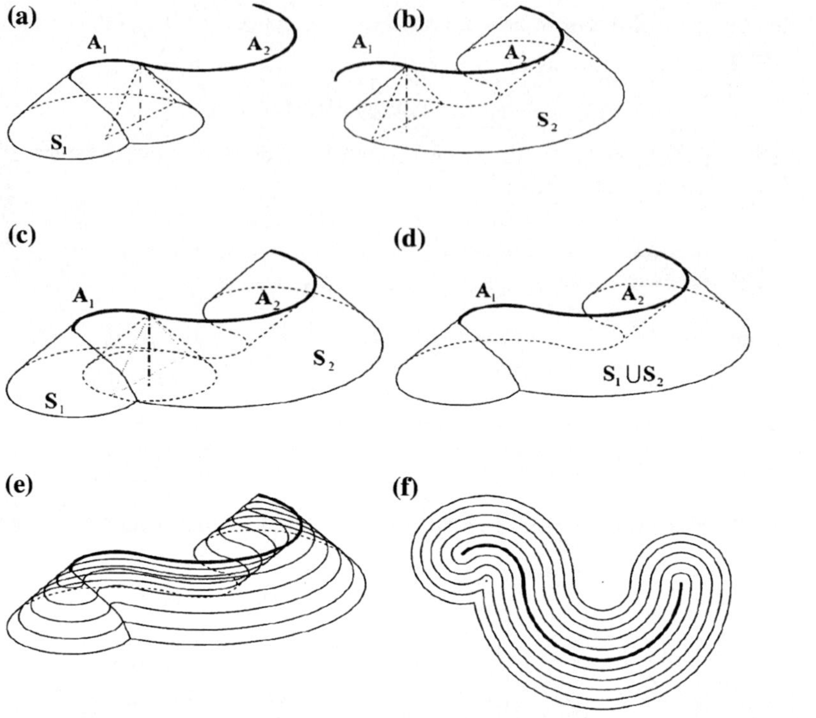

Fig. 7 Generating the offset curve by distance function **(a)** Distance function for *arc* A_1. **(b)** Distance function for *arc* A_2. **(c)** Distance function of both *arc*. **(d**)Resultant distance function. **(e)** Sectioning the distance function. **(f)** Projecting the family family of *offset curves*

6 Discussion

6.1 Offset Curve Representation

Most of the existing approaches to offset curve representation compute the exact offset curve points and then approximate the curves by interpolation using polynomials or rational curves. Furthermore, these approaches are iterative processes. The approximation error range is calculated and compared with the required tolerance. If the tolerance is outside the error range, the progenitor curve is subdivided and the process is repeated until the tolerance falls within the error range.

The proposed approach approximates the progenitor curve by a set of arcs and then obtains the exact offset curves. Hence, the offset curve is composed of a set of arcs with G^1 continuity. Since each arc can be represented by a rational Bezier curve, the offset curve can be represented by a nonuniform B-spline curve. The excess knots are removed to reduce the number of control points [14].

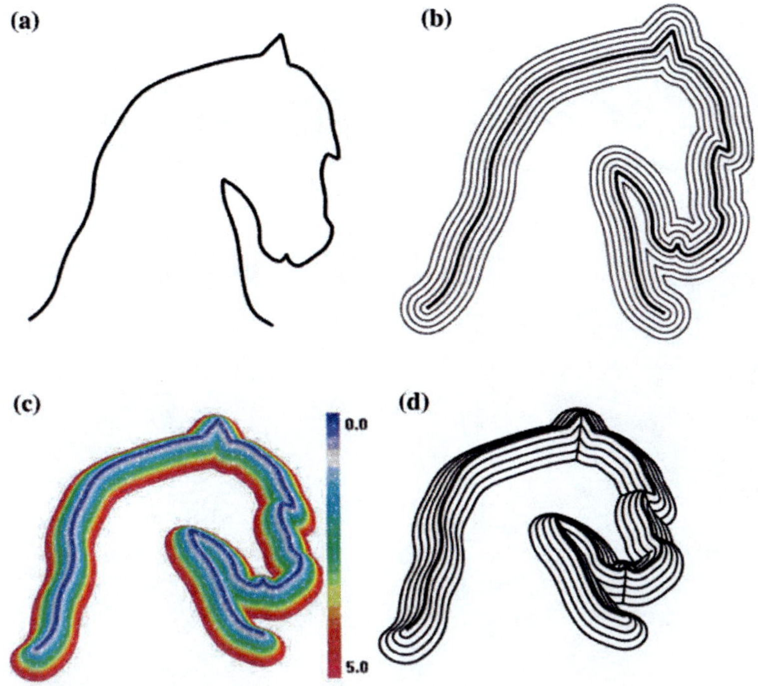

Fig. 8 The offset of an open curve . (**a**) An *open curve.* (**b**) The *offset curves.* (**c**) The distance function. (**d**) Solid geometry with sections

6.2 Self-Intersection

The changes of topology between the progenitor curve and the offset curve are mainly due to the local and global self-intersections of offset curves. The common approach to dealing with these self-intersections is to detect these changes and trim them away. However, self-intersection detections are nontrivial and increase the complexity of the algorithm. Using set theory and solid modeling, the proposed implementation eliminates the detection and trimming process in offset curve computation.

Plotting the distance function of a curve is a major step in computing the offset curve. It is based on the fact that a planar curve consists of an infinite number of points, which can be expressed as $\mathbf{E} = \cup_i\{\mathbf{P}_i\}$ $(\forall i)$. The distance function of a point, $\mathbf{P}_i$, is given as ∂K_i in Eq. (6); however, the distance function of a curve is $\partial(\cup_i K_i)$ instead of $\cup_i(\partial K_i)$. This set operation arrangement will automatically eliminate the self-intersection of the offset curve and results in a simple offset curve computation algorithm. It can be seen that all the local and global self-intersections of the offset curves shown in Figs. 5b, 6f, 7b, and 8b are eliminated

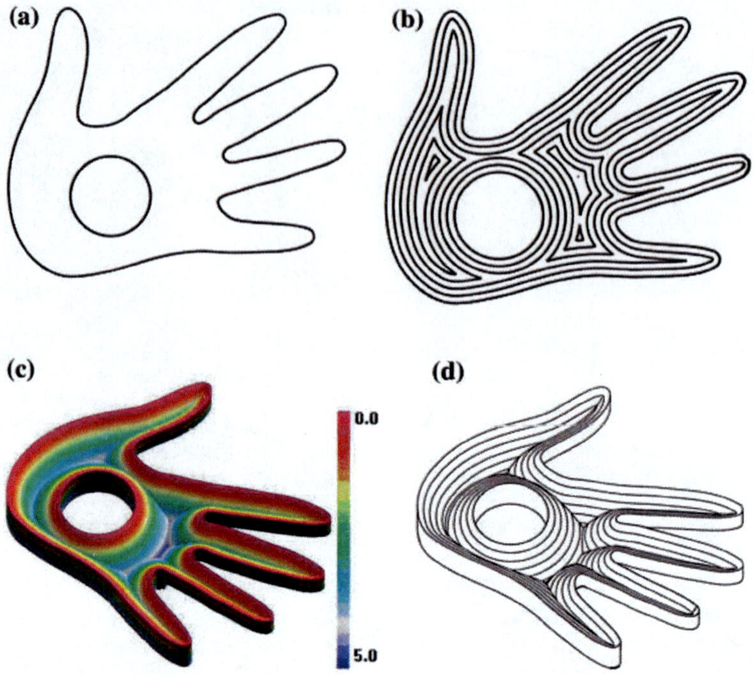

Fig. 9 The offset of a *closed curve*. (**a**) A *closed curve* with an island. (**b**) The inward *offset curves*. (**c**) The distance function. (**d**) Solid geometry with sections

even though no self-intersection detection and trimming routines are found in the algorithm.

6.3 Error Control

Most of the offset curve representations using polynomial or rational functions are approximated by interpolating the sample points from the offset curve Equation (1). An error function is defined to measure the deviations [1, 7–9]. The errors between the approximated offset curve and the exact offset curve are computed. If the error is larger than the required tolerance, the offset curve will be re-interpolated with more sample points to reduce the error. This is a timely iterative process, as errors are not directly controlled. Furthermore, the error function for deviation measurement is dependent upon the offset curve computation method and must be correctly defined; otherwise, the approximation errors may be either over- or under-estimated. Alternatively, an error bound [20] is employed to estimate the errors rather than computing the actual deviation.

The method proposed here uses a different approach. Instead of approximating the offset curve, it approximates the progenitor curve; the exact offset of the approximated progenitor curve is then computed. This approach has direct control over the errors between the progenitor curve and its bi-arc approximation [12, 16]. The implementation of this proposed algorithm employs the robust algorithm provided by Piegl and Tiller [16]. The examples presented here were all created using NURBS curves, and were approximated by a set of bi-arcs with a tolerance of 10^{-6} mm. The curve offset computation performance is very fast and the calculation can be done almost in real time with the required tolerance.

7 Summary

This chapter presented a method to compute offset curves. The method can be summarized as follows:

1. The progenitor curve is reformatted by the bi-arc approximation.
2. A distance function yielding the minimum distance from a point to the progenitor curve approximation is plotted by using set theory and a union Boolean operation.
3. The offset curves are obtained by an intersection Boolean operation.

The major advantage of this approach is its coverage of both self-intersection and curve representation issues. The algorithm is simple, requiring no search and trimming operation. The Boolean set operations (union and intersection) will automatically handle the topological changes. The offset curves are expressed as a set of arcs with first-order continuity. These arcs can be reformulated to a NURBS representation with the excess knots removed.

Additionally, an approach to compute the distance between a point and a curve, as well as to carry out solid modeling, was introduced by applying the set theory and Boolean operation. This distance computation method will be useful for medial axis transformation and Voronoi diagram computation.

References

1. Ahn YJ, Kim YS, Shin Y (2004) Approximation of circular arcs and offset curves by Bezier curves of high degree. J Comput Appl Math 167:405–416
2. Coquillart S (1987) Computing offset of B-splines curves. Comput-Aid Des 19:305–309
3. Farouki RT, Sakkalis T (1990) Pythagorean hodographs. IBM J Res Dev 34:736–752
4. Held M (1991) On the computational geometry of pocket machining. Springer, Berlin
5. Klass R (1988) An offset spline approximation for plane cubic splines. Comput-Aid Des 20:33–40
6. Lai YL, Wu JSS, Hung JP, Chen JH (2006) A simple method for invalid loops removal of planar offset curves. Int J Adv Manuf Tech 27:1153–1162

7. Lee IK, Kim MS, Elber G (1996) Planar curve offset based on circle approximation. Comput-Aid Des 28:617–630
8. Lee IK, Kim MS, Elber G (1997) New approximation methods of planar offset and convolution curves. In: Strasser W, Klein R, Rau R (eds) Geometric modelling: theory and practice. Springer, Berlin
9. Lee IK, Kim MS, Elber G (1998) Polynomial approximation of Minkowski sum boundary curves. Graph Model Image Process 60:136–165
10. Li W, Martin RR (2011) Global topological changes of offset domains. In: Eighth international symposium on voronoi diagram in science and engineering. Qingdao
11. Ma YS (1994) The generation of tool paths in 21/2-D milling. PhD thesis, University of Manchester, Manchester
12. Park H (2004) Error-bounded biarc approximation of planar curves. Comput-Aid Des 36:1241–1251
13. Park SC, Choi BK (2001) Uncut free pocketing tool-paths generation using pair-wise offset algorithm. Comput-Aid Des 33:739–746
14. Piegl LA, Tiller W (1997) The NURBS book. Springer, Berlin
15. Piegl LA, Tiller W (1999) Computing offsets of NURBS curves and surfaces. Comput-Aid Des 31:147–156
16. Piegl LA, Tiller W (2002) Biarc approximation of NURBS curves. Comput-Aid Des 34:807–814
17. Seong JK, Elber G, Kim MS (2006) Trimming local and global self-intersections in offset curves/surfaces using distance maps. Comput-Aid Des 38:183–193
18. Seong JK, Johnson DE, Elber G, Cohen E (2010) Critical point analysis using domain lifting for fast geometry queries. Comput-Aid Des 42:613–624
19. Tiller W, Hanson EG (1984) Offsets of two-dimensional profiles. IEEE Comput Graph Appl 4:36–46
20. Zhao HY, Wang GJ (2007) Error analysis of reparameterization based approaches for curve offsetting. Comput-Aid Des 39:142–148
21. Zhao HY, Wang GJ (2009) Offset approximation based on reparameterization the path of a moving point along the base circle. Appl Math: J Chin Univ 24:431–442

Feature Transformation from Configuration of Open-Loop Mechanisms into Linkages with a Case Study

Abiy Wubneh, C. K. Au and Y.-S. Ma

1 Introduction

This chapter proposes a method for feature synthesis of mechanisms and manipulators from user specifications based on a hybrid approach employing both neural network and optimization techniques. The mechanism design modeling problem with the lack of solution convergence observed in optimization is addressed by using a neural network method to generate reliable initial solutions. This chapter also discusses a module by which the validation of prescribed precision configuration points is evaluated. An excavator arm mechanism is used as a case study to test and validate the method. The necessary training data for the neural network is generated primarily through the use of forward kinematics equations, while the proposed method is analyzed using dimensional data collected from existing products.

Existing products are frequently modeled as a type of assembly features [9]. They can be redesigned and customized to meet specific operational needs and increase efficiency. Such customizable and yet conceptually proven products are commonly used to perform atypical tasks under space constraints, such as specialized manipulators. These products can be developed as a cluster of instances of a generic product because of their inherent common engineering principles. The generic product model are modeled in the form of assembly features. In most cases, the design objective can be achieved by adopting a different set of configuration parameter values based on a generic product model of the existing design features using the same design procedures developed during the initial design. Such well-defined assembly features, whose parameters can be assigned

A. Wubneh · Y.-S. Ma (✉)
Department of Mechanical Engineering, University of Alberta,
Edmonton, AB T6G 2G8, Canada
e-mail: yongsheng.ma@ualberta.ca

C. K. Au
Engineering, University of Waikato, Waikato, New Zealand

Y.-S. Ma (ed.), *Semantic Modeling and Interoperability in Product and Process Engineering*, Springer Series in Advanced Manufacturing,
DOI: 10.1007/978-1-4471-5073-2_9,

with different values, enable the product configuring mechanism to achieve increasing versatility and to address customization needs. However, specifying an effective and valid design feature data set for those existing feature models is difficult given the expected complicated and interdependent constraints [8].

For certain machinery equipment, such as excavators, the final spatial access envelope diagrams of the overall assemblies, which are referred to here as product *specification features*, are the basis of customer evaluation of the dimensional specifications. This kind of specification feature can be appreciated as a subtype of the assembly design feature as defined by Ma et al. [9]. Unlike those well-defined modular mechanical products, such as the mold bases used in the plastic molding industry [9], the conceptual design of manipulator products usually starts with a set of target specification features, i.e., the envelope diagrams or prescribed paths and motions identified by end users that need to be achieved by the overall mechanism.

There is another type of feature that has to be defined in this chapter: the design *configuration feature*, which is a product-level assembly feature [9] that represents the design intent with characteristic dimensions, geometry, patterns, and other associative relations to interfacing components. In the context of typical manipulator mechanism design, the configuration features are realized by materializing the *component design features* [10] involved. The process of design realization involves a feature dimensional synthesis phase in which engineers analyze relations among component design features to compose a workable and satisfactory product configuration that is governed by the aforementioned configuration feature. Of course, the final product design has to be completed by focusing on determining individual component feature dimensions and their related constraints, as well as material application patterns, in order to meet the configuration feature requirements after the mechanism is assembled. This chapter covers the design transformation process from the specification features to the configuration features. The remaining further design processes will be covered in the next chapter.

With reference to the context of manipulator design, a single pose (position and orientation) of the end effectors of the manipulator, or a valid instance of the specification feature, is defined by customers with their application demands; to translate the specification features into a set of configuration design features, the transformation process needs to use those known values of the manipulator's linkage dimensions together with the joint parameters to go through trials of inverse kinematic fitting. Typically, the specified poses (or instances of specification features) can only be verified and eventually confirmed by using kinematics procedures with the assumed (or known) linkage dimensions in the manipulator, or in other words, the related component design features.

However, the challenge arises when the customer requires and defines a set of expected configuration poses: what is the method to transform such input into those materialized linkage configuration features? In such a case, instead of solving a forward kinematics problem (direct configuration feature development) from the known components, the nature of the (which is common manipulator design) problems require solving kinematics problems in reverse based on a set of given specification feature instances. In other words, with the predefined specification feature instances defined in the form of access envelopes, the configuration definition of the mechanism needs to be inversely calculated.

Multiple configuration solutions to such inverse kinematics problems usually exist. Obviously, with a set of required feature behavior instances (poses) to be targeted in the form of an access envelope space or, even more critically, an access path, the calculation for the mechanism dimensions becomes very complicated due to the combined constraints of kinematics and the existence of multiple solutions. Finding these solutions is already a challenge, but evaluating or validating them is even more difficult. For a single pose problem, the existing methods are manageable with reasonable effort. However, a multi-pose problem, which requires that the calculated linkage dimensions and joint variables fully satisfy a set of configurations or path parameters, makes the inverse kinematics approach difficult to implement.

Thus far most researchers have tried to solve the inverse kinematics and optimization problems by using a computational workspace searching method, but the optimization results are not reliable due to the fact that there are multiple solutions. To obtain the necessary convergence toward the expected range of a solution, researchers need initial suggestions as the input of the searching procedures. This initial solution requirement creates considerable challenges when trying to automate the conceptual design process and implement it using computer programs. There are dense correlations among the configuration feature parameters; arbitrary values cannot be assumed in their places when solving the system equations. Failure to use an appropriate starting parameter vector may produce mathematically accurate but physically impossible solutions.

The two remaining tasks are formulating a set of parametric geometric relationships for the specification feature of a manipulator, which has to be associated with typical linkage configuration feature parameters, and searching a workable solution, which needs an optimization technique.

This chapter proposes a method by which feature dimensional synthesis for manipulator mechanisms is performed based on the end user's specification parameter input. The implementation of this method requires a vector of initial suggestions of linkage configuration parameters that has to be close to the expected solution. A smart neutral network procedure is used to generate the feasible initial suggestions of the linkage parameters. A case study of an excavator arm mechanism is carried out and the results are promising. The algorithm has been implemented in MATLAB, a numerical analysis software tool produced by MathWorks.

2 Background of Relevant Research

Optimizing mechanism dimensions is a well-known design problem in the field of robotics and machinery, and a broadly studied research area. However, due to the multiple members of a mechanism and the dependencies among them that are constrained by kinematics, the problem is complicated. For example, Wu et al.

[13] formulated the kinematic constraints of closed chain mechanism as a mapping from Cartesian space to a higher dimensional projective space called image space by representing planar displacements with planar quaternions. The researchers pointed out that the use of this method enables one to reduce the problem of dimensional synthesis by determining algebraic parameters that define the image spaces. Computational simplification was achieved by transforming kinematic equations into geometric constraints. Dealing with the geometric parameters of the constraint manifold instead of the mechanism parameters provides ease and flexibility due to the decoupled nature of the relationships.

Optimization techniques are usually applied to solve the mechanism linkage dimensions). For example, a procedure of synthesizing the linkage dimensions of a four-bar spherical linkage mechanism was proposed by Alizade and Kilit [1] The procedure used a polynomial approximation to transform 5 nonlinear equations into 15 linear equations and solve five design parameters. The objective of this study was to determine the dimensions of a spherical four-bar linkage mechanism by linearizing a set of nonlinear equations. The requirement for the mechanism was that it will be able to trace five precision points in space. The minimum deviation area (MDA) was proposed as a constraint criterion to select the most appropriate solution. The result of this investigation was tested by plotting the path of the mechanism against the prescribed precision points using AutoCAD 2000.

Jensen and Hansen [6] have suggested a method by which dimensional synthesis for both planar and spatial mechanisms are accomplished by taking the problem of non-assembly into consideration. The method makes use of a gradient-based optimization algorithm. Analytic calculation of sensitivities is performed by direct differentiation. The problem was mathematically formulated as a standard optimization problem with inequality to take the non-assembly nature of the problem into account. The Newton–Raphson method, due to its rapid convergence property, is used in the minimization of the kinematic constraints. Saddle's point and steepest descent methods were used to verify the direction of convergence and stability of the minimization method, respectively.

Kinematic synthesis of redundant serial manipulators has become the focus of research for Singla et al. [11]. They used an augmented Lagrangian optimization technique to determine optimum dimensions for a redundant serial manipulator. The algorithm was used for its robustness in identifying feasible solution ranges effectively. The formulation of the problem was based on the minimization of the positional error subject to the constraints of avoiding manipulator collisions with either external obstacles or its own links.

The workspace boundary definition can be more complicated. Laribi et al. [7] discussed an optimization technique used for determining the linkage dimensions)of a DELTA parallel robot for a prescribed workspace. The technique uses a genetic algorithm to minimize an objective function developed by writing expressions for the end effector location, based on a concept called the power of the point. The dimensions of the robots were calculated by minimizing a volume created by three intersecting surfaces that contain the prescribed cubic workspace. A penalty function screened out infeasible and select feasible solutions from the available solution domain. Zhao et al. [15] used a similar approach, but the prescribed workspace was represented by a cylinder contained inside the minimum workspace volume and enclosed by the manipulator movement boundary surfaces. An optimization-based dimensional synthesis procedure was suggested to determine dimensional parameters for the design of a 2-UPS-PU parallel manipulator. The researchers used a cylindrical coordinate system when formulating the kinematic relationships, including the forward and inverse kinematics of the manipulator together with the Jacobian matrix for force and velocity analysis.

However, it has been identified that the multiple numbers of possible solutions is the primary disadvantage of analytical solutions methods. Gao et al. [4] reported that for their six degree of freedom (DOF) parallel manipulator, the traditional optimization techniques in the areas of dimensional synthesis lack the badly needed convergence property in their solutions when it is used for handling a larger number of geometric variables and complex objective functions. Non-traditional optimization methods need to be explored in order to address the problems of convergence uncertainties and limitations, on a maximum number of precision point problems solved using optimization and analytical techniques.

Gao et al. [4] also used generic algorithms and artificial neural networks (ANNs) as tools to deal with the optimization of the manipulator's stiffness and dexterity based on kinematic analysis procedures. Levenberg–Marquardt and standard back propagation algorithms were used in the neural network to approximate stiffness and dexterity analytical solutions. Because of the large numbers of variables included in the analysis, they have used two different approaches for the optimizations: Single Objective Optimizations (SOOs) and Multiple Objective Optimizations (MOOs) multiple objective optimization (MOO). With the first approach, the two objectives, stiffness and dexterity, were investigated separately; with the second approach, they were investigated together to understand their combined effect. Both approaches proved to be compatible.

It is worth pointing out, as Vasiliu and Yannou [12] did in their work, that "the absence of continuity between different morphologies prohibited and discouraged the use of interpolation techniques" in such a problem. Vasiliu and Yannou also proposed the use of ANNs. The ANN designed to be used for the synthesis application takes in the prescribed path and motion as an input and gives out the linkage parameters as an output. Erkaya and Uzmay [2] aimed to overcome problems arising from joint clearances in a four-bar mechanism. They used neural networks to characterize the clearances and the mechanism, and genetic algorithms to optimize them with the path and transmission angle errors used as part of the objective function. The clearances were represented by high stiffness and weightless links to make them suitable to be studied under rigid motion considerations but without affecting the overall inertial property of the mechanism. ANN procedures were also used by Hasan et al. [5] to study the relationship between the joint variables and the position and orientation of the end effector of a six DOF robot. The study was motivated by the fact that the use of ANN does not require an explicit knowledge of the physics behind the mechanism. The network was trained by the use of real-time data collected by sensors mounted on the robot. Designed with an input layer of six neurons for three Cartesian location coordinates and three linear velocity components, the network was used to establish a mapping pattern between the input and output. The project mainly focused on finding the kinematic Jacobian solutions.

The advantage of using ANN is that it does not require any details of the mathematical and engineering knowledge involved [5]; it is thus suited to a wide range of similar applications. It was suggested that as long as there is sufficient data for training purposes, the ANN can be used to predict the Jacobian kinematics of other configurations without the need to learn and understand the explicit robot philosophies. Modifications and changes in existing robot structures can always be addressed by training the ANN with a new set of data reflecting the new modifications.

The problems and shortcoming associated with using ANN are also discussed in Hasan et al. [5]. The first challenge discussed is the difficulty of selecting the appropriate network architecture, activation functions, and bias weights. The other problem discussed is the difficulty and impracticality of collecting large amounts of data for the purpose of training the neural network.

As an alternative to using the ANN approach, some researchers are more interested in simulation and spatial configuration performance analysis of manipulators. Their work is motivated by the need to understand the manipulators' performance under certain environmental constraints. Frimpong and Li [3], for example, modeled and simulated a hydraulic shovel to investigate its kinematics and spatial configurations when the shovel is deployed in a constrained mining environment. Denavit-Hartenberg homogeneous coordinate transformation techniques were used to represent the relative orientations and configurations of adjacent links as well as the overall assembly. Forward kinematics of the machine was investigated as a five-linkage manipulator. After formulating the kinematics equations the manipulator was modeled in 3D and was simulated using the MSC ADAMS simulation software for selected time steps.

Therefore, as suggested by Vasiliu and Yannou [12], the requirement of a large number of data for training ANN can be addressed by simulation of the paths for a number of given sets of linkage parameters. The ANN can be trained using the simulated data in reverse, i.e., that for the given sets of mechanism parameters, the information of the access paths were determined. The other important point discussed in Vasiliu and Yannou's work is that neural networks perform well only within the data range they were trained with. Normalization of parameters during the utilization phase of the network is needed to bring the input values to the known range of the training set.

The constraints imposed for manufacturing the products usually dictates the capacities and efficiencies of the machineries. The general design and modification requirements can sometimes be achieved by merely redesigning an existing mechanism out of a different set of existing product sales materials.

Remote operability of hydraulic excavators, initiated due to operational safety and hazard issues, has recently become the focus of some researchers. The task of controlling the motion of excavator arm mechanisms has been attempted by various remote control mechanisms. The method developed by Yoon and Manurung [14] is based on mapping the angular joint displacements of the human arm joints to that of the excavator arm joints. Their work is motivated by the need to include intuitivism into the control system.

3 The Proposed Hybrid Approach

3.1 Overall Concept Description

Most optimization techniques usually require a very good initial solution to be defined in order to produce sound solutions. One of the objectives of this chapter is to introduce a feature-based system by which a set of initial solutions that are reasonably close to the actual solution can be generated. Optimization techniques, when applied to the problems of dimensional synthesis of prescribed precision points, commonly encounter the difficulty of giving reasonable and practical results. There are two reasons for this: first is the proximity of the goal solution to the predefined initial solution; second is the compatibility or feasibility of the prescribed precision points. This is to say that prescription of unrealistic and ambitious specifications most likely produce, if the search converges to a solution at all, mathematically sound but physically inapplicable solutions.

The hybrid method proposed in this paper can be summarized by the flowchart as shown in Fig. 1. It is the objective of this chapter to introduce a new method in which a well-trained artificial neural network (ANN) tool is used to generate a set of high-quality initial solution suggestions for mechanism parameters based on user specifications, while optimization techniques are used to finally synthesize the necessary dimensions. The hybrid method attempts to jointly employ optimization and neural network procedures to synthesize the linkage design mechanisms' feature dimensions and further map them to the real manipulators. User specifications are also qualified with the checking of their priorities and ranges acceptable as the prescribed input values. The individual modules and procedures are explained in detail in the following subsection.

3.2 Synthesis and Validation Procedure

The proposed method can be divided into the following stages: (1) ANN training; (2) input parameter validation; (3) system testing; (4) initial solution generation; (5) mechanism parameter synthesis; (6) result verification; and (7) random system check. To make full use of the neural network's advantage, its inner transformation matrices must first be trained to reflect the intricate nature of input and output relations.

3.2.1 Artificial Neural Network Training

Essentially, the purpose of training the ANN is to build a database that will be used to generate the feasible suggestions of the initial mechanism parameters according to new configuration specifications. The first step is to collect the training data. Ideally, such training data can be obtained from existing similar product information catalogues, usually in the form of product families, because the relevant data from that channel is proven workable with both input and output sets. As shown in Fig. 1, the proposed method makes use of such data as indicated by the top job block. Unfortunately, although these real product data sets are quite useful for training the ANN, the number of available data sets is never sufficient. To find a solution for the shortage of training data, forward mechanism simulation can be utilized to create as many input/output data sets as required [5]. Note that the generation of such simulation data is necessary bccausc thc available data is usually insufficient to serve the training purpose and the extra effort of collecting additional real product data is prohibitively costly.

In the case of the data generation process, the specification feature parameters which define the total workspace of the mechanism assembly will be generated from the given set of linkage configuration feature dimensions using forward kinematic equations. This is a mapping process in which the mechanism design feature parameters (linkage dimensions) are mapped to the specification feature terms, i.e., the envelope configuration parameters of the workspace or the working path in the case of a planar mechanism.

When training the ANN, both the existing real product feature data sets and the generated data sets will be used in reverse: the existing specification feature parameters are used as the input data for the training while the mechanism configuration feature parameters are used as the target output data. Note that most of the training data sets can be generated from the "artificial" forward mechanism behavior algorithm as used by Laribi et al. [7], which had provided a satisfactory outcome. In addition, real product data sets are collected from the market, and play a more important role in incorporating the industrial optimization factors into the ANN module. Those overall industrial design factors are embedded implicitly in real products on top of engineering mathematical solutions.

Since the ANN is expected to be effectively used only for those parameters lying within the ranges of its training data [7], to make the training data more generically useful, unification of the input vector as well as the output vector during the training cycles should be assured. Similarly, during the application of the trained neural network, the input and output for the new dimensional parameters have to be scaled or normalized to make sure they lie within the training ranges.

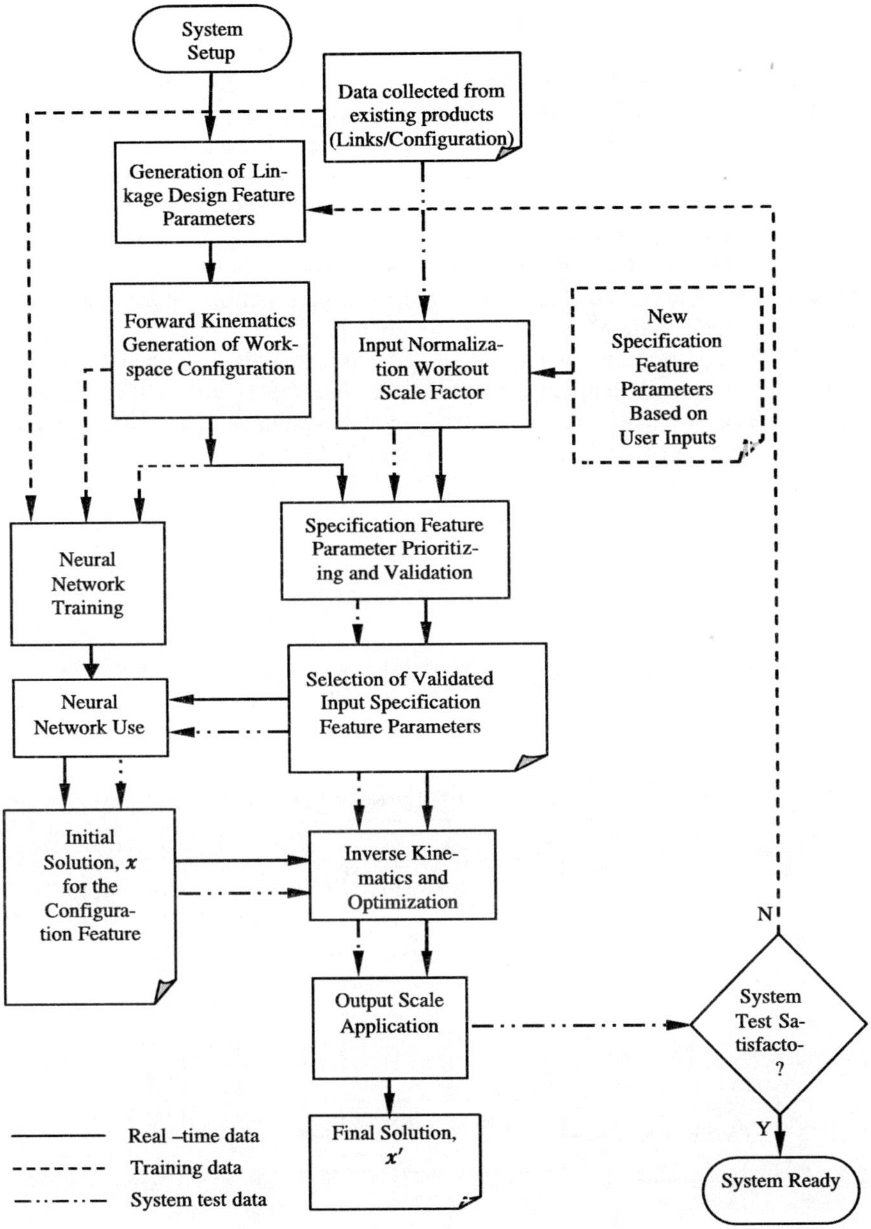

Fig. 1 Dimensional synthesis procedure

3.2.2 Input Specification Feature Parameter Validation

In addition to the training of the ANN, to search for a feasible mechanism parameter solution from a given set of configuration parameters it is necessary that the configuration parameter values be compatible with each other and their practical coexistence be feasible. If this condition is not met, the results of the analysis may be inapplicable. Figure 2 shows the procedure adopted to validate input configuration parameters. It is worth noting that the term *validation* is used here only to describe the applicability of a prescribed parameter set to a particular machine or manipulator configuration. The validation is performed by determining whether the configuration's given multiple input parameters, after being scaled or normalized, lie within the relative ranges established by the collected and generated data. The ranges derived from collected data are based on the results of statistical analysis of all the real product models available. The ranges derived from ANN-generated data are to be discussed in Sect. 4.3, which addresses the implementation algorithm.

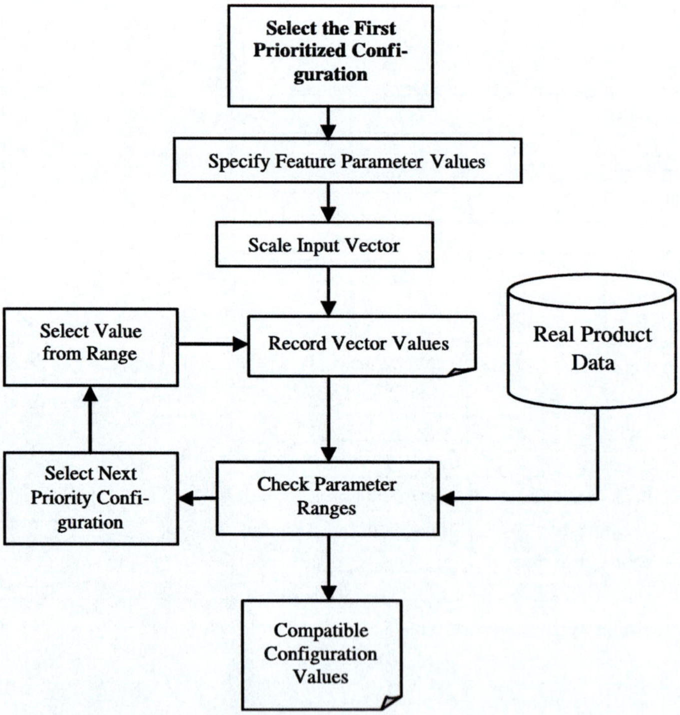

Fig. 2 Configuration prioritizing and selection

3.2.3 System Testing

To validate the overall procedure, real product specification feature data sets are to be used again for testing purposes, as shown by the step "Selection of Validated Input specification Feature Parameters" in Fig. 1. To test the system's reliability, which is different from the ANN training process, the real product configuration parameters are fed into a trained ANN module to generate initial suggestions of the linkage configuration feature dimensions. Then, together with envelope specification feature parameters, the initial linkage configuration feature dimensions are used as the seeding vector to search for the goal vector of the targeted mechanism configuration dimensions. Then the output goal vector is compared with the real product mechanism dimension vectors. Theoretically, the system output deviations should be well within the specified tolerance of the system's accuracy requirements. Note that the real product data sets are only a relatively small portion of the overall ANN training data sets. If the system does not meet the accuracy expectations, then more training data sets are required from both channels (as discussed previously).

3.3 Application of the Smart Design Feature Transformation

3.3.1 Initial Inverse Kinematic Solution Generation for Application

After the ANN has been successfully configured, it should be ready for application. At the beginning of the application design stage, the validated specification feature parameters are passed to the ANN module to generate initial design solutions. The initial solutions will then be used by the appropriate optimization procedure to refine and derive the goal solutions.

3.3.2 Mechanism Configuration Feature Dimension Synthesis

The dimension synthesis, which is specific to the nature of the mechanism in question, is carried out through optimization algorithms. The case that follows in Sect. 4 is a study of an excavator arm linkage system, and includes details of the algorithms. Ultimately, the resulting mechanism configuration feature parameter solutions in this research must be scaled back to the original ratio before being further processed.

3.4 Results Validation

The optimization results are to be validated before they are adopted in the design and displayed in an appropriate CAD context. This straightforward procedure is to apply forward mechanism simulation and check the envelope space or path details against the specifications. If the results are not satisfactory, a troubleshooting procedure must be carried out. As discussed in the following case study, the reported research results have so far been satisfactory with a limited number of tests; the troubleshooting method was therefore not further explored.

3.5 Random System Validation Check

To measure and validate the performance of the system, randomly selected configuration parameter data sets from the existing products' database can be selected and the corresponding mechanism configuration feature parameters generated using the proposed method. The results can be cross-checked against the actual dimensions and the efficiency of the method will be determined.

4 Case Study

4.1 Excavator Case Representation

In the conceptual process of designing an excavator, translating the access specification parameters (prescribed points or an envelope path) into linear dimensions of the arm mechanism represents the first stage. To do this, the boom, stick, and buckets of the planar mechanism are represented by linear linkages, and other elements, such as hydraulic cylinders and bucket transition four-bar linkages, are left out of consideration at this stage (see Fig. 3). These three links, connected in boom-stick-bucket sequence, are positioned and oriented in different poses such that their final configurations pass through the input specifications. Figures 3a and 3b show the traditional catalogue specification dimensions $(S_1, S_2, \ldots S_5)$, listed in Table 1, and the representation of the mechanism by linear elements (l_1, l_2, l_3, β), respectively.

Hence, the design process involves determining a set of individual linkage dimensions) for the excavator arm mechanism (listed in Table 2) so that when they are connected to each other and conform to the overall mechanism, they will satisfy the working-range requirements.

Unlike forward kinematic problems in which the location and other properties of the end effector are to be calculated based on different joint variables and linkage dimensional inputs, this problem involves determining the joint variables

(a)

(b)

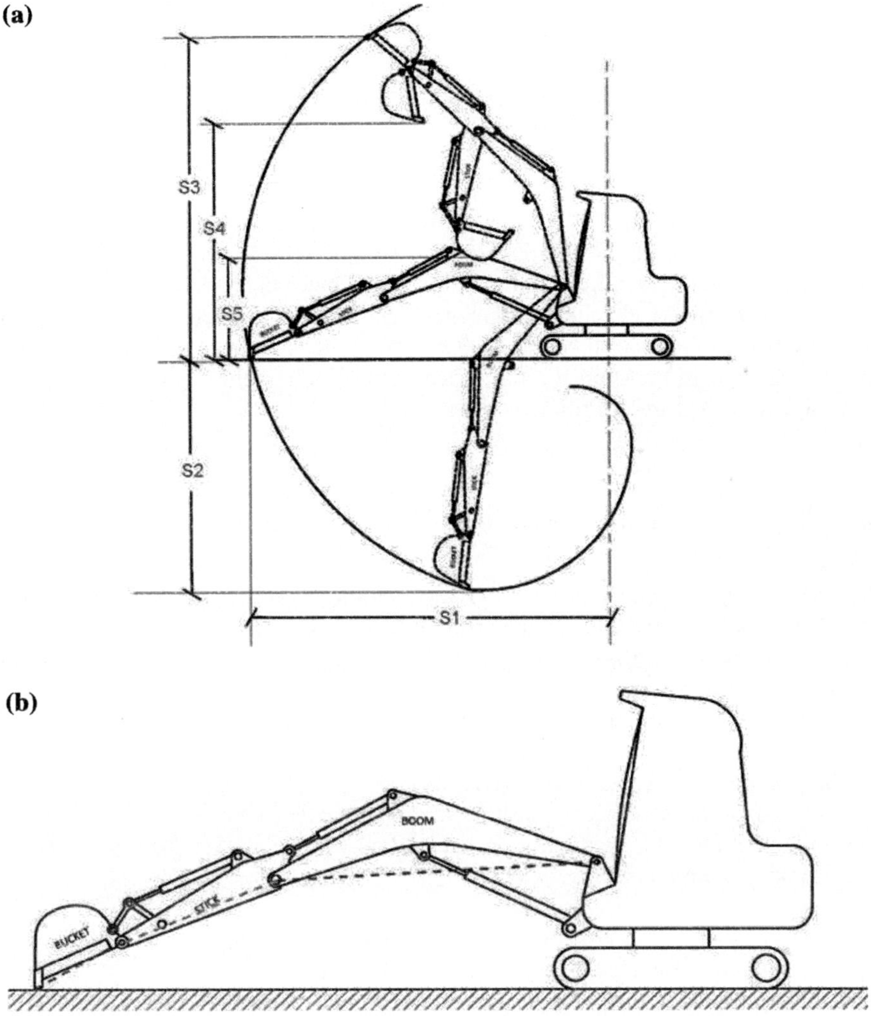

Fig. 3 An example excavator arm configuration. **a** Typical commercial work-range specifications. **b** Linear arm elements

Table 1 Hydraulic excavator workspace configuration parameters

*S*1	Maximum reach at ground level
*S*2	Maximum digging depth
*S*3	Maximum cutting height
*S*4	Maximum loading height
*S*5	Minimum loading height
*S*6	Maximum depth cut at level bottom
*S*7	Maximum vertical wall digging depth

Table 2 Mechanism linkage dimensions

*l*1	Hinge to hinge boom length
*l*2	Stick length
*l*3	Hinge to tip bucket length
β	Boom deflection angle

and linkage dimensions) for a given set of end effector configurations (bucket in this case). In forward kinematics or direct configuration analysis, the task is usually to determine the final configuration of the mechanism based on a given set of joint variables and linkage dimensions); this is a relatively simple and straightforward process, since the analysis usually leads to a unique solution. The inverse process in question, on the other hand, is relatively complex due to the availability of multiple solutions.

4.2 Data Generation for Neural Network Training

The main purpose of this task is to generate configuration and linkage parameter data sets to be used for training the proposed ANN. The ANN will be used in later stages to narrow down and select a physically viable set of linkage parameters to be used as initial solutions. This is entirely a forward kinematic procedure in which each final vector of configuration parameters, *S*, is determined from a given set of linkage dimensions) and joint variables, *L*.

Here $\boldsymbol{S} = (S_1, S_2, \ldots, S_5)$; $\boldsymbol{L} = (l_1, l_2, l_3, \beta)$.

The following subsections describe the mathematical model used for working out the envelope path configuration parameters $(S_1, S_2, \ldots, S_5)$ from the mechanism linkage parameters, (l_1, l_2, l_3, β).

4.2.1 Maximum Reach-Out at Ground Level (S_1)

The position of the bucket tip is calculated using forward kinematic methods. The individual rotational and linear transformation matrices are formulated using the Denavit-Hartenberg convention.

By applying the Law of Cosine to Fig. 4, the following mathematical relationship is formulated:

$$c^2 = (l_2 + l_3)^2 + l_1^2 - 2l_1(l_2 + l_3)\cos(180 - \beta) \tag{1}$$

$$c^2 = (l_2 + l_3)^2 + l_1^2 + 2l_1(l_2 + l_3)\cos\beta \tag{2}$$

$$c = \sqrt{(l_2 + l_3)^2 + l_1^2 + 2l_1(l_2 + l_3)\cos\beta} \tag{3}$$

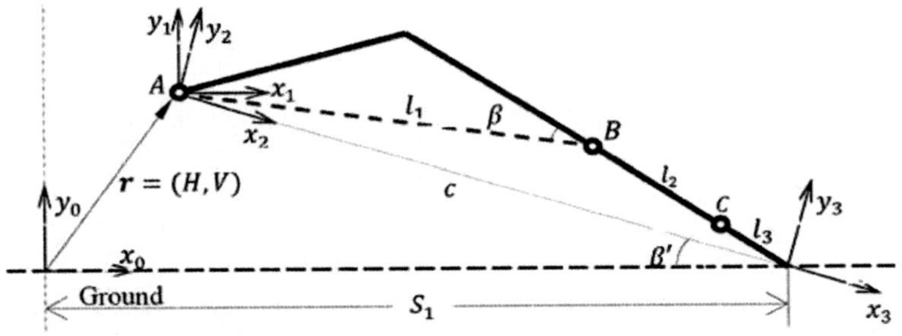

Fig. 4 Maximum out-reach at ground level

$$\sin\beta' = \frac{V}{c} = \frac{V}{\sqrt{(l_2+l_3)^2 + l_1^2 + 2l_1(l_2+l_3)\cos\beta}} \tag{4}$$

$$\beta' = \sin^{-1}\left(\frac{V}{c}\right) = \sin^{-1}\left(\frac{V}{\sqrt{(l_2+l_3)^2 + l_1^2 + 2l_1(l_2+l_3)\cos\beta}}\right) \tag{5}$$

$$(S_1 - H)^2 = (l_2+l_3)^2 + l_1^2 + 2l_1(l_2+l_3)\cos\beta - V^2 \tag{6}$$

$$(l_2+l_3)^2 + l_1^2 + 2l_1(l_2+l_3)\cos\beta - V^2 - (S_1 - H)^2 = 0 \tag{7}$$

The sequence of frame of reference translation from the origin to a frame located at the tip of the bucket is represented by the homogeneous transformation

$$A = T_{x'H}T_{y'V}R_{z'-\beta}R_{z'-\beta'} \tag{8}$$

where

$T_{x'H}$ Linear displacement in the positive x direction with H value
$T_{y'V}$ Linear displacement in the positive y direction with V value
$R_{z'-\beta'}$ Rotation about the z axis by angular value of $-\beta'$
$T_{x'c}$ Linear displacement in the positive x direction by a value of c.

The rotation sequences of Eq. 8, when represented by the corresponding matrices, take the following form:

$$A_{S1} = \begin{bmatrix} 1 & 0 & 0 & H \\ 0 & 1 & 0 & 0 \\ 0 & 0 & 1 & 0 \\ 0 & 0 & 0 & 1 \end{bmatrix} \begin{bmatrix} 1 & 0 & 0 & 0 \\ 0 & 1 & 0 & V \\ 0 & 0 & 1 & 0 \\ 0 & 0 & 0 & 1 \end{bmatrix} \begin{bmatrix} \cos(-\beta') & -\sin(-\beta') & 0 & 0 \\ \sin(-\beta') & \cos(-\beta') & 0 & 0 \\ 0 & 0 & 1 & 0 \\ 0 & 0 & 0 & 1 \end{bmatrix} \begin{bmatrix} 1 & 0 & 0 & c \\ 0 & 1 & 0 & 0 \\ 0 & 0 & 1 & 0 \\ 0 & 0 & 0 & 1 \end{bmatrix} \tag{9}$$

The resulting homogenous transformation matrix is then given by

$$A_{s1} = \begin{bmatrix} \cos\beta' & \sin\beta' & 0 & H + c\cos\beta' \\ -\sin\beta' & \cos\beta' & 0 & V - c\sin\beta' \\ 0 & 0 & 1 & 0 \\ 0 & 0 & 0 & 1 \end{bmatrix} \quad (10)$$

The value of the *maximum out-reach at ground level* is then extracted from the above homogenous transformation matrix. The expression in cell $(1,4)$ is the value of the x coordinate of the bucket tip from the origin of the fixed reference frame, which in this case is the same as the value of the maximum reach-out at ground level, $S1$.

$$S_1 = |A_{S1}(1,4)| = |H + c\cos\beta'| \quad (11)$$

4.2.2 Maximum Digging Depth (S_2)

The maximum digging depth requires the definition of angle α_2, measured from the vertical to indicate the lower limit of the boom angular displacement around the base hinge. For a given value of this limiting angle, the maximum digging depth is expressed mathematically using the Denavit-Hartenberg convention:

Again, by using Law of Cosine,

$$l_1^2 = b^2 + b^2 - 2b^2\cos(180 - 2\beta) \quad (12)$$

where b is the length of each side of the boom. For the purpose of simplification, they are assumed to be of equal length in this development.

$$l_1 = 2b\cos\beta \quad (13)$$

Referring to Fig. 5,

$$S_2 = l_1\cos\alpha_2 + l_2 + l_3 - V \quad (14)$$

$$S_2 = l_1\cos\alpha_2 + l_2 + l_3 - V \quad (15)$$

$$l_1\cos\alpha_2 + l_2 + l_3 - V - S_2 = 0 \quad (16)$$

The homogeneous transformation sequence in this case is given by

$$A_{S2} = T_{x'H}T_{y'V}R_{z'(\alpha_2)}T_{y'-l1}R_{z'-\beta}T_{y'-b} \quad (17)$$

Fig. 5 Maximum digging depth

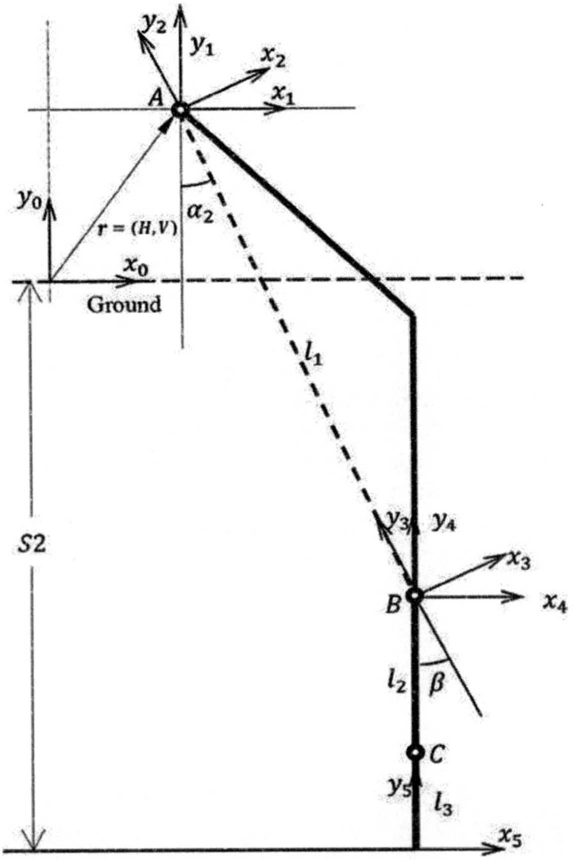

$$A_{S2} = \begin{bmatrix} 1 & 0 & 0 & H \\ 0 & 1 & 0 & 0 \\ 0 & 0 & 1 & 0 \\ 0 & 0 & 0 & 1 \end{bmatrix} \begin{bmatrix} 1 & 0 & 0 & 0 \\ 0 & 1 & 0 & V \\ 0 & 0 & 1 & 0 \\ 0 & 0 & 0 & 1 \end{bmatrix} \begin{bmatrix} \cos(\alpha_2) & -\sin(\alpha_2) & 0 & 0 \\ \sin(\alpha_2) & \cos(\alpha_2) & 0 & 0 \\ 0 & 0 & 1 & 0 \\ 0 & 0 & 0 & 1 \end{bmatrix} \begin{bmatrix} 1 & 0 & 0 & 0 \\ 0 & 1 & 0 & -l1 \\ 0 & 0 & 1 & 0 \\ 0 & 0 & 0 & 1 \end{bmatrix}$$

$$\cdots \begin{bmatrix} \cos(-\beta) & -\sin(\beta) & 0 & 0 \\ \sin(-\beta) & \cos(-\beta) & 0 & 0 \\ 0 & 0 & 1 & 0 \\ 0 & 0 & 0 & 1 \end{bmatrix} \begin{bmatrix} 1 & 0 & 0 & 0 \\ 0 & 1 & 0 & -b \\ 0 & 0 & 1 & 0 \\ 0 & 0 & 0 & 1 \end{bmatrix} \quad (18)$$

The resulting homogeneous transformation matrix takes the form of

$$A_{S2} = \begin{bmatrix} \cos(\alpha_2-\beta) & -\sin(\alpha_2-\beta) & 0 & H + l1\sin(\alpha_2) + b\sin(\alpha_2-\beta) \\ \sin(\alpha_2-\beta) & \cos(\alpha_2-\beta) & 0 & V - l1\cos(\alpha_2) - b\cos(\alpha_2-\beta) \\ 0 & 0 & 1 & 0 \\ 0 & 0 & 0 & 1 \end{bmatrix} \tag{19}$$

The cell in this matrix representing the maximum digging depth is the y displacement in cell $(2, 1)$.

$$S_2 = |A_{S2}(2,1)| = |V - l1\cos(\alpha_2) - b\cos(\alpha_2-\beta)| \tag{20}$$

4.2.3 Maximum Cutting Height (S_3)

For a given value of the upper angular limit of the boom rotation, α_1, the procedure for the maximum cutting height expression formulation follows a procedure similar to the maximum digging depth calculation.

Referring to Fig. 6, the following relationship is developed for the maximum cutting height configuration:

$$H_2 = l_1 \cos\theta \tag{21}$$

where θ in this case is given by

$$\theta = (\alpha_1 - \beta) \tag{22}$$

$$H_2 = l_1\cos(\alpha_1 - \beta) \tag{23}$$

$$H_3 = l_2\cos(\theta - \beta) \tag{24}$$

$$H_3 = l_2\cos(\alpha_1 - 2\beta) \tag{25}$$

$$H_4 = l_3\cos(\theta - \beta + \alpha_{\mathrm{bu}}) \tag{26}$$

$$S_3 = V + H_2 + H_3 + H_4 \tag{27}$$

$$S_3 = V + l_1\cos(\alpha_1 - \beta) + l_2\cos(\alpha_1 - 2\beta) + l_3\cos(\alpha_1 - 2\beta + \alpha_{\mathrm{bu}}) \tag{28}$$

$$l_1\cos(\alpha_1 - \beta) + l_2\cos(\alpha_1 - 2\beta) + l_3\cos(\alpha_1 - 2\beta + \alpha_{bu}) + V - S_3 = 0 \tag{29}$$

The homogenous coordinate transformation sequence for this configuration is given by

$$A_{S3} = T_{x'H}T_{y'V}R_{z'(\alpha_1-\beta)}T_{y'l1}R_{z'-\beta}, T_{y'b} \tag{30}$$

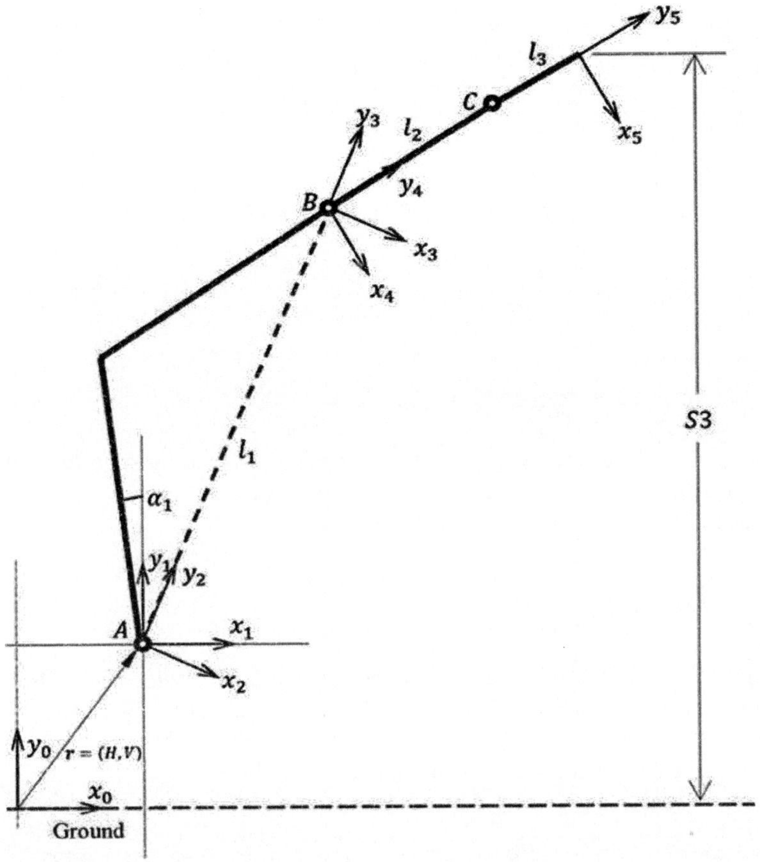

Fig. 6 Maximum cutting height

$$A_{S3} = \begin{bmatrix} 1 & 0 & 0 & H \\ 0 & 1 & 0 & 0 \\ 0 & 0 & 1 & 0 \\ 0 & 0 & 0 & 1 \end{bmatrix} \begin{bmatrix} 1 & 0 & 0 & 0 \\ 0 & 1 & 0 & V \\ 0 & 0 & 1 & 0 \\ 0 & 0 & 0 & 1 \end{bmatrix} \begin{bmatrix} \cos(\alpha_1 - \beta) & -\sin(\alpha_1 - \beta) & 0 & 0 \\ \sin(\alpha_1 - \beta) & \cos(\alpha_1 - \beta) & 0 & 0 \\ 0 & 0 & 1 & 0 \\ 0 & 0 & 0 & 1 \end{bmatrix}$$
$$\begin{bmatrix} 1 & 0 & 0 & 0 \\ 0 & 1 & 0 & l1 \\ 0 & 0 & 1 & 0 \\ 0 & 0 & 0 & 1 \end{bmatrix} \begin{bmatrix} \cos(-\beta) & -\sin(-\beta) & 0 & 0 \\ \sin(-\beta) & \cos(-\beta) & 0 & 0 \\ 0 & 0 & 1 & 0 \\ 0 & 0 & 0 & 1 \end{bmatrix} \begin{bmatrix} 1 & 0 & 0 & 0 \\ 0 & 1 & 0 & b \\ 0 & 0 & 1 & 0 \\ 0 & 0 & 0 & 1 \end{bmatrix} \tag{31}$$

$$A_{S3} = \begin{bmatrix} \cos(\alpha_1 - 2\beta) & -\sin(\alpha_1 - 2\beta) & 0 & H - b\sin(\alpha_1 - 2\beta) - l1\sin(\alpha_1 - \beta) \\ \sin(\alpha_1 - 2\beta) & \cos(\alpha_1 - 2\beta) & 0 & V + b\cos(2\beta - \alpha_1) + l1\cos(\alpha_1 - \beta) \\ 0 & 0 & 1 & 0 \\ 0 & 0 & 0 & 1 \end{bmatrix} \tag{32}$$

The y displacement component of this matrix represents the maximum cutting height.

$$S_3 = |A_{S3}(2,4)| = |V + b\cos(2\beta - \alpha_1) + l1\cos(\alpha_1 - \beta)| \tag{33}$$

4.2.4 Maximum Loading Height (S_4)

The vertical position assumed by l_3 in Fig. 7 is represented by slightly modifying the expression developed for maximum cutting height and ignoring the orientation angle of the last frame of reference, as follows:

$$S_4 = V + H_2 + H_3 - l_3 \tag{34}$$

$$l_1\cos(\alpha_1 - \beta) + l_2\cos(\alpha_1 - 2\beta) - l_3 + V - S_4 = 0 \tag{35}$$

The expression for maximum cutting height is modified with minor changes to make it fit this configuration. The last linear coordinate translation in this case is limited to $l2$ instead of $(b = l_2 + l_3)$. The bucket length l_3 is further deducted from the y displacement component of the matrix.

The final result is given by the following matrix:

$$A_{S3} = \begin{bmatrix} \cos(\alpha_1 - 2\beta) & -\sin(\alpha_1 - 2\beta) & 0 & H - l2\sin(\alpha_1 - 2\beta) - l1\sin(\alpha_1 - \beta) \\ \sin(\alpha_1 - 2\beta) & \cos(\alpha_1 - 2\beta) & 0 & V - l3 + l2\cos(2\beta - \alpha_1) + l1\cos(\alpha_1 - \beta) \\ 0 & 0 & 1 & 0 \\ 0 & 0 & 0 & 1 \end{bmatrix} \tag{36}$$

$$S_4 = |A_{S3}(2,4)| = |V - l_3 + l_2\cos(2\beta - \alpha_1) + l_1\cos(\alpha_1 - \beta)| \tag{37}$$

4.2.5 Minimum Loading Height (S_5)

Following a similar procedure gives an expression for the homogeneous transformation matrix of the minimum cutting height configuration, as represented by Fig. 8.

$$S_5 = V + H_2 - l_2 - l_3 \tag{38}$$

$$S_5 = V + l_1\cos(\alpha_2 - \beta) - l_2 - l_3 \tag{39}$$

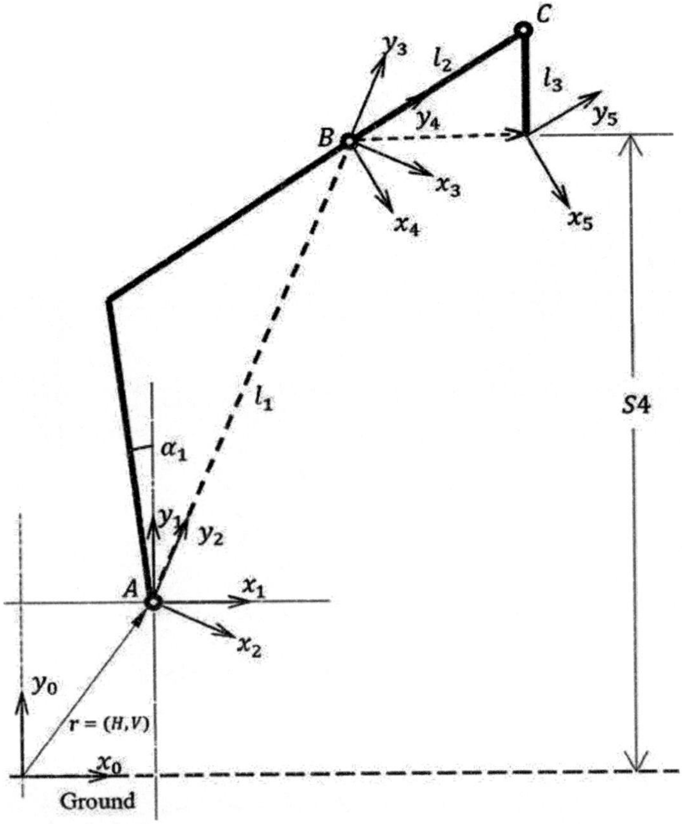

Fig. 7 Maximum loading height

$$V + l_1 \cos(\alpha_1 - \beta) - l_2 - l_3 - S_5 = 0 \tag{40}$$

$$A_{S3} = \begin{bmatrix} \cos(\alpha_1 - 2\beta) & -\sin(\alpha_1 - 2\beta) & 0 & H - l1 \sin(\alpha_1 - \beta) \\ \sin(\alpha_1 - 2\beta) & \cos(\alpha_1 - 2\beta) & 0 & V + l1 \cos(\alpha_1 - \beta) - l2 - l3 \\ 0 & 0 & 1 & 0 \\ 0 & 0 & 0 & 1 \end{bmatrix} \tag{41}$$

$$S_5 = |A_{S3}(2,4)| = |V + l1 \cos(\alpha_1 - \beta) - l2 - l3| \tag{42}$$

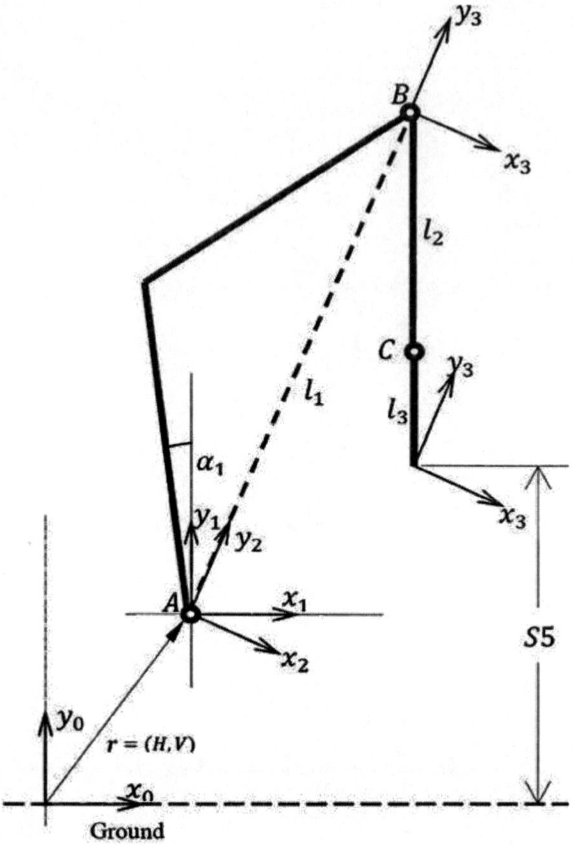

Fig. 8 Minimum loading height

4.3 Generation of Training Data

The required training data is generated by mapping the configuration parameter for a set of mechanism dimension parameters. MATLAB is used to implement this task.

$$\begin{bmatrix} l_1 \\ l_2 \\ l_3 \\ \beta \end{bmatrix} \rightarrow \begin{bmatrix} S_1 \\ S_2 \\ S_3 \\ S_4 \\ S_5 \end{bmatrix} \tag{43}$$

4.4 Neural Network Training

Since the ANN is needed to fulfill the purpose of preliminary inverse kinematic analysis, the output data generated from the forward simulation, ***S***, will be used as the input data for its training, while the linkage parameters vector ***L***, is the target data. Since the values of the configuration parameters depend also on the overall dimensions of the vehicle on which they are mounted, constant values for the ***xx*** and *yy* coordinates of the base hinge, H and V, are used in the analysis.

Accordingly, as shown in Fig. 9, a two-layer feed forward ANN is designed to map seven input configuration parameters to four target parameters. The ANN has one hidden layer with twenty neurons and one output layer with four neurons. The network is trained using the *Levenberg–Marquardt* back propagation algorithm. Sigmoid activation functions are used for the first layer and linear one-to-one activation functions for the output layer. The neural network is implemented using the neural network toolbox of the MATLAB programming language.

Given any one of the configuration parameters, $S_1, S_2, \ldots, S_5$, the developed method identifies possible ranges of the other four configuration parameters based on the data generated in the previous section. Since the data is generated by simulating specific ranges of the linkage dimensions), this method scales input configuration parameters to make sure they lie within the available data range. Selected output ranges by this method are scaled back to the original before being displayed for the user.

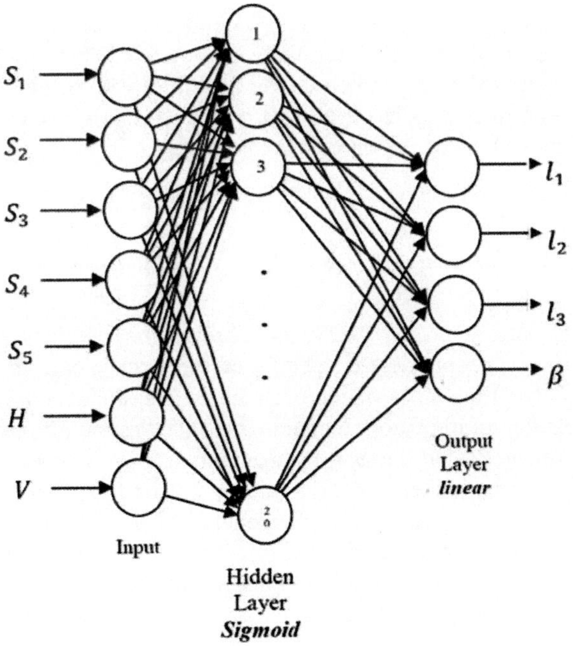

Fig. 9 Architecture of the neural network

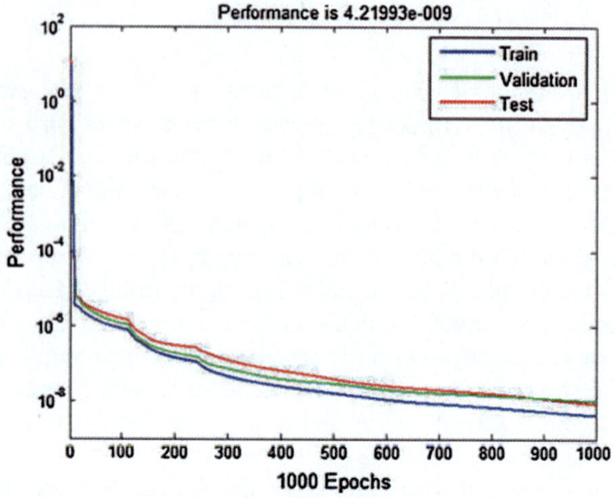

Fig. 10 Performance of the neural network

The method implemented using a MATLAB program called *f_Parameter_Sorter* provides an option for the user to select a configuration parameter with which to begin and the sequence of upcoming selections. This option allows the flexibility to prioritize the operational configurations as needed. Once the first item is entered for the first choice of the configuration parameter, four different compatible configuration parameter ranges will be suggested for the others.

This process will be repeated on the remaining four parameters by selecting which configuration parameter to prioritize and picking its value from the range provided. The result of this second operation modifies the ranges of compatible values of the remaining three parameters. This process is repeated until all configuration parameters are assigned valid values. Figure 10 shows the convergence performance of the ANN training cycles, while Fig. 11 shows the standard ANN algorithm regression chart.

4.5 Solving for Linkage Configuration Feature Parameters

Equations 11, 20, 33, 37, and 42 relate the specification values S_1, S_2, S_3, S_4, and S_5 to the geometric dimensions of the excavator arm mechanism l_1, l_2, l_3, and β. Given the values of the other constants, these nonlinear equations can be solved using optimization techniques to determine the optimum linear and angular dimensions of the arm mechanism.

Since buckets are available as standard parts, the calculation of this algorithm focuses on determining the lengths of the boom and the stick together with the boom deflection angle, i.e., l_1, l_2, and β. The selection of the bucket is made based on the initial solution suggested by the ANN. To determine the above three unknown variables, a combination of three of the above nonlinear equations is

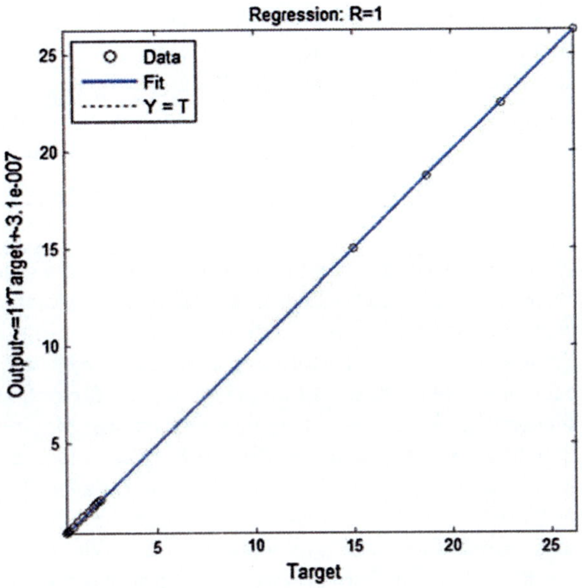

Fig. 11 Regression result

solved using a MATLAB function, *fsolve()*, which employs the power of the *trust-region-reflective* algorithm

$$F(\boldsymbol{X}, \boldsymbol{S}) = \boldsymbol{0} \tag{44}$$

where $\boldsymbol{X}$ and $\boldsymbol{S}$ are vectors of unknown mechanism dimension variables and input configuration specification parameters.

$$\boldsymbol{X} = \begin{bmatrix} l_1 \\ l_2 \\ \beta \end{bmatrix} \quad \boldsymbol{S} = \begin{bmatrix} S_1 \\ S_2 \\ S_3 \\ S_4 \\ S_5 \end{bmatrix} \tag{45}$$

Considering the maximum reach-out at ground level, maximum cutting height, and maximum loading height, the vector of equations will be formulated as follows:

$$\begin{bmatrix} l_1^2 + (l_2 + l_3)^2 + 2l_1(l_2 + l_3)\cos\beta - V^2 - (S_1 - H)^2 \\ l_1\cos(\alpha_2 - \beta) + l_2\cos(\alpha_2 - 2\beta) + l_3\cos(\theta - \beta + \alpha_{bu}) + V - S_3 \\ l_1\cos(\alpha_2 - \beta) + l_2\cos(\alpha_2 - 2\beta) - l_3 + V - S_4 \end{bmatrix} = 0 \tag{46}$$

The *trust-region-reflective* algorithm used to find the solution requires an initial solution to be defined as a starting point. The accuracy of the output for this particular problem greatly depends on the closeness of the initial solution to the actual solution. This is the stage where the suggested initial solution by the neural network is used. It is also expected that at this stage the viability of the initial input parameters, $S_1, S_2, \cdots, S_5$, is confirmed by the use of valid ranges developed according to the procedure discussed previously.

4.6 Case Study Analysis Results and Discussion

In all, ten existing excavator product configuration data sets were collected; their contents are given in Table 3. A total of 1,296 forward simulation data sets were generated and they were used to train the ANN module developed with MATLAB. To test the system performance, the ten product configuration envelope path parameters were then fed into the ANN, and the output of the ANN, i.e., the initial suggestions for the downstream optimization module, was presented in Table 4 (left half). For the sake of comparison, the solutions generated after the optimization process are also listed in Table 4 (right half).

Clearly, the ANN module has served the purpose of providing useful initial suggestions that enabled the optimization module to find feasible solutions for the given mechanism. Furthermore, Table 5 shows the comparison results between the solutions and the original real product data obtained for the ten existing configurations. The average errors for linear dimensions are pretty close, i.e., within 10 %, but the angular β shows a bigger difference from the original dimension: about 24 %. The deviations of these errors are relatively small. Therefore, we can conclude that the proposed method is feasible and the results show a good agreement with the testing input data set. The method can be further improved by fine-tuning the optimization algorithms and the boundary conditions as well as by using more realistic product data sets for ANN training.

Table 3 System testing data collected from the existing products (units: cm/degree)

	Configuration					Mechanism dimensions				Vehicle	
Product	S1	S2	S3	S4	S5	11	12	13	β	H	V
1	359	183	344	226	107	174.1	88.2	51.9	24.5	63	75
2	413	252	384	271	109	205.9	102	61.1	25	68	86
3	412	260	359	246	111	201.2	99.4	67.9	28	74	93
4	435	228	422	283	106	203.3	105.2	64.9	30	78	90
5	409	248	385	267	125	201.3	99.5	61.5	25	66	84
6	372	208	371	257	110	171.4	89.1	58.2	24	77	82
7	352	196	331	235	92	159	86.3	49.6	22	77	71
8	345	203	338	238	99	165.1	88.4	49	20	64	73
9	332	184	335	238	104	163.4	83.8	47.5	24	55	71
10	415	254	368	272	110	204.9	102	63.1	25.76	68	81

Table 4 The initial and final solutions generated from the system

Product	ANN initial solution (m)				Optimization final solution (m)			
	l1	l2	l3	β	l1	l2	l3	β
1	1.747	0.688	0.634	17.28	2.09	0.93	0.634	34.97
2	1.9697	0.9599	0.6556	19.4858	2.0022	0.9547	0.6556	30.0233
3	2.0453	0.8433	0.7019	26.524	2.0282	0.8581	0.7019	35.2464
4	1.8104	0.7914	0.783	10.7072	1.6981	0.8382	0.783	18.6694
5	1.7703	0.7381	0.6662	15.6772	2.0433	0.8323	0.6662	30.1607
6	1.6806	0.805	0.6103	13.2475	1.9883	0.9117	0.6103	25.8929
7	1.5389	0.8259	0.5279	15.2362	2.0148	0.9988	0.5279	30.014
8	1.6275	0.9462	0.5138	12.3882	1.97	0.9848	0.5138	32.8084
9	1.6105	0.909	0.4687	14.3593	2.1144	0.9756	0.4687	32.3521
10	1.9589	1.0773	0.6004	24.8409	1.879	0.926	0.6004	31.9021

Table 5 Accuracy statistics of the system results

Dimensions	Average error (%)	Unbiased standard deviation	Root mean square error (RMSE)
l_1	8.627	0.1569	0.1489
l_2	1.4641	0.1356	0.1286
l_3	7.1778	0.085	0.0806
β	23.858	0.2652	0.2516

5 Conclusion

In this chapter, a hybrid feature transformation method from specification feature instances to mechanism configuration features was presented. The method uses ANN and optimization tools to solve feature-based dimension synthesis problems. The ANN, in order to reflect the mapping relations between accessing the envelope path and the linkage lengths in an excavator arm case study, needs to be trained before becoming usable. A mix of real product data sets from those existing product families and the generated data sets from forward kinematic simulation calculation methods are used for ANN training purposes. The forward data generation method is used to solve the problem of a shortage in real product data, and to produce enough "artificial" training data. The results of the analysis show a satisfactory estimation of the initial solutions based on the ANN model. For a set of existing product configurations, after testing the system on the whole cycle and searching for the final solutions with the optimization module, it can be concluded that the method is feasible and the results are promising, although more research analysis and evaluation are required. While this research used an excavator arm mechanism for the case study, the proposed method is not a product-dependent approach. Potentially, this hybrid method can be used for many other mechanism design processes as well.

Acknowledgments The authors would like to thank the Natural Sciences and Engineering Research Council of Canada (NSERC) for its Discovery grant support (No. 355454-09) and the University of Alberta for its NSERC GRF (G121140079) grant.

References

1. Alizade RI, Kilit O (2005) Analytical synthesis of function generating spherical four-bar mechanism for the five precision points. Mech Mach Theor 40:863–878
2. Erkaya S, Uzmay I (2008) A neural-genetic (NN–GA) approach for optimizing mechanisms having joints with clearance. Multibody Syst Dyn 20:69–83
3. Frimpong S, Li Y (2005) Virtual prototype simulation of hydraulic shovel kinematics for spatial characterization in surface mining operations. Int J Surf Min, Reclam Environ 19:238–250
4. Gao Z, Zhang D, Ge Y (2010) Design optimization of a spatial six degree-of-freedom parallel manipulator based on artificial intelligence approaches. Robot Comput Integr Manuf 26:180–189
5. Hasan AT, Hamouda AMS, Ismail N (2009) Trajectory tracking for a serial robot manipulator passing through singular configurations based on the adaptive kinematics Jacobian method, vol 223. In: Proceedings of the institution of mechanical engineers, Part I: journal of systems and control engineering, pp 393–415
6. Jensen OF, Hansen JM (2006) Dimensional synthesis of spatial mechanisms and the problem of non-assembly. Multibody Syst Dyn 15:107–133
7. Laribi MA, Romdhane L, Zeghloul S (2007) Analysis and dimensional synthesis of the DELTA robot for a prescribed workspace. Mech Mach Theor 42:859–870
8. Li YL, Zhao W, Ma YS (2011) A strategy for resolving evolutionary performance analysis coupling at the early stages of complex engineering design. J Eng Des 22:603–626
9. Ma YS, Britton GA, Tor SB, Jin LY (2007) Associative assembly design features: concept, implementation and application. Int J Adv Manuf Tech 32:434–444
10. Shah JJ, Mantyla M (1995) Parametric and feature-based CAD/CAM concepts, techniques and applications. Wiley-Interscience, New York
11. Singla E, Tripathi S, Rakesh V (2010) Dimensional synthesis of kinematically redundant serial manipulators for cluttered environments. Robotics Auton Syst 58:585–595
12. Vasiliu A, Yannou B (2001) Dimensional synthesis of planar mechanisms using neural networks: application to path generator linkages. Mech Mach Theor 36:299–310
13. Wu J, Purwar A, Ge QJ (2010) Interactive dimensional synthesis and motion design of planar 6R single-loop closed chains via constraint manifold modification. J Mech Robot 2:031012
14. Yoon J, Manurung A (2010) Development of an intuitive user interface for a hydraulic backhoe. Automat Constr 19:779–790
15. Zhao Y, Tang Y, Zhao Y (2008) Dimensional synthesis and analysis of the 2-UPS-PU parallel manipulator. 1st international conference on intelligent robotics and applications, ICIRA 2008

Feature-Based Mechanism Design

Abiy Wubneh and Y.-S. Ma

1 Introduction

This chapter presents a feature-based concept design method for variational mechanisms. The proposed method integrates dimensional synthesis, mechanical design, and CAD generation with advanced feature technology with minimal designer intervention. Extended feature definitions are used in this research to create a smooth data flow connection between different engineering tools and applications. An excavator arm mechanism is used as a case study to demonstrate the proposed design method. The design of the given mechanism is carried out based on its digging mode configurations as introduced in the previous chapter.

The design process of multi-component products which are subject to frequent changes and modification is very complex due to the large amount of data involved. Dimensions and parameters defined by customers at the initial stages of the design process are used by latter design stages during a product lifecycle. This situation is further complicated when different parts of the product are designed by people located in various geographic locations. Changes and modifications evoked by one department are not reflected on components being developed by other departments without considerable time and resources. Constraint definition and management is just one area affected by the chosen data management system adopted in the design process.

Traditional CAD systems lack effective means for the implementation of effective knowledge-driven design procedures, especially for embedding engineering semantic patterns and the subsequent evolvement in CAD-supported design processes. This has forced researchers to look into possible ways of enriching information in the traditional CAD models.

A. Wubneh · Y.-S. Ma (✉)
Department of Mechanical Engineering, University of Alberta,
Edmonton, AB T6G 2G8, Canada
e-mail: yongsheng.ma@ualberta.ca

Y.-S. Ma (ed.), *Semantic Modeling and Interoperability in Product and Process Engineering*, Springer Series in Advanced Manufacturing,
DOI: 10.1007/978-1-4471-5073-2_10,

Features were introduced as a means to address the engineering semantic pattern modeling need. Features are basically object data structures containing a range of information and service functions required to fully describe and manipulate the shape and related semantic aspects of the engineering pattern. They have been used to deal with the most commonly used data elements including model geometries, materials, manufacturing methods, tolerances, and machining procedures. Recently, more complicated and sophisticated features have been defined to cover previously overlooked but critical aspects of the design process.

This research is motivated by the advancement of feature definitions and their potential applications in the areas of intelligent design automation and integration. The intention is to extend the use of feature-based modeling concepts to include mechanism design intents and constraints.

2 Statement of Problem

Traditional design systems, including CAD and CAE tools, have very limited capability in terms of storing rich-information with data format that can be accessed by different phases of the product development cycles. This limitation directly affects the choices of the design methodologies and the necessary data communication mechanisms. The information required by different stages of product design process has to be transferred in a effective manner to achieve maximum automation and efficiency. Attention is now shifted to developing a method by which the pertinent design information will be stored in a consistent and reusable data structure that is integrated with the product's CAD models. Thus, the previously fragmented design data files and those corresponding product CAD models have to be combined into a single product data model. In fact, modern product models are no longer merely a collection of CAD geometric entities. Customizable and advanced semantic definitions called features have long been used to integrate different engineering data structures with strong associativity to engineering thinking patterns and semantics.

The conceptual design process of variational mechanism, which is the focus of this research, is known for its various fragmented modules running on some sets of common data. Design process and knowledge development is iterative in nature. Each design cycle generates a set of data or a data model to be used in the creation of the next phase model of the product being designed. For example, since the product will have different inertia properties after its 3D embodiment, the next design cycle will have to consider the new properties (both physical and geometric) before commencing the next cycle. Effective automation of this process requires the systematic creation of CAD models in a very consistent manner.

Associative relationships between different features of a mechanism design model have to be systematically constrained to ensure the final generated model has comprehensive physical and engineering descriptions. A design procedure equipped with methods which can satisfy these requirements will earn itself an important place in advancing the industrial application.

3 Objectives

The objective of this research is to propose a feature-based conceptual design automation scheme specific to variational mechanism products. This scheme will attempt to use the capabilities of features in accommodating for both geometrical and non-geometrical types of data into a CAD system. It aims to utilize features to bridge the current automation gap in the conceptual design cycle and to further investigate features' applicability in terms of embedding conceptual design rules for complex part shapes and structures development.

The applicability of the proposed methods will be demonstrated using the design procedure of an excavator mechanism as a case study. This chapter addresses feature-based product design aspects of product configuration, linkage optimization, and their programming implementation.

4 Scope of Study

This research is proposing a conceptual product design automation method with the integration of feature-based CAD model generation via APIs C++ and MATLAB programming tools. The case study mentioned includes the basic generation and optimization design calculations of an excavator arm mechanism. Only the digging operational conditions are considered for the design purpose. The design has been carried out mainly taking the strength requirements (working stresses) into consideration in a numerical approach (design for strength). Other design aspects, such as dynamic behavior and finite element analysis (FEA), are beyond the scope of this work due to the resource constraints.

5 Literature Review

This chapter reviews research works and publications of other scholars which are relevant to the objective of this research. The overall organization of this section is targeted to cover the following major topics:

- Design automation and integration
- Parametric and feature-based CAD modeling
- Reverse engineering and knowledge fusion in product development.

5.1 Parametric and Feature-Based CAD Modeling

Several methods and procedures have been developed to automate and increase the efficiency of CAD modeling processes. The depth of data embedded in the CAD modes greatly depends on the techniques employed to carry out the process [22]. Parametric modeling, among others, has become one of the most rapidly growing and commonly adopted methods of product modeling in the manufacturing industries. Much research effort has been focused on modifying the *Standardized Exchange of Product* (STEP) format, which contains only geometric information, to accommodate for additional part-specific parameterized information [15].

This approach takes the traditional CAD geometry modeling method a step further by enforcing permanent geometric constraints among members of CAD objects and features. This system has known limitations in terms of validity in change and modification implementations. Basak and Gulesin's study [1] suggested and demonstrated a method in which parametric modeling was used in coordination with feature-based and knowledge fusion approaches to increase its robustness in the areas constraint validation. This method also used standard feature libraries to investigate the practicality of proposed part manufacturing techniques by the designers.

Programming through the application programming interfaces (APIs) of existing commercial CAD packages provides designers with more flexibility to embed the design intent into the CAD models [9]. In their approach, Myung and Han [13] took design unit and functionality features into consideration for the configuration and CAD modeling of selected products. The work of Wang et al. [25] proposed a modeling environment integration method by which interactive parametric design modification was demonstrated using a CAD-linked virtual environment.

The success of parametric modeling greatly depends on the consistency and preservation of topology of CAD features used in the creation of the part being modeled. The defined parent–child relationships have to be consistently validated in order to apply this method in the design process of customizable products. The use of explicit relationship was suggested by Van et al. [22] to increase user control and add sophistication to the modeling process.

Although parametric modeling techniques are widely used in today's design and manufacturing industries for facilitating CAD modeling processes, their use has been limited to geometric attributes. Incorporation of additional sets of information such as product material, method of manufacturing, assembly procedures, and design intents to the CAD models have been the focus of several recent research works.

Features, which are basically data structures, have been used to attach additional information to the CAD models. The type of information ranges from purely geometric to non-geometric parameters. Traditional features represented only those attributes related to the geometry of the part. Recently, new types of feature definitions [20] have been introduced to embed other non-geometric aspects of the product/part being designed with the CAD models.

The employment of parametric and feature-based modeling techniques has been proven to contribute significantly the implementation of an integrated and automated product design procedure [26]. The interest of manufacturers in reducing the time to market and costs associated with the design process has motivated researchers [27] to investigate features in greater depth. The power of feature-based modeling methods was coupled with the concepts of reverse engineering techniques [27] to embed design intents and other constraints into existing products retrieved by CAD scanning techniques (reverse engineering). By employing this method, manufacturers will reduce the time required to re-fabricate a given existing product with different materials and modified design constraints.

The data structures of features can handle more than one type of information. As discussed earlier, information pertinent to product development such as conceptual design intents, geometric constraints, non-geometric parameters, and manufacturing and assembly procedures can be embedded into the CAD model of the product by manipulating its feature (data structure). The extent to which this information can be exploited mainly depends on the feature definition and the level of organization and communication architectures of the network [26]. Ter et al. [20] discussed this issue in their work and proposed a high level abstract unified feature-based approach by categorization and generalization of conceptual data.

The traditional definition of a feature, which used to be merely a description of the shapes and geometries of the CAD models, has been extended to cover assembly design features and other vital aspects in regard to manufacturing and concurrent engineering [3]. Associative relationships, both geometric and non-geometric, between various parameters of two or more members of an assembly were discussed by Ma et al. [9]. This ability opens the door for design automation of frequently updated and modified products. Design customization of products can benefit from the inclusion of design intents, constraints, and assembly hierarchy data [2] on the CAD files. Incorporating rules and constraints in the CAD files in the form of features requires the definition of a new set of features. By treating a feature more like a data structure than a geometric parameter description, associative relationships between parts that were not previously considered can be defined. Additionally, the feature definition is extended to cover information pertinent to component mating conditions and interfaces within an assembly.

5.2 Design Automation and Integration

The implementation of a collaborative product development process requires a large amount of data to be transferred between applications used by different designers working toward a single product [14]. Change and modifications are part of a two-way process in this approach. Ma and his colleagues defined a data structure (feature) called *operation* in an effort to address the need to communicate data at the feature level [11]. An associative fine-grain product repository mechanism with a four-layer information scheme was proposed and demonstrated by the team for this purpose. The method was proposed with consideration for the probability of using different applications and platforms due to the interdisciplinary nature of the product design process.

Features, which have a higher level of semantic organization than the simple geometric aspects of a product, are currently being used to create the link and bridge the gap in terms of the amount and detail of information needed to be shared by CAD and CAM systems [12, 18]. Concurrent and collaborative engineering product development processes require the implementation of an effective change propagation and constraint management mechanism to handle the flow of data between various development stages. In their work, Ma et al. [10] proposed a unified feature approach method for constraint management and change propagation to be used in a collaborative and concurrent environment. The algorithm developed uses the JTMS-based dependency network. The data model was categorized under constraint-based associativity and shares entity references. Lourenco et al. [8], in a slightly different approach, investigated a method of interactive manipulation of feature parameters. They proposed a solver-driven algorithm for optimization of geometric constraints using non-application specific constraint solvers.

Excavator arm mechanisms have been investigated from different research perspectives. Solazzi discusses [19] the advantages and quantitative analysis of performance improvements achieved by redesigning an existing hydraulic arm mechanism with a different material. Yoon and Manurung [28] investigated the development of a control system by which the operations of an excavator arm mechanism are controlled by the operator's own arm movements. Sensors attached at different joint locations of an operator's arm are used to map the arm joint displacements in the mechanism's motion.

The development of new design automation procedures [16, 21] together with existing mechanical simulation tools such as SimMechanics of MATLAB® and MSc ADAMS® have given researchers the opportunity to fully impose explicit constraints when investigating mechanisms and manipulators' kinematic and dynamic responses [23]. The forward and inverse kinematics analyses involved in the design of mechanisms and manipulators benefit from the implementation of parametric and feature-based modeling approaches [23]. Work space configurations of manipulators and their dynamic responses require frequent changes and fine-tuning initial parameters which can be easily implemented with the use of appropriate feature definitions.

5.3 Reverse Engineering and Knowledge Fusion

Reverse engineering (RE) techniques had been used to extract shapes and geometries from existing products [4, 5]. The results of RE procedures usually poorly represent the design logic used to create the parts. The gap between RE techniques and the requirement of embedding design intents into the CAD files of products retrieved using this method was discussed by Durupt et al. [4, 5].

The traditional RE tools allow creating the geometries of existing products but lack the capability of embedding the design intents. The method proposed in Durupt's work suggested procedures to integrate RE tools with knowledge fusion techniques. Similarly, Li et al. suggested the use of knowledge-driven graphical partitioning approaches to include design intents in the RE scan results [7, 24].

Topological approaches have recently become more popular for their ability to generate 3D free shape models based on finite element concepts. However, like that of the RE techniques, there is much research to be done before it will be possible to smoothly extract simple CAD models from these shapes. The work of Larsen and Jensen [6] focuses on the investigation of methodologies to extract parametric models out of topologically optimized 3D shapes.

6 The Proposed Approach

Product modeling involves creating and combining individual basic semantic entities called features and further generating part geometries. A feature is a data structure with members of geometric elements and associative relations. The ability to create relationships between the data members of different features allows controlling part dimensions parametrically. With this modeling approach, constraints can be imposed on the geometric entities defining the features. The data from the features can be easily accessed and modified, making this method ideal for managing change propagations and modifications in the design process. In addition to geometric parameters, these features can be designed to store other design entities such as part material specifications and manufacturing methods. The fact that features are basically data structures allows them to play an important role in the automation processes of conceptual design cycles.

In this chapter, we will propose and discuss two methods to be used in the implementation of feature-based CAD modeling techniques in the development and design of automation processes of variational mechanisms.

6.1 General Design Automation Method

The design of mechanisms and products that are subject to frequent changes and modifications involves several application-dependent processes utilizing a set of common data. The given specifications, standards, and design requirements may be changed at any time during the development process. These changes can be evoked by customers as well as due to newly arising engineering requirements. Without a system to address these changes efficiently, the costs associated with the modifications could undermine the product need.

This section proposes a method by which such changes and design intent modifications are handled in a very cost-effective and timely manner using a feature-based approach to reduce the CAD modeling and the overall design cycle times. By employing commercially available programming and feature-based 3D modeling tools, it is possible to create a reliable automation procedure which accommodates for inevitable changes and modifications in the design process.

6.2 The Proposed Design Procedure

The following flowchart summarizes the general automation procedure proposed for this purpose. The area of data communications between different programming and modeling tools are beyond the scope of this chapter. The authors will instead focus on the intended communication, which is achieved through the use of neutral text data files written and updated by program codes developed for this purpose.

In Fig. 1, the design process begins with input information in the form of user specifications. This input, together with additional engineering rules and intents, is used by the *Kinematic Analysis* algorithm (discussed in the next section) to synthesize the linear dimensions of the mechanism. These dimensions will then be used as the skeleton assembly model in an initial kinematic and dynamic analysis. This analysis results in the identification of forces and moment reactions between contacting joints and bodies. The output of the *Dynamic Analysis and Simulation* module will be used to establish the free body diagrams (FBDs) and mass-acceleration diagrams (MADs) which will be used during the design and optimization phases.

Results obtained in these stages, together with the initial input specification values, will be used in the design of linkages and members of the mechanism. One or more applicable optimization criteria can be used in order to determine a set of optimum cross-dimensional parameters for the machine elements. In addition to the abovementioned considerations in the design and optimization phases, several additional factors may be considered (depending on the type of product) such as design codes, standards, assumption, and safety factors.

Based on dimensional data determined by previous processes, the 3D models of the mechanism components will be modeled using feature-based techniques. Application Programming Interfaces (API) of most commonly used modeling platforms can be used for this purpose. The choice of the programming and modeling tools depends on the compatibility of tools and the familiarity of the personnel using them.

For this case study, the programming component was carried out in C++ programming using Visual Studio 2008®. The final 3D models were generated from the codes using the UG NX 7.5 modeling software. These models, preferably assembled, will be exported back to the *Dynamic Analysis and Simulation* block to take the effects of their newly created dimensions (inertia effects) into consideration. This first stage loop will be repeated until a stopping criterion is met.

The strength and deformation of parts and models passing this screening stage can be further examined using FEA techniques. In the event these components fail to meet the qualification criteria set for the FEA stage, the entire iteration can be restarted with modified input parameters to address the shortcomings.

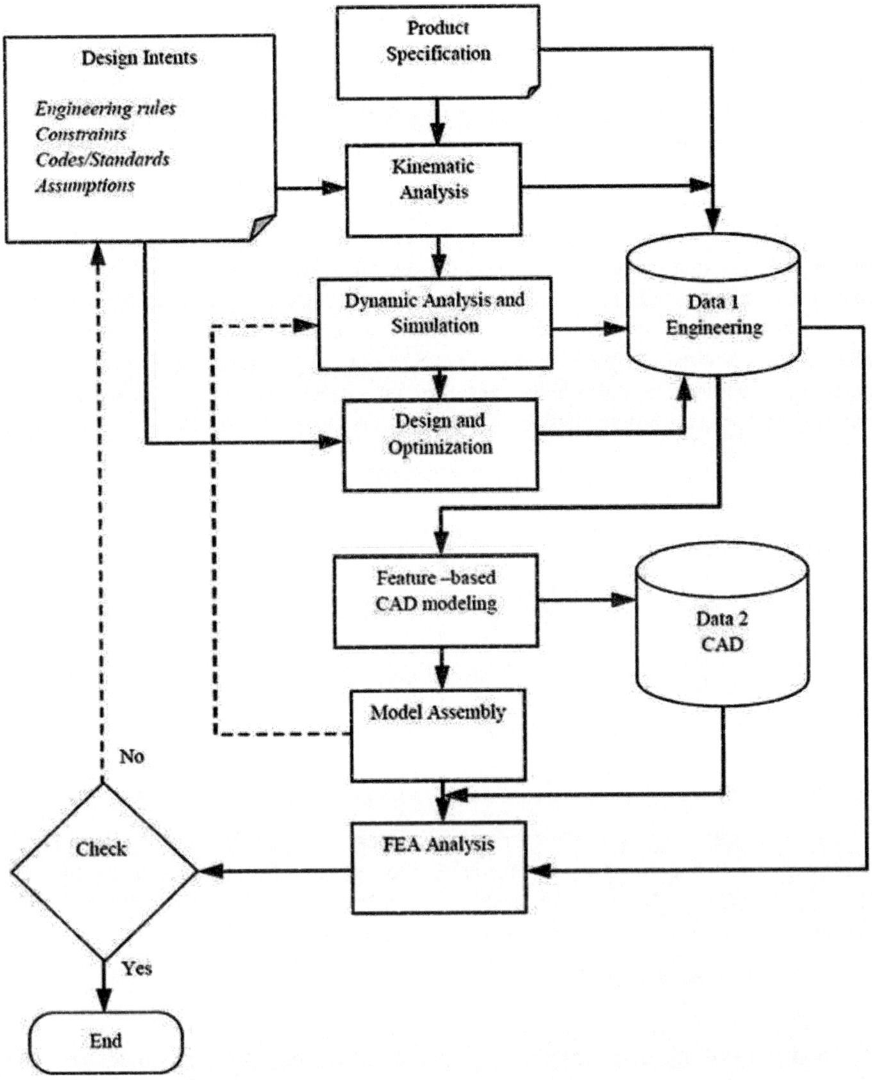

Fig. 1 Design process modules and data communication routes

7 Features and Data Structures

Concurrent engineering and product development processes involve the participation of personnel with different engineering and technical backgrounds. In most cases these individuals work from within different departments, requiring an efficient mechanism for smooth information transfer among them.

Any information, whether in the form of initial input or generated data, has a very good chance of being used by more than one function module or application. In addition, a series of design data for a particular product family needs to be stored in a systematic repository database due to its potential for future use in the areas of knowledge fusion, as well as a training source for artificial neural network applications.

Data structures, implemented by using object-oriented programming tools, address these needs. The *Product Specification* input shown in Fig. 1 needs to be organized systematically and its scope needs to be defined as "global" or "local" in order to establish its accessibility by individual program modules. This is done by defining a data structure and instantiating its object. The following is an example of a class defined in MATLAB®. The data structure for handling a particular problem is defined by creating an object instance of this class and entering values to its data members.

```
classdef Product_Specification_c
    properties
        Title = 'Specification Parameters'
        T_1 = 'Geometric Spec.'
        G1 = 0;
        G2 = 100;
        Gn = 0;
        T_2 = 'Material Spec'
        Modulus_Elasticity = 210e9;
        Poisson_ratio=0.3;
    end
end
```

For example, a data structure for a new product model called Product_Spec_2010 is created by instantiating the above definition and using the following command. (Note: Neither this particular example data structure nor its values are real; they are used here only for explanatory purposes.)

```
global Product_model_2010
Product_model_2010 = Product_Specification_c
```

The values of this data structure are updated using the following object-oriented programming syntax:

```
Product_model_2010.G1: new value
Product_model_2010.G2: new value
Product_model_2010.Gn: new value
Product_model_2010.Modulus_Elasticity: new value
Product_model_2010.Poisson_ratio: new value
```

The collection and input methods of the individual entities for the data structure greatly depend on the convenience and applicability to a particular problem. Initial values can be assigned during the definition of the data structure, or they can be updated afterward using both the command line and graphical user interfaces (GUI).

The following is a real example data structure taken from the excavator arm mechanism case study.

```
classdefc_Spec_Data_SI
  properties
      Title = 'Commercial Specifications and Vehicle Dimensions'
      Maximum_Reachout_at_Ground_Level_S1 = 0;
      Maximum_Digging_Depth_S2 = 0;
      Maximum_Cutting_Height_S3 = 0;
      Maximum_Loading_Height_S4 = 0;
      Minimum_Loaidng_Height_S3 = 0;
      Horizontal_Distance_H = 0;
      Vertical_Distance_V = 0;
      Vehicle_Weight = 5000;
  end
end
```

An object of this structure, SpcDat, instantiated and completed with its own values takes the form:

```
SpcDat = c_Spec_Data_SI
 Properties:
      Title: 'Commercial Specifications and Vehicle Dimensions'
   Maximum_Reachout_at_Ground_Level_S1: 5.6700
         Maximum_Cutting_Height_S3: 3.7248
         Maximum_Loading_Height_S4: 1.3521
           Horizontal_Distance_H: 0.9857
            Vertical_Distance_V: 1.2300
               Vehicle_Weight: 5000
```

The set of data generated within the *Product Specification* (PS) module is used directly by the Kinematic Analysis (KA) module when calculating the linear dimensions of the mechanism or manipulator. The KA, in turn, generates its own data structure and makes it available for use by downstream functional modules.

The number of programming applications and tools involved in the system dictate the number of databases involved. If there are two or more programming tools running on different platforms involved, it may be required to devise a mechanism by which their respective databases are able to communicate with each other.

In Fig. 1, it is assumed that the programming environment used for kinematic analysis and dimensional synthesis is different from the one employed by the API of the CAD modeling application, as this is the usual case. This is a very common practice since MATLAB® and Maple are usually used for engineering design calculations and optimizations processes while C#, C++, and VB are used for programming CAD with the API tools. However, all of these tools are expected to operate based on a common set of data models and parameters produced during the initial phase of the conceptual design cycle. Accordingly, *Data 1* and *Data 2* in Fig. 1 are communicated by neutral intermediate text data files. Similar data structures need to be defined from within the other programming applications involved to read and import the data exported by other applications. These definitions do not have to be an exact copy of the previous one as long as the necessary parameters are imported. Defining all corresponding data structures consistently avoids confusion and facilitates better data management.

The concept of feature has been investigated in great depth in recent decades to address emerging product development and manufacturing needs and challenges. Originally, the term feature was used to refer only to the geometrical aspects of a model such as slots, holes, and chamfers. Since the product development process involves much more than geometric entities, researchers have sought ways of embedding more information into the CAD models. Consequently, today's features have a much broader definition; both geometric and non-geometric information are embedded in the model, aiding in rapid and reliable product information transfer. Some of the many features developed in the work reported here include:

- Coordinate system (CSYS) features: Used in the creation of relative and absolute CSYS
- Skeleton functional features: Used in the development of skeleton product profiles
- Embodiment features: Features responsible for creation of 3D geometries
- Sheet body features
- Solid body features
- Curve features.

8 Linkage Geometry Design

8.1 Homogeneous Coordinate Transformation

The output of the SimMechanics simulation provides only joint forces and motions expressed in the global reference frame. These global generalized joint forces have to be transformed into and expressed in separate coordinate systems local to the links or frame members under investigation. In order to achieve this, three coordinate transformation matrices are developed. The *boom*, due to its geometrical deflection, has two sides and requires two different matrices to express forces in frames local to these sides. Since the *stick* has a straight axis, it needs only a single transformation matrix for reference frame manipulation.

The first step in this process is to identify stationary angles which, together with the linear dimensions, help to fully define the geometry of the boom and stick parts. This is followed by defining variable angles responsible for the operational configuration of the arm mechanism—in this case, the digging operation (Fig. 2).

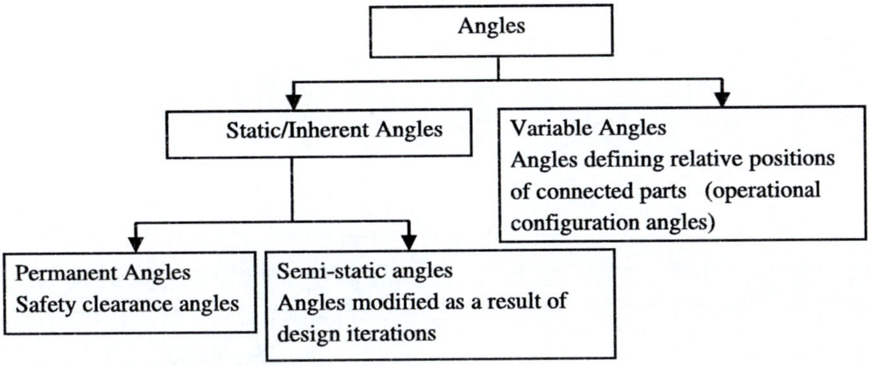

Fig. 2 Classification of angles

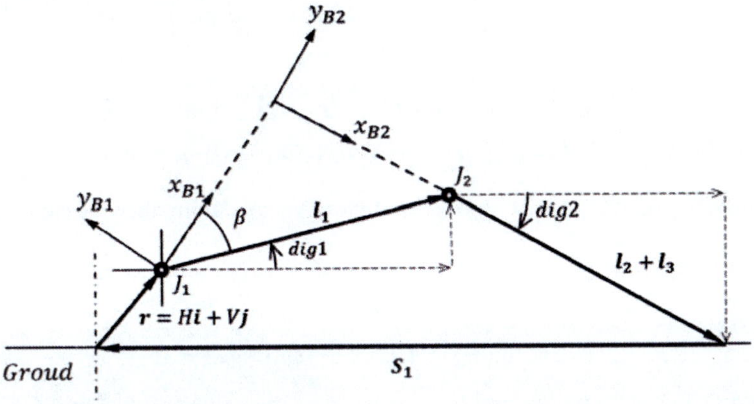

Fig. 3 Operational configuration angles

As shown in Figs. 3 and 4, angle dig1 and dig2 define the orientations of the boom and the stick with respect to the ground during a digging operation. Unlike static angles which are always assigned positive values, variable angles are direction-sensitive.

8.2 Boom Geometries

Referring to Fig. 3, expressions for variable angles dig1 and dig2 can be formulated as follows.

Summing the components of vectors along the horizontal direction gives:

$$\sum V_x = S_1 - H \tag{1}$$

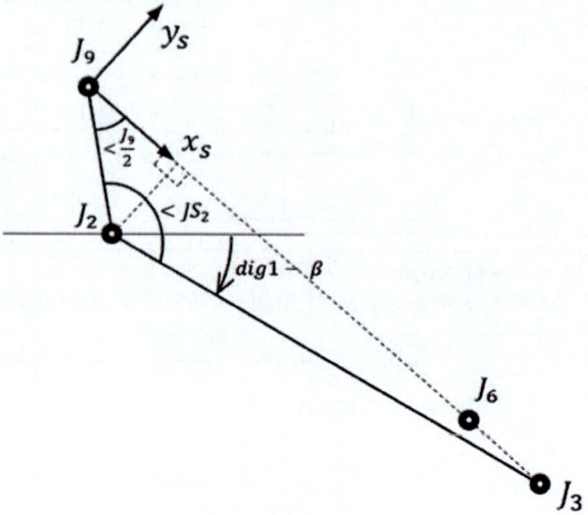

Fig. 4 Stick structural angles

$$l_1 \cos(\text{dig1}) + l_2 \cos(\text{dig2}) + l_3 \cos(\text{dig2}) = S_1 - H \tag{2}$$

$$l_1 \cos(\text{dig1}) + (l_2 + l_3) \cos(\text{dig2}) - S_1 + H = 0 \tag{3}$$

Similarly, summing the components of these vectors along the vertical direction gives another expression.

$$\sum V_y = 0 \tag{4}$$

$$V + l_1 \sin(\text{dig1}) + (l_2 + l_3) \sin(\text{dig2}) = 0 \tag{5}$$

Solving Eqs. (3) and (5) simultaneously gives the values of dig1 and dig2. The homogeneous coordinate transformation matrices for the first and second side of the boom are then derived using these calculated angles.

Boom Rotation Matrix I, RB_1

$$\text{RB}_1 = \begin{bmatrix} \cos(\text{dig1} + \beta) & -\sin(\text{dig1} + \beta) & 0 \\ \sin(\text{dig1} + \beta) & \cos(\text{dig1} + \beta) & 0 \\ 0 & 0 & 1 \end{bmatrix} \tag{6}$$

Boom Rotation Matrix II, RB_2

$$\text{RB}_2 = \begin{bmatrix} \cos(\text{dig1} - \beta) & -\sin(\text{dig1} - \beta) & 0 \\ \sin(\text{dig1} - \beta) & \cos(\text{dig1} - \beta) & 0 \\ 0 & 0 & 1 \end{bmatrix} \tag{7}$$

8.3 Stick Rotation Matrix

Since the axis of the stick is not parallel to the axis of the second section of the boom, the transformation matrix developed for the second part of the boom cannot be directly used to transform global forces into the frame of reference local to the stick.

The rotation matrix for the stick is formulated by carefully observing the subsequent chain of angular transformations starting from joint $J1$.

$$\tan(J9_L) = \frac{h_{\text{stick}}}{h_{\text{tail}}} \tag{8}$$

$$J9_L = \tan^{-1}\left[\frac{h_{\text{stick}}}{h_{\text{tail}}}\right] \tag{9}$$

$$J9_U = J9L \tag{10}$$

$$J9 = J9_U + J9_L = 2 \times J9_L \tag{11}$$

$$TS_2 = \sqrt{l_2^2 - h_{\text{stick}}^2} \tag{12}$$

$$J2_l = \tan^{-1}\left[\frac{h_{\text{tail}}}{h_{\text{stick}}}\right] \tag{13}$$

$$J2_r = \tan^{-1}\left[\frac{\text{TS}_2}{h_{\text{stick}}}\right] \tag{14}$$

$$J2 = J2_l + J2_r \tag{15}$$

$$J3_l = \tan^{-1}\left[\frac{h_{\text{stick}}}{\text{TS}_2}\right] \tag{16}$$

$$J3_U = J3_L \tag{17}$$

$$J3 = J3_L + J3_U = 2 \times J3_L \tag{18}$$

$$R_{Z,(\beta+\text{dig1})} \rightarrow R_{Z,-2\beta} \rightarrow R_{Z,J2} \rightarrow R_{Z,(J9_L-180)} \tag{19}$$

$$\begin{aligned} \text{net angle of rotation} &= \beta + \text{dig1} - 2\beta + J2 + J9_L - 180 \\ &= \text{dig1} - \beta + J2 + J9_L - 180 \end{aligned} \tag{20}$$

The stick rotation matrix RS is then given by the expression:

$$\text{RS} = \begin{bmatrix} \cos(\text{dig1} - \beta + J2 + J9_L - 180) & -\sin(\text{dig1} - \beta + J2 + J9_L - 180) & 0 \\ \sin(\text{dig1} - \beta + J2 + J9_L - 180) & \cos(\text{dig1} - \beta + J2 + J9_L - 180) & 0 \\ 0 & 0 & 1 \end{bmatrix} \tag{21}$$

8.4 Transition Four-Bar Dimensional Synthesis

The major linear dimensions defining the working range of the overall mechanism have already been synthesized using the method developed in the previous chapter. For the purpose of making the mechanism complete, the transition four-bar mechanism's dimensions have to be synthesized for the given dimensions of standard buckets and sticks. For a given dimension of the bucket b_o, the length of the other two linkages of the transition four-bar mechanism can be calculated as follows.

As shown in Fig. 5, there are two configurations of the four-bar linkage mechanism resulting in a phenomenon called "*mechanism lock.*" These two angular positions are considered to be the upper and lower range limits within which the bucket operates.

In Fig. 5, γ_{ec} is the eccentricity angle necessary to prevent the mechanism from self-locking. γ_{s1} is the minimum angle between the back of the bucket and the stick. The value of γ_{s1} depends on the safety clearance angle necessary to avoid physical contact between the links. The value of this angle is subject to change during the life of the design cycle, reflecting changes in the dimensions of the contacting parts as a result of cyclic modifications.

γ_{s2} serves the same purpose as γ_{s1} but on the opposite end of the bucket's angular displacement range. Limiting factors in this case include direct contact between mechanical components and volumetric allowance for extra bulk material when loading the bucket. In addition to b_o and b_1, these two angles are assumed to be known to evaluate the lengths of links b_2 and b_3.

The critical values of b_2 and b_3, i.e., those that result in mechanical locking of the mechanism, are determined from the following two simplified geometries corresponding to the two cases as shown in Figs. 6 and 7.

Applying Law of Cosines to the geometry of Fig. 6 we get

$$b_3^2 = b_o^2 + (b_1 + b_2)^2 - 2b_0(b_1 + b_2)\cos\gamma_{s1} \tag{22}$$

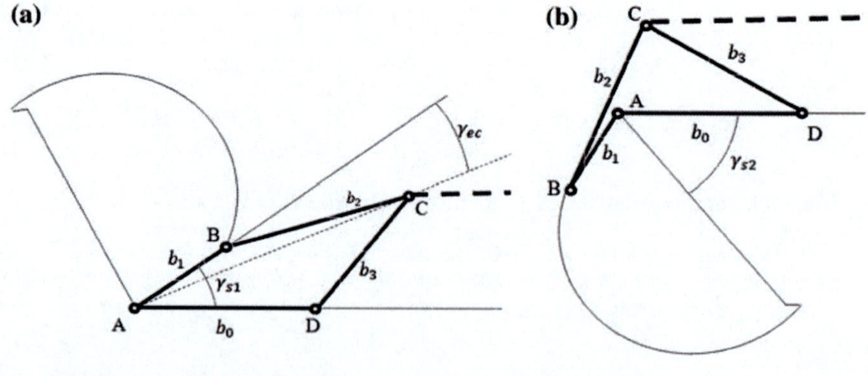

Fig. 5 Transition four-bar work ranges

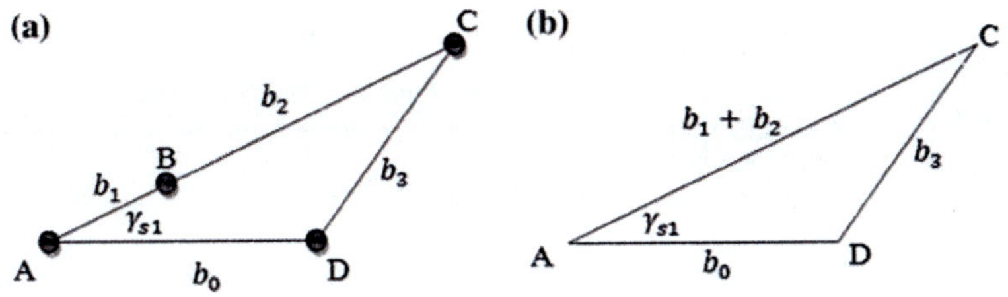

Fig. 6 Upper mechanism locking configuration

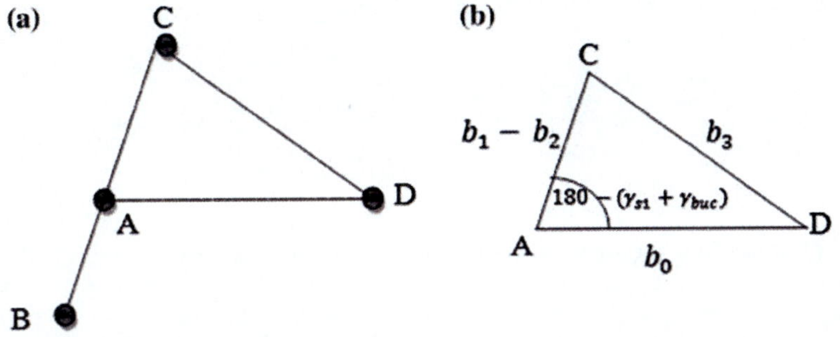

Fig. 7 Lower mechanism locking configuration

Similarly referring to Fig. 7

$$b_3^2 = b_o^2 + (b_2 - b_1)^2 + 2b_0(b_2 - b_1)\cos(\gamma_{s2} + \gamma_{bk}) \tag{23}$$

Solving Eqs. (22) and (23) simultaneously gives the values of b_2 and b_3.

This calculation is implemented using the custom MATLAB® function *f_Four-bar_Solver* in the main program.

9 Stress and Strength Calculations

9.1 Basic Stresses Involved

The expressions for the various stresses considered in this section are developed based on the assumptions and procedures outlined by Shigley and Mischke [17].

9.1.1 Transverse Shear Stress, τ_b

Shear stress, τ, as a result of shear force and bending moment is derived from the relation (Fig. 8)

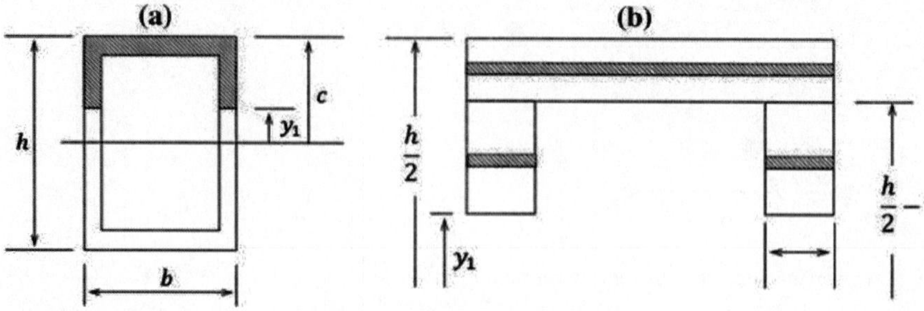

Fig. 8 Cross-sectional area under transverse shear stress

$$V = \frac{dM}{dx} \tag{24}$$

$$\tau = \frac{VQ}{Ib} \tag{25}$$

where Q is the first moment of area about the neutral axis given by:

$$Q = \int_{y1}^{C} y\,dA \tag{26}$$

Q for the given cross-sectional geometry computed by dividing the area into two sections.

Case 1

For $0 < |y_1| < \left(\frac{h}{2} - t\right)$

$$Q = \int_{y_1}^{h/2} y\,dA = \int_{y_1}^{\frac{h}{2}-t} y\,dA_1 + \int_{\frac{h}{2}-t}^{\frac{h}{2}} y\,dA_2 \tag{27}$$

where $dA_1 = 2t\,dy$ and $dA_2 = b\,dy$

$$Q_1 = \int_{y_1}^{\frac{h}{2}-t} 2ty\,dy + \int_{\left(\frac{h}{2}-t\right)}^{\frac{h}{2}} by\,dy \tag{28}$$

$$Q_1 = t\left(\frac{h}{2} - t\right)^2 - ty_1^2 + \frac{bt(h-t)}{2} \tag{29}$$

Case 2

For $\left(\frac{h}{2} - t\right) \le |y_1| \le \frac{h}{2}$

$$Q_2 = \int_{y_1}^{h/2} y\,dA_2 = \int_{y_1}^{h/2} by\,dy \tag{30}$$

$$Q_2 = \frac{b}{8}\left(h^2 - 4y_1^2\right) \tag{31}$$

The second moment of area of the entire cross section,I, is given by

$$Q_{\max} = t\left(\frac{h}{2} - t\right)^2 + \frac{bt(h-t)}{2} \tag{32}$$

$$I = \frac{bh^3}{12} - \frac{(b-2t)(h-2t)^3}{12} \tag{33}$$

$$\tau_{\max} = \frac{VQ_{\max}}{Ib} \tag{34}$$

$$b_{\max} = 2t \tag{35}$$

$$\tau_b = \frac{V\left[t\left(\frac{h}{2} - t\right)^2 + \frac{bt(h-t)}{2}\right]}{2t\left[\frac{bh^3}{12} - \frac{(b-2t)(h-2t)^3}{12}\right]} \tag{36}$$

9.1.2 Torsional Stress, τ_{tor}

The equations expressing torsional stresses in the cross-sectional areas are developed based on the assumptions and procedures outlined by Shigley and Mischke [17].

Area of torsion, A_{tor}, is given by the following expression (Fig. 9).

$$A_{\mathrm{tor}} = \left(b - 2\frac{t}{2}\right)\left(h - 2\frac{t}{2}\right) \tag{37}$$

$$A_{\mathrm{tor}} = (b-t)(h-t) \tag{38}$$

The torsional stress, τ_{tor}, is given by

$$\tau_{\mathrm{tor}} = \frac{T}{2(b-t)(h-t)t} \tag{39}$$

where T, the torque creating the torsional stress, is recorded at every joint during the simulation of the mechanism in the SimMechanics environment.

9.1.3 Direct Stress, σ_{dx}

Area of direct stress distribution is given by the expression

$$A_{\mathrm{dx}} = 2[bt + t(h-2t)] \tag{40}$$

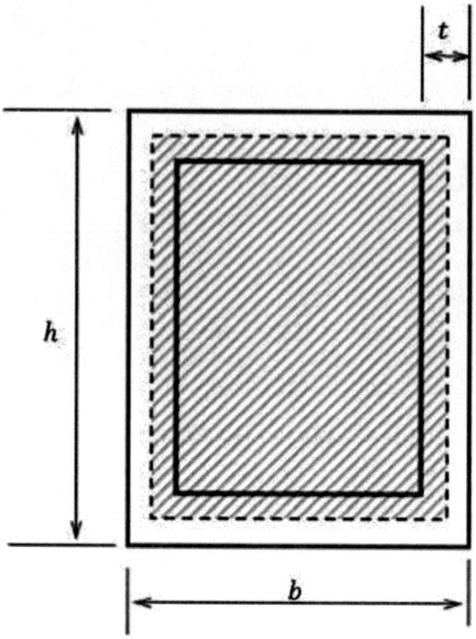

Fig. 9 Torsional cross-sectional area

For a given axial longitudinal force, F_{ax}, the direct stress is calculated by using the formula

$$\sigma_{\mathrm{dx}} = \frac{F_{ax}}{A_{\mathrm{dx}}} \tag{41}$$

$$\sigma_{\mathrm{dx}} = \frac{F_{ax}}{2[bt + t(h - 2t)]} \tag{42}$$

9.1.4 Bending Stresses, σ_{zx}

Bending stress, σ_{zx}, due to bending moment acting about the z-axis, M_z, is given by:

$$\sigma_{zx} = -\frac{M_z y}{I_{zz}}\left(-\frac{h}{2} \leq y \leq \frac{h}{2}\right) \tag{43}$$

where $I_{zz}$is the second moment of area of the cross section about the centroidal z-axis and it is given by:

$$I_{zz} = \frac{bh^3 - (b - 2t)(h - 2t)^3}{12} \tag{44}$$

This stress reaches its maximum value when at the outer most boundaries of the cross section, i.e., when:

$$y = \pm \frac{h}{2} \tag{45}$$

$$\sigma_{zx(\max)} = -\frac{6M_z h}{bh^3 - (b-2t)(h-2t)^3} \tag{46}$$

In a similar manner, bending stress, σ_{yx}, due to moment acting about the y-axis is given by:

$$\sigma_{yx} = -\frac{M_y z}{I_{yy}}\left(-\frac{b}{2} \le z \le \frac{b}{2}\right) \tag{47}$$

where I_{yy} in this case is the second moment of the cross-sectional area about the centroidal y-axis

$$I_{yy} = \frac{hb^3 - (h-2t)(b-2t)^3}{12} \tag{48}$$

$$\sigma_{yx(\max)} = -\frac{6M_z b}{hb^3 - (h-2t)(b-2t)^3} \tag{49}$$

Superposition of the effects of these two bending stresses and the direct stress gives the maximum value of stress in the x-direction.

$$\sigma_{xx(\max)} = \sigma_{dx} + \sigma_{yx(\max)} + \sigma_{zx(\max)} \tag{50}$$

9.2 Pin Design

9.2.1 Methods of Pin Failure

1. Localized contact stress
2. Failure of pin due to double shear
3. Failure of pin due to bending moment.

9.2.2 Contact Stresses

The interaction between the pin and the casing is modeled by a cylinder-plane contact instead of cylinder–cylinder contact. The resulting magnitudes of Hertz's contact stresses will have relatively higher values than if they were calculated using the cylinder–cylinder assumption because of the reduced contact area (Fig. 10).

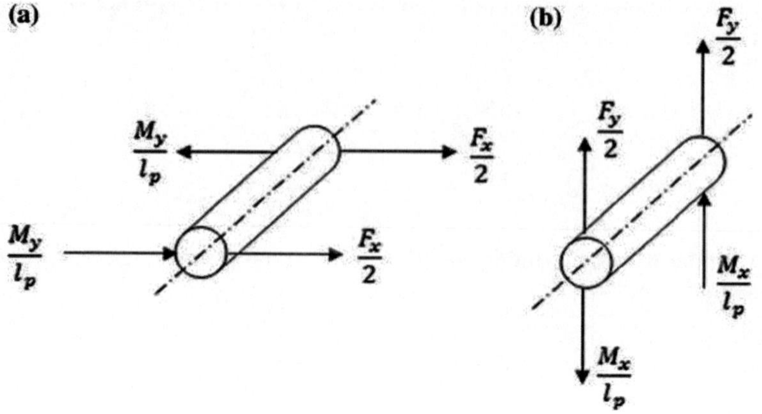

Fig. 10 Pin loads in global vertical and horizontal planes

where

d_1 Diameter of pin
d_2 Diameter of base hole
ν_1 Poisson's Ratio of the pin material
ν_2 Poisson's Ratio of the base material
E_1 Young's Modulus of Elasticity of the pin material
E_2 Young's Modulus of Elasticity of the base material.

The total force exerted on the first end of the pin is given by:

$$F_p = \begin{bmatrix} \frac{F_x}{2} + \frac{M_y}{l_p} \\ \frac{F_y}{2} + \frac{M_x}{l_p} \\ 0 \end{bmatrix} \tag{51}$$

where d_p and l_p are the pin diameter and its effective length, respectively.
The magnitude of this force, F_{pin}, is:

$$F_p = \sqrt{\left(\frac{F_x}{2} + \frac{M_y}{l_p}\right)^2 + \left(\frac{F_y}{2} + \frac{M_x}{l_p}\right)^2} \tag{52}$$

9.2.3 Hertz's Contact Stress

The maximum stress due to contact between the surfaces of the pin and the base is calculated using the Hertz's method. The contact zone between these two surfaces is approximated by a rectangular region. The width of this region, commonly known as "contact half width" is calculated by using the formula:

$$b = k_b\sqrt{F_p l_p} \tag{53}$$

$$\text{where } k_b = \sqrt{\frac{2}{\pi t}\frac{\left(\frac{1-v_1^2}{E_1}+\frac{1-v_2^2}{E_2}\right)}{\frac{1}{\mathrm{d}_1}+\frac{1}{\mathrm{d}_2}}} \tag{54}$$

Since the casing holes are modeled by a plane for the purpose of making the design compatible with higher stresses, the reciprocal term containing d_2 will be approximated by zero. The resulting expression takes the form:

$$k_b = \sqrt{\frac{2d_1\left(\frac{1-v_1^2}{E_1}+\frac{1-v_2^2}{E_2}\right)}{\pi t}} \tag{55}$$

The maximum contact pressure is then written as a function of the length (l_p) and diameter (d_p) of the pin as follows:

$$P_{\max}(l_p, d_p) = \frac{2F_p l_p}{\pi b t} \tag{56}$$

The resulting principal stresses in the pin are given by the relations

$$\sigma_x = -2v_1\frac{2F_p l_p}{\pi b t}\left[\sqrt{1+\left(\frac{z}{b}\right)^2}-\left|\frac{z}{b}\right|\right] \tag{57}$$

$$\sigma_y = -\frac{2F_p l_p}{\pi b t}\left[\frac{1+2\left(\frac{z}{b}\right)^2}{\sqrt{1+\left(\frac{z}{b}\right)^2}}-2\left|\frac{z}{b}\right|\right] \tag{58}$$

$$\sigma_z = -\frac{2F_p l_p}{\pi b t}\frac{1}{\sqrt{1+2\left(\frac{z}{b}\right)^2}} \tag{59}$$

To calculate the maximum value of stress from one of these three equations, these stress are computed at the critical section $z/b = 0.786$

$$\sigma_x = -1.944v_1\frac{F_p l_p}{\pi b t} \tag{60}$$

$$\sigma_y = -0.371\frac{F_p l_p}{\pi b t} \tag{61}$$

$$\sigma_z = -1.572\frac{F_p l_p}{\pi b t} \tag{62}$$

The allowable contact stress is generally taken as the minimum of $\sigma_y/4$ or $\sigma_u/6$.

9.2.4 Failure of Pin Due to Double Shear

The total shear area of each pin is represented by the equation (Fig. 11)

$$A_{sh} = 2\frac{\pi d_p^2}{4} \tag{63}$$

Direct shear stress on the pin is calculated by dividing the shearing force by the total area. An equation for this stress is derived in terms of the pin length and the pin diameter so that it can be used in calculating an optimum values for these two pin dimensions.

$$\tau_{sh} = \frac{2|F_p|}{A_{sh}} \tag{64}$$

$$\tau_{sh} = \frac{4|F_p|}{\pi d_p^2} \tag{65}$$

9.2.5 Failure of Pin Due to Bending Moment

Referring to Fig. 12 for the loading distribution, the maximum bending moment at the mid-span of the pin is given by the expression:

$$t_1 = t + e_{b1} \tag{66}$$

$$t_2 = t + e_{b2} \tag{67}$$

$$M_{max} = F_p\left(\frac{t_1}{3} + t_2 + \frac{l_p}{2}\right) - F_p\left(\frac{t_2}{2} + \frac{l_p}{2}\right) \tag{68}$$

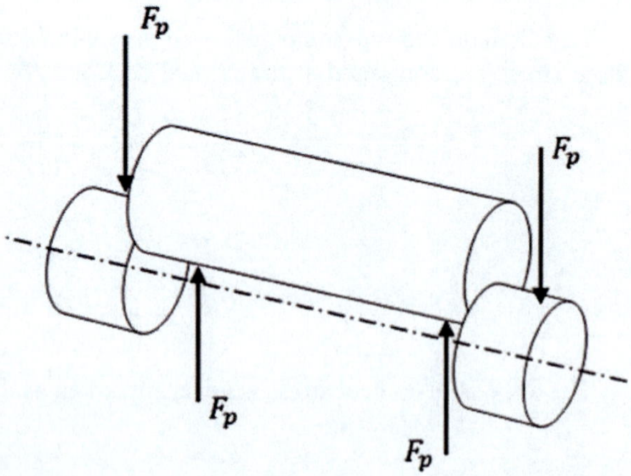

Fig. 11 Pin under double shear

where e_{b1} and e_{b2} are base reinforcement extensions.

The maximum bending stress is then given by substituting the above expression into the general formula

$$\sigma_b = \frac{M_{\max}}{z} \tag{69}$$

$$\sigma_b = \frac{F_p\left[\frac{t_1}{3} + \frac{t_2}{2}\right]}{\frac{\pi}{32} d_p^3} \tag{70}$$

10 Mechanism Dynamics Analysis with Virtual Mechanism Modeling and Simulation

The design process of linkages and members in the mechanism requires the identification of generalized reaction forces at each joint. These forces will be used in the free body diagrams (FBD) and mass-acceleration diagrams (MAD) during the design stages. This task is implemented by using SimMechanics®, the mechanical simulation and analysis tool available in MATLAB® software.

10.1 Simulation Setup

Now that the major linear dimensions of the mechanism are identified, the next step is determining the reaction forces and moments experienced by each joint as a result of the digging operation. To do this, a skeleton mechanism is constructed

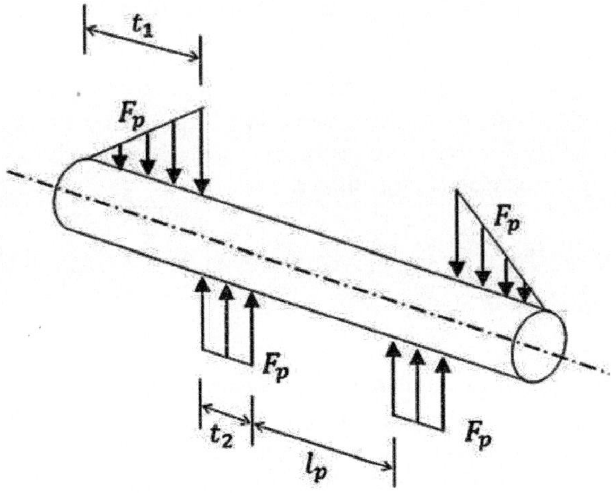

Fig. 12 Bending load distribution on pins

using the previously calculated linear dimensions in the SimMechanics modeling environment of MATLAB®. Using a skeleton mechanism provides the flexibility of calculating interaction forces and exporting the results into the workspace for further processing. Furthermore, during latter stages of the design cycles, the weightless links can be substituted by their 3D counterparts and the simulation can be re-run to consider the inertia effects.

When using this tool, the mechanism under investigation is constructed by connecting linear linkages of prescribed inertia properties with virtual mechanical joints available in the joint library. The virtual weld joint is used in the event it is required to model bending, splitting, or merging mechanical members. For this approach to be viable in the automation of design processes, two general requirements have to be met. The first is that all linear dimensions need to be defined either numerically or symbolically. The second is that all forms of mating constraints between connecting members have to be imposed by the use of joints and limits on joint parameters.

Although it is possible to assign inertia properties to these bodies in the initial stages as mentioned above, the procedure adopted in this case study uses a different approach, deemed more suitable for cyclic modifications. The first round of the simulations starts with weightless and high-stiffness rigid bodies; further simulations will be carried out with updated 3D solid bodies produced as a result of previous design cycles.

The necessary kinetic and kinematic joint variables registered during the simulations are extracted to MATLAB® workspace using the Simulink® scope readers. SimMechanics Link®, another useful tool to import and export CAD solid models into and out of the SimMchanics® simulation environment, supports only SolidWorks® and Pro/E®, but not UG NX. To overcome this issue, SolidWorks is used as an intermediate file transfer tool in exporting the 3D models to SimMechanics.

10.2 Simulink Model Construction

Figure 13 shows the major components of an excavator arm mechanism while Fig. 14 shows the representation of the same mechanism by linear components. The latter model is constructed in SimMechanics environment using the standard rigid body and joint library.

10.3 Boom Construction

The boom as shown in Fig. 15 has two sides: BS_1 and BS_2. It is hinged to the vehicle body with joint J1 and to the stick with joint J2. Joints J10 and J11 are connecting points for hydraulic cylinders C2 and C1, respectively. The deflection of the structure by an angle 2β is modeled by welding the two linear sides.

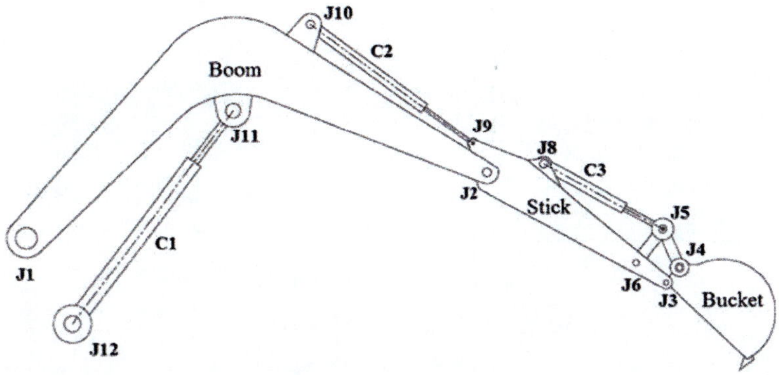

Fig. 13 Typical excavator arm mechanism

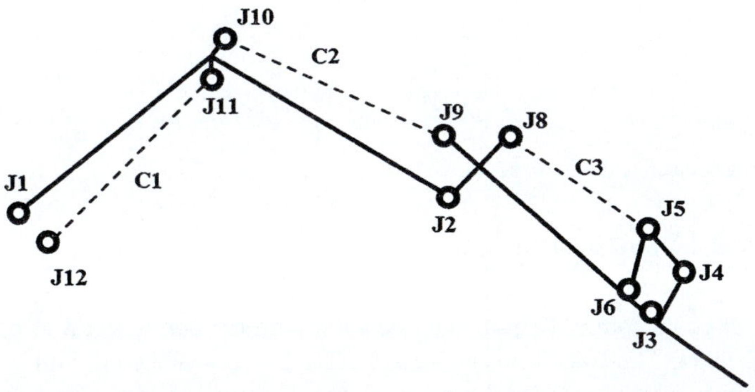

Fig. 14 Skeleton representation of excavator arm mechanism

Fig. 15 Skeleton representation of boom structure

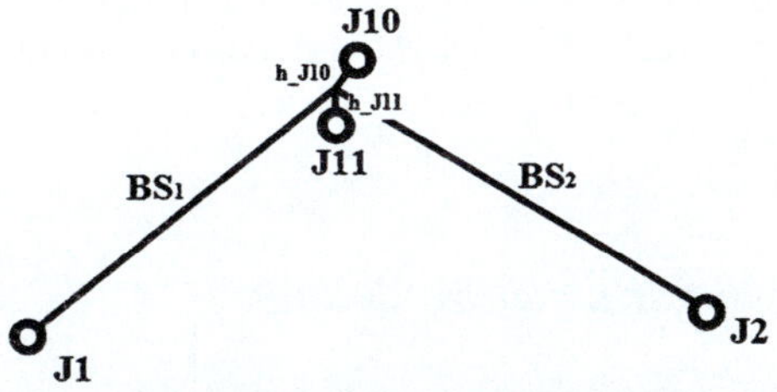

The two hinges for the hydraulic cylinders, **J10** and **J11**, are located at the end of the extension sticks *h_J10* and *h_J11* to represent for an initial thickness of the boom. The values of these lengths are updated at the end of each conceptual design cycle from the cross-sectional calculation results. This is done because the hinge locations are assumed to be on the surfaces of the boom which are subject to modification at the end of each conceptual design cycle (Fig. 16).

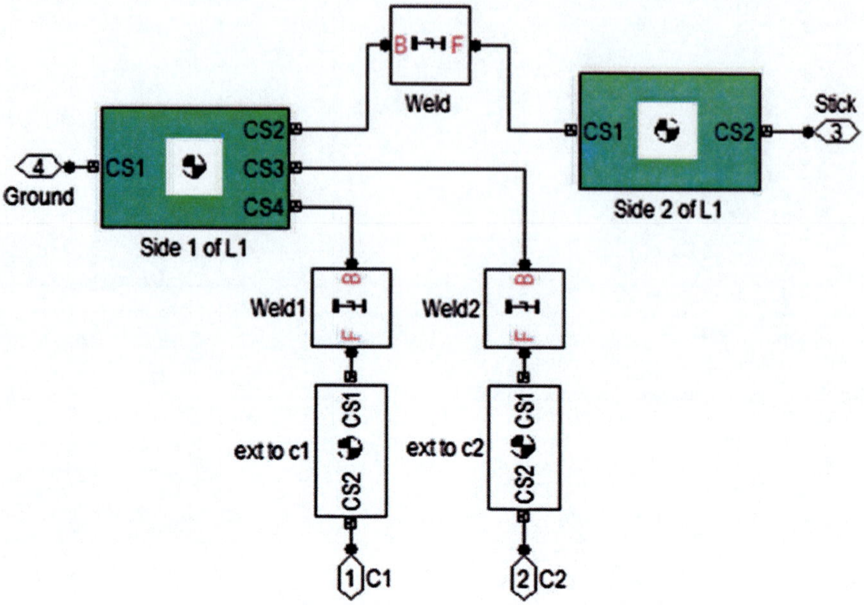

Fig. 16 SimMechanics model of boom

10.4 Stick Construction

The stick has four joints: J2 connects the stick with the boom while J3 connects it with the bucket. The transition four-bar mechanism is connected with the stick at joint J6 with revolute joint. Joint J8 is the connection point for the second hydraulic cylinder, C2.

Again in this case, the final distances of hinges J2 and J8 from the longitudinal axis of the stick, h_J2 and h_J8, are determined based on the final 3D dimensions of the stick. To start the simulation, however, initial values are assigned for these dimensions. These parameters will be updated at the end of each cycle (Figs. 17, 18).

10.5 Bucket Modeling

The bucket, which is the follower link of the transition four-bar mechanism, has only two joints: J3 and J4 to connect it to the stick and the coupler link of the four-bar, respectively (Figs. 19, 20).

The application point of the ground reaction force is selected in such a way that the overall mechanism will be subject to severe loading conditions. Twisting movements about the *x*-axis and bending movements about the *y*- and *z*- axes register maximum readings when the digging force is applied on the bucket at a

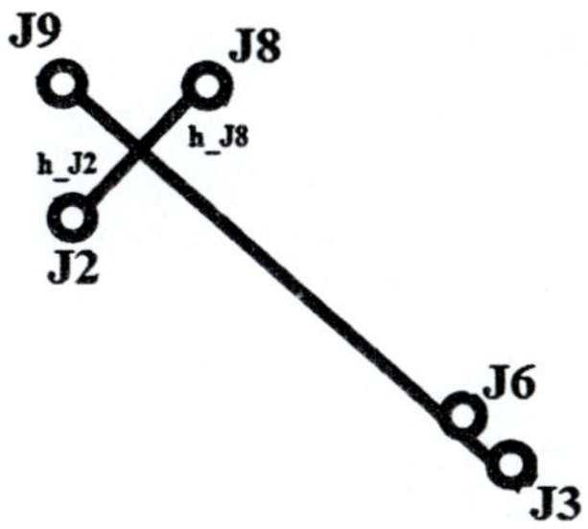

Fig. 17 Skeleton representation of stick

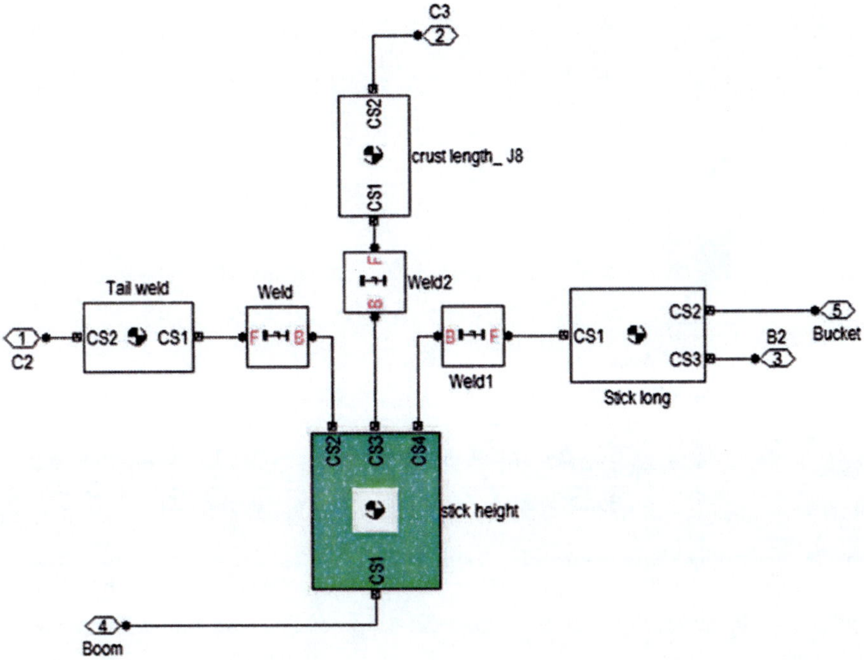

Fig. 18 SimMechanics model of stick

point furthest from the *x*-axis. An eccentricity loading distance of half the width of the bucket is introduced for this purpose as shown in Fig. 21.

10.6 Hydraulic Cylinders

Hydraulic cylinders are represented by simple weightless rigid links solely for the purpose of keeping the mechanism rigid. The forces registered at the opposite ends of these links can be used to determine the required hydraulic capacity of the

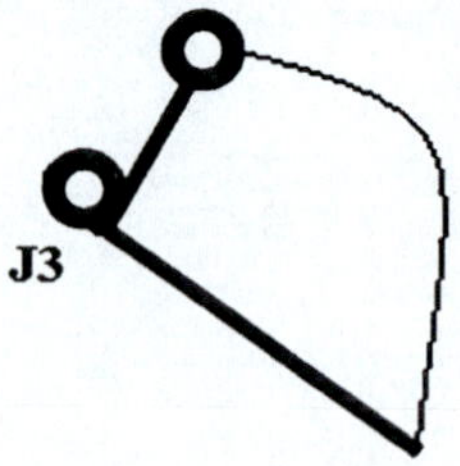

Fig. 19 Bucket schematics

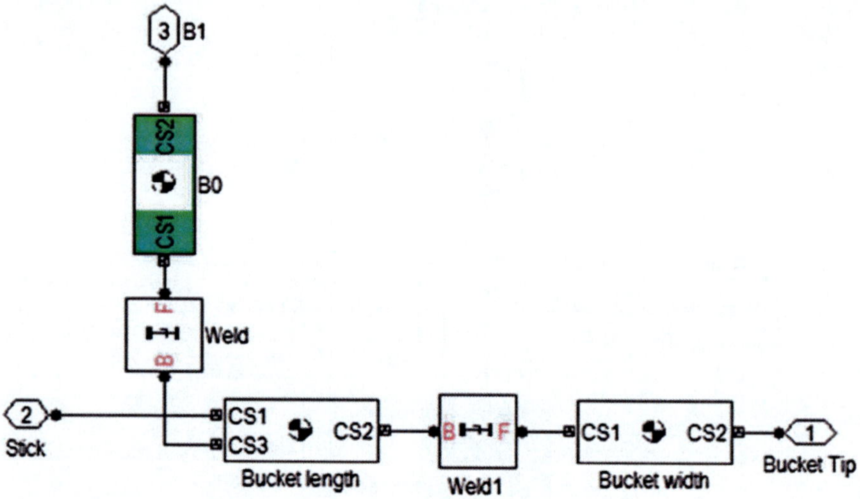

Fig. 20 Bucket SimMechanics model

cylinders. However, this task is beyond the scope of this research and will not be discussed here.

10.7 Transition Four-Bar Linkages

The two remaining linkages in the transition four-bar mechanism are represented by simple blocks with joints at both ends. The actual total number of revolute joints in the mechanism is 11. However, in the SimMechanics model one more joint needs to be introduced due to the location of the hydraulic force application point on the driving link of the transition four-bar. Two coaxial joints in the real mechanism are required to be modeled by combining them into a single joint. Because this point is chosen to be at the connection point of the driving and

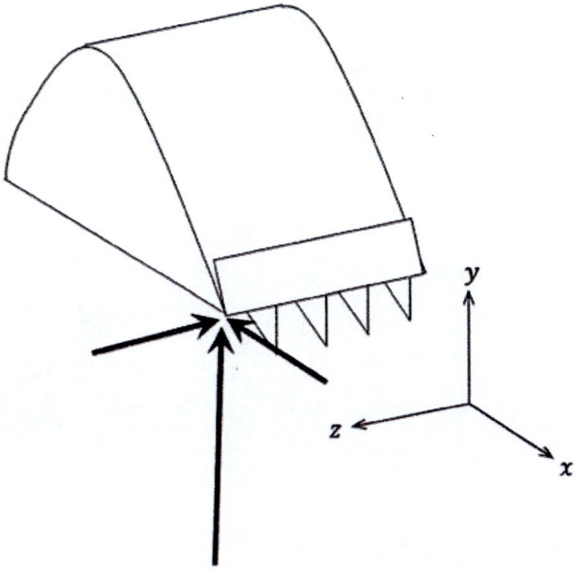

Fig. 21 Location of application point of digging force

coupler links, an additional redundant hinge joint was required for creating a three-branch connection. This representation will not have any negative effect on the final outcome of the analysis.

The above SimMechanics sub-models are assembled and simulation environment parameters are defined. Figure 22 below shows the final assembled SimMechanics Model of the excavator arm mechanism in the real-time simulation window.

The simulation for this model is run for two seconds at the digging orientation. Scopes in the model such as $SD1, SD2, \ldots, SD12$ register and export joint variable data to the MATLAB® workspace in vector form. Because of the selection of the digging mode for the design purpose, it was not necessary to define angular displacement and speed limits.

Figure 23 shows the assembled SimMechanics diagram of the excavator arm mechanism. The digging force is represented by a constant vector and is applied at the left tip of the bucket.

11 Example Simulation Analysis

An example problem is investigated in this section to demonstrate the applicability of the proposed methods. The methods and the necessary engineering rules and design intents are programmed and implemented in MATLAB®. This demonstration will be carried out in two stages, which are designed to show the operational procedures of the two proposed methods and their corresponding results.

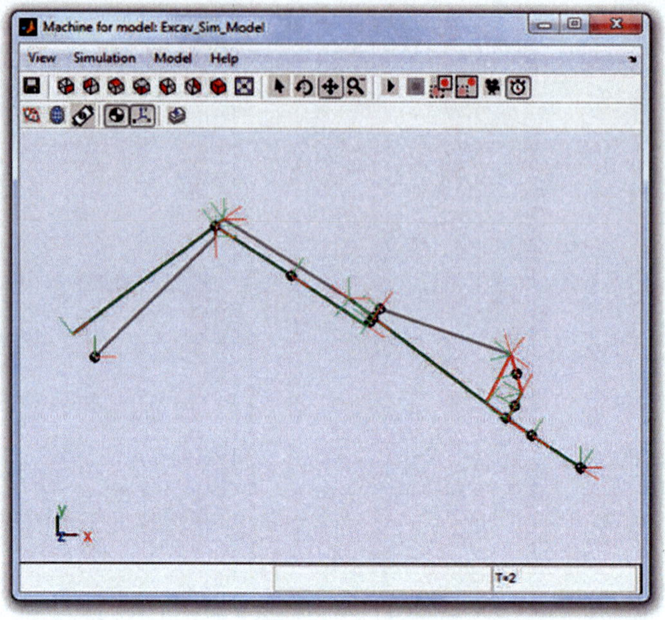

Fig. 22 Real-time simulation environment

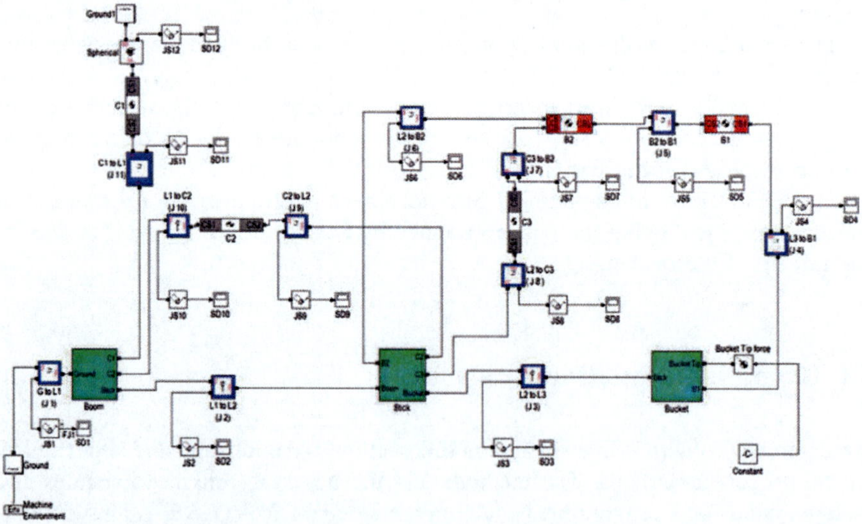

Fig. 23 SimMechanics diagram for an excavator arm mechanism

The first stage deals with the application of the hybrid ANN-Optimization technique in the process of dimensional synthesis of mechanisms. In the second section, the calculations involved and the generated results in the areas of optimizations of cross-sectional dimensions of the boom and the stick will be discussed. The input for the module handling the dimensional synthesis is a set of required output configurations of the excavator arm mechanism. As discussed before, these values are checked for compatibility to ensure the feasibility of their co-existence.

Input Problem:

SpcDat = c_Spec_Data_SI

Properties:
Title: 'Commercial Specifications and Vehicle Dimensions'
Maximum_Reachout_at_Ground_Level_S1: 5.6700
Maximum_Cutting_Height_S3: 3.7248
Maximum_Loading_Height_S4: 1.3521
Horizontal_Distance_H: 0.9857
Vertical_Distance_V: 1.2300
Vehicle_Weight: 5000

Once the linear dimensions of the overall mechanism are calculated, the next task will be calculating the optimum cross-sectional dimensions of the boom and the stick. This process is started by defining the necessary geometric and non-geometric constraints. In addition to these constraints, initial values for some variables and iteration-dependent dimensions are also initiated at this stage.

Bucket geometric properties:

BuckGeo= c_Bucket_Geo_SI

Properties:
Title: 'Bucket Geometries and Dimensions'
Bucket_Length_l3: 0.8112
Bucket_Width_BW: 0.4867
Bucket_Height_b0: 0.2839
Bucket_Pin_Width_bw: 0.2434
Bucket_Angle_teta_bucket: 95
Bulk_Volume_Clearance_Angle: 40
Maximum_Upward_Bucket_Open_Limit_Angle: 35

Dimensional Constraints:

Dimensional_Constraints = c_Dimensional_Constraints_SI

Properties:
Title: 'Structural Dimensions Constraints'

Minimum_Plate_Thickness:	*0.0070*
Maximum_Plate_Thickness:	*0.0200*
Minimum_Base_Dimension:	*0.1000*
Maximum_Base_Dimension:	*0.5000*
Minimum_Boom_and_Stick_Height:	*0.0100*
Maximum_Boom_Height:	*0.5000*
Maximum_Stick_Height:	*0.5000*
Extension_of_Boom_Pin_Reinforcement:	*0.0140*
Extension_of_Stick_Pin_Reinforcement:	*0.0140*

Material Properties:

MaterProp = c_Material_Properties_SI

Properties:
Title: 'Material Selection and Properties'
Prpertiy: 'Poisons E YS_kpsiTS_kpsiYS_MPaTS_MPaElong_2% Area_% BHN'

Pin_Material:	*[0.3000 210 234 260 1612 1791 12 43 498]*
Base_Material:	*[0.3000 210 26 47 179 324 28 50 95]*
Allowable_Stress_in_Pin:	*447750000*
Poison_Ratio_Pin:	*0.3000*
Youngs_Modulus_Pin:	*2.1000e+011*
Allowable_Stress_in_Base:	*81000000*
Poison_Ratio_Base:	*0.3000*
Youngs_Modulus_Base:	*2.1000e+011*
Safety_Factor_Pin:	*1.1500*
Safety_Factor_Boom:	*1.1500*
Safety_Factor_Stick:	*1.1500*
Safety_Factor_Linkages:	*1.1500*

Initial Variable Linkage Imitation:

InitialParam = c_Initial_Parameters_SI

Properties:
Title: 'Initial Values for Variable Dimensions'

Distance_to_J2_and_J8_on_Stick:	*0.1000*
Distance_to_J10_on_Boom:	*0.1000*
Distance_to_J11_on_Boom:	*0.1000*

Linkage Geometries:

LinkDims= c_LinkDims_SI

Properties:

Title: 'Boom and Stick Dimensions'

Boom_Shortcut_Length_l1:	*2.7126*
Boom_Deflection_Angle_betta:	*35.6055*
Side_Length_of_Boom_T:	*1.6682*
Stick_Length_l2:	*1.5616*
Stick_Angle_J2:	*156.9267*
J2_left:	*70.5982*
J2_right:	*86.3285*
Stick_Angle_J8:	*156.9267*
J8_left:	*70.5982*
J8_right:	*86.3285*
Stick_Angle_J9:	*38.8037*
J9_up:	*19.4018*
J9_lower:	*19.4018*
Stick_Angle_J3:	*7.3429*
J3_up:	*3.6715*
J3_lower:	*3.6715*
Distance_to_J2_and_J8_on_Stick:	*0.1000*
Distance_to_J10_on_Boom:	*0.1000*
Distance_to_J11_on_Boom:	*0.1000*
Stick_Tail_Length:	*0.2839*
Stick_Forward_Length:	*1.5584*

Transition Four-bar Dimensions:

FourbarDims = c_Fourbar_Solver_SI

Properties:

Title: 'Fourbar Linkage Dimensions'

Fourbar_Link_b0:	*0.2839*
Fourbar_Link_b1:	*0.3692*
Fourbar_Link_b2:	*0.4461*
Fourbar_Link_b3:	*0.2434*

Operational Configuration Matrices and Variables:

OperConfig = c_Operational_Configuration_SI

Properties:
Title: 'Configuration Parameters and Rotational Matrices'

Boom_opertating_angle_dig1:	*1.8619*
Boom_Rotational_Matrix_Sec1_RB1:	*[3x3 double]*
Boom_Rotational_Matrix_Sec2_RB2:	*[3x3 double]*
Stick_Rotational_Matrix_RS:	*[3x3 double]*
Fourbar_teta_1:	*72.8748*
Fourbar_teta_2:	*-36.7587*
Fourbar_teta_3:	*278.6715*

Generalized Joint Forces and Moments:

Joint_Forces = c_Joint_Forces_SI

Properties:
Title: 'Generalized Forces on Joints'
JointForces: [12x7 double]
FORCES: ''
F1: [3x1 double]
F2: [3x1 double]
F3: [3x1 double]
F4: [3x1 double]
F5: [3x1 double]
F6: [3x1 double]
F7: [3x1 double]
F8: [3x1 double]
F9: [3x1 double]
F10: [3x1 double]
F11: [3x1 double]
F12: [3x1 double]
MOMENTS: ''
M1: [3x1 double]
M2: [3x1 double]
M3: [3x1 double]
M4: [3x1 double]
M5: [3x1 double]
M6: [3x1 double]
M7: [3x1 double]
M8: [3x1 double]
M9: [3x1 double]
M10: [3x1 double]
M11: [3x1 double]
M12: [3x1 double]

Force and moment values can be extracted by calling the members of the data structure as follows:

Joint_Forces.F5 $= 1.0e + 004$ **(0.5644, -1.9908, 0)*

12 Feature-Based CAD Embodiment

The Application Programming Interface (API) open platform is used to write program codes and generate the feature-based 3D CAD parts of the boom and stick in NX. The programming part is implemented using Visual Studio 2008® C ++.

The results of the engineering design calculations carried out in previous sections using MATLAB® and SimMechanics® needed to be imported in a systematic manner to be used in the generation of the CAD models. Additionally, the process of importing and exporting data was required to be performed without direct manual involvement. This was accomplished by creating intermediate sets of text data files to bridge the gap.

At the end of the engineering design cycle calculations, a MATLAB® program is used to create or update a set of.dat* text files containing the necessary input dimensions and parameters data structures. The MATLAB® program writes/updates these files and stores them in specific directories created for this purpose. These files will automatically be accessed by the API C ++ code during the generation of the CAD models. Similar to the exporting command in MATLAB®, a C ++ program is developed, which is responsible for reading the values of this files and storing them in the internal memory.

The locations of the shared directories were determined with consideration for the possibility of different tasks being performed on different systems. A free Internet file storage service was used to create a common directory shared by two computers involved in this research. In practical industrial applications, this approach lends itself to the implementation of efficient collaborative project, as it provides the flexibility of assigning different tasks to different engineers working in different geographical locations.

Classes and their corresponding objects are instantiated and used to effectively handle the data imported. Most of these data were used in the program more than once and adopting an object-oriented programming approach proved helpful in managing the data. The following lines show a class for handling custom datum coordinate system (CSYS) creating function parameters.

```
struct DATUM_CSYS_DATA{
    double  offset_x;
    double  offset_y;
    double  offset_z;
    double  angle_x;
    double  angle_y;
    double  angle_z;
    bool    transform_sequence;
    int     rotation_sequence[2];};
```

12.1 Reusability of Functions

All the functions developed for this project were created with the C ++ API functions provided in NX open documentations. Direct application of the basic functions to this research was found to be very difficult and time consuming due to the need to specifically define most initializing parameters unrelated to the objective of this work.

An effort was undertaken to generalize most of the developed functions and ensure their reusability. Based on the basic C ++ API functions, customized functions were developed by incorporating additional procedures to bring user intuitivism while simplifying the definitions of input and output arguments. More than 40 functions were developed and used in the creation of the boom and the stick CAD files. The following are lists of some of the tasks these functions are responsible for.

- Reading external data files
- Creating new part model files in specified directories
- Creation of datum CSYS (Absolute and Relative)
- Creation of datum planes
- Extraction of datum planes/axis out of datum CSYS
- Creation of geometric objects such as points, lines, arcs, and B-spline curves from data points.

12.2 Boom Modeling

The modeling of the boom part is initialized by creating a blank NX.prt file using the function

_Create_New_Part_File(char file_path[UF_CFI_MAX_FILE_NAME_SIZE])

After creating a blank CAD modeling environment, the next step was to properly position user-defined CSYS features for the purpose of simplicity in additional features and object creation. The relative angular orientations and offset distances between consecutive CSYSs were represented by instantiating an object of the class DATUM_CSYS_DATA. In addition to relative linear displacements and angular orientations, these objects also define the coordinate transformation sequences (Fig. 24).

In the case study most of the CSYSs were defined and located at the joint locations for the purpose of simplifying creation of joint associative features such as hinges. The custom functions used for this purpose are:

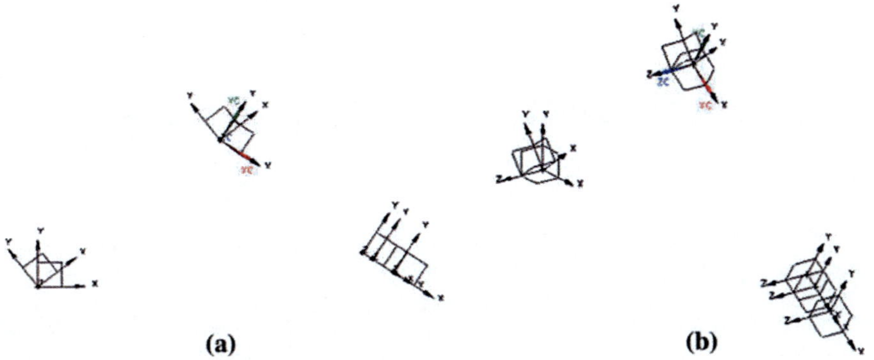

Fig. 24 Boom coordinate system (CSYS) features

- *_CSYS_origin_and_direction(void)*
- *_CSYS_offset(tag_t referece_datum_CSYS,*
const double linear_offset[3],
const double angular_offset[3],
bool operation_sequence)

The optimization result vectors for the two sides of the boom exported from MATLAB® were saved. These vectors define point coordinates of the top left edge of the boom. B-spline curves representing each side of the boom were created from these data points by importing within their respective CSYS (Fig. 25).

Since these edges are symmetrical about the local *x–y* and *x–z* planes, the curves defining the lower right edges of the boom are created by reflecting the existing curves about their *x–z* planes within the local CSYS (Fig. 26).

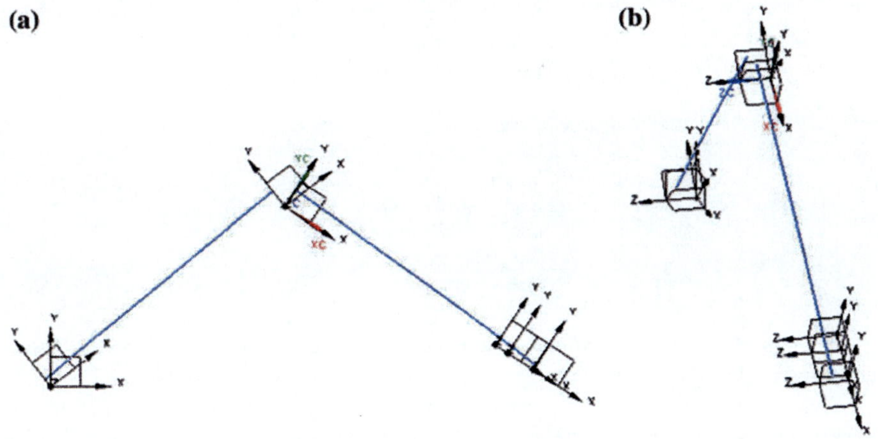

Fig. 25 Boom B-spine curve features

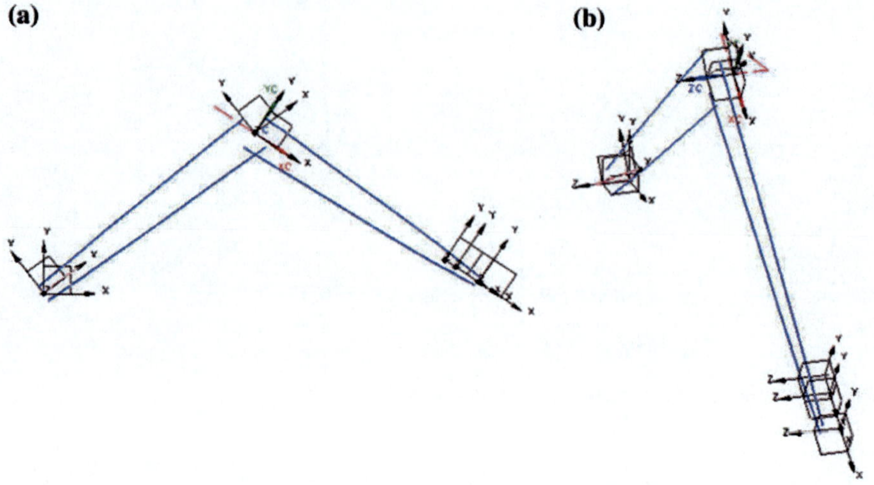

Fig. 26 Evolvement of features

The function used for this purpose is:

_Mirror_a_Curve_through_a_Plane(tag_t Curve_Tag, tag_t Plane_ Tag)

The curves are modified by trimming and bridging operations to create a joined curve feature. The following functions are used to for these operations.

```
tag_t _Trim_Curve_by_Datum_Plane(
        tag_t    Curve_Tag,
        tag_t    Datum_Plane_Tag,
        int      which_end);

tag_t _Trim_Curve_by_Curves(
        tag_t    Target_curve_Tag,
        tag_t    Tool_curvel,
        tag_t    Tool_curve2);

tag_t _Bridge_Curves(
        tag_t    Curve_1_Tag,
        tag_t    Curve_2_Tag,
        int      Reversel,
        int      Reverse2,
        int      parl,
        int      par2);
```

The end of the boom at joints J1 and J2 are closed by arcs tangent to the upper and lower edge curves and centered at the origins of the CSYSs (Fig. 27).

This closed curve feature represents the side wall of the boom. To create the upper and lower floors of the boom it is required to create other sets of curve features defining the boundaries of the surfaces. The above modified closed curve,

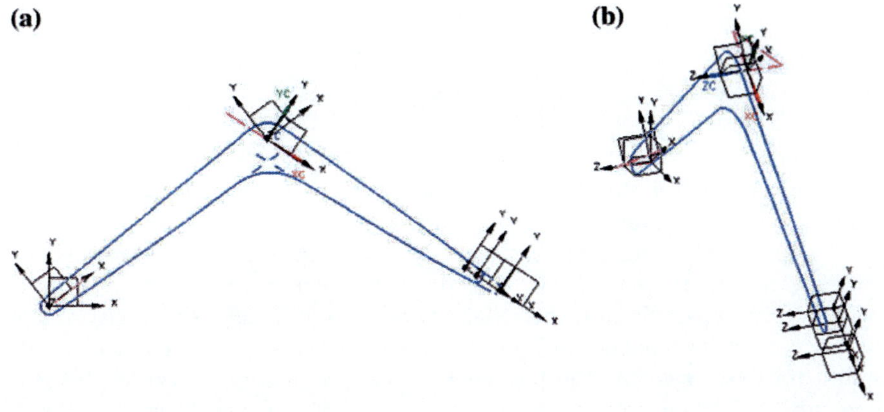

Fig. 27 Joined curve features

shown by the green line in Fig. 28, is projected onto the vertical *x*–*y* plane to form a guideline to be used for surface sweeping operation together with the existing one. This projected curve will serve the purpose of defining the right section of the boom as seen from the—*x* directions.

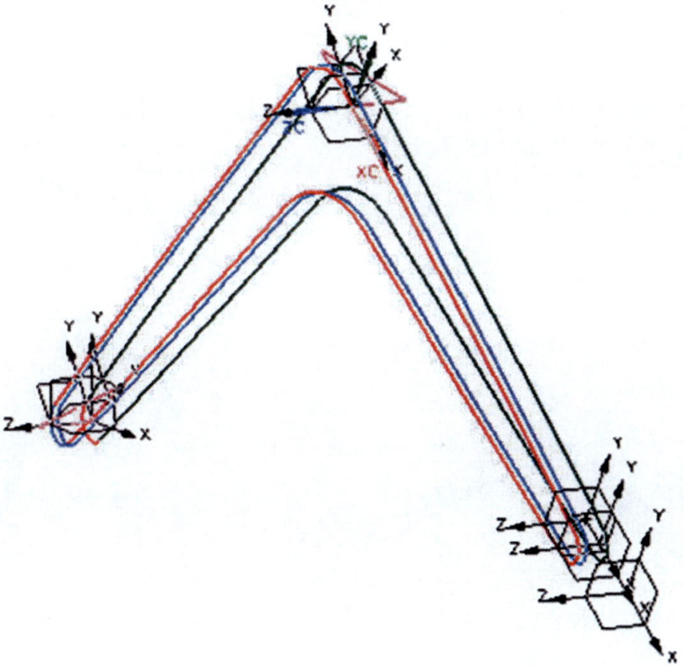

Fig. 28 Embodiment features

tag_t _Create_Projected_Curve(
tag_t curve_tag,
tag_t Datum_CSYS_tag,
int Plane_Num)

The third closed curve, colored green in Fig. 28, is created in a very similar procedure as that of the above curve but with an offset value added in the *z* direction to accommodate for a welding space.

The top and bottom floor surface features of the boom are generated by sweeping a linear curve guided by the red and the green curves. To avoid potential modeling errors associated with availability of multiple solutions for a given input to the responsible function, this process was carried out in two stages. The side wall surface was created from bounding curves. The functions used for this purpose are:

*tag_t _Join_Curves(tag_t *curves,int n)*
tag_t _Point_from_Spline(tag_t curve_Tag, int line_end)
tag_t _Lines_from_two_points(tag_t point1, tag_t point2)
tag_t _SWEEP_2_guides(tag_t Guide_s1, tag_t Guide_s2, tag_t Section)
tag_t _BPLANE(tag_t Curve_String[2])

Hinge joint features are created by sketching and extruding their profiles. The boom has two types of hinge joints: one that passes through the plate walls and one that is welded externally to the boom structure.

Joint J1 is constructed by introducing a hollow cylindrical feature of appropriate dimensions to the boom structure while joints J2, J10, and J12 are constructed from scratch by sketching and extruding their profile (Fig. 29).

Functions used for this purpose include:

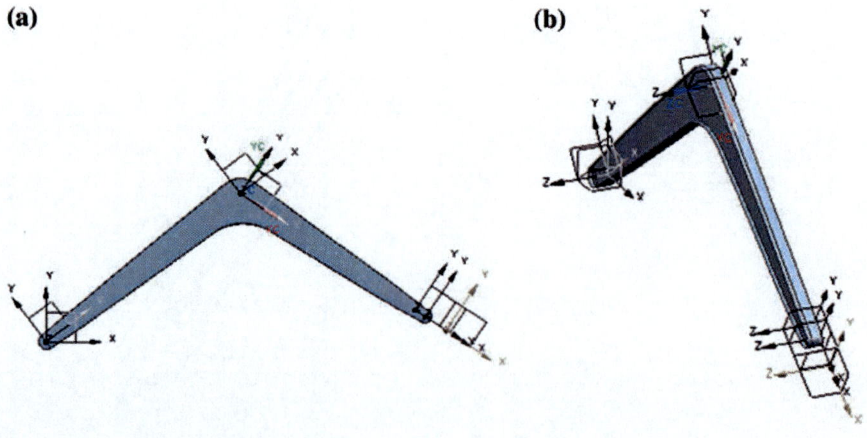

Fig. 29 Sheet body features

```
tag_t   _SKETCHES_J2_adopter(
        char    name[30],
        tag_t   Refrence_CSYS,
        int     Plane_num,
        int     Axis_num)

        tag_t   _ARC_on_sketch(tag_t Reference_CSYS)

        tag_t   _ARC_Center_Radius(tag_t Reference_CSYS,
        int     Plane_num,
        double  radius,
        double  arc_center[3],
        double  start_ang,
        double  end_ang)

tag_t _ARC_Point_Point_Radius(
        tag_t   Reference_CSYS,
        int     Plane_num,
        double  radius,
        double  arc_center[3],
        tag_t   p1,
        tag_t   p2)

tag_t _Extrude(tag_t Connected_Curve, char* limit[2])

tag_t _SKETCH_J11(
        tag_t   Reference_CSYS,
        int     Plane_num,
        tag_t   line_Tag,
        tag_t   bridge_tag)

tag_t _SKETCH_J12(
        tag_t   Reference_CSYS,
        int     Plane_num,
        tag_t   Curve_Tag);
```

The sheet bodies are thickened and the joint sketches are extruded with initial and final limits to create the final solid body features. After the necessary modification on the joint solid features, the left side of the solid boom is created by merging the individual solids with each other (Figs. 30, 31).

Custom functions used for these operations include:

```
tag_t THICKEN_Sheet(tag_t sheet_body_tag)
tag_t _UNITE_SOLIDS(tag_t Target_Solid, tag_t Tool_Solid)
```

This left side of the boom is mirrored about the x–y plane to create the other half of the boom. The original and the newly created halves are then merged to create the final boom solid body shown by Fig. 32.

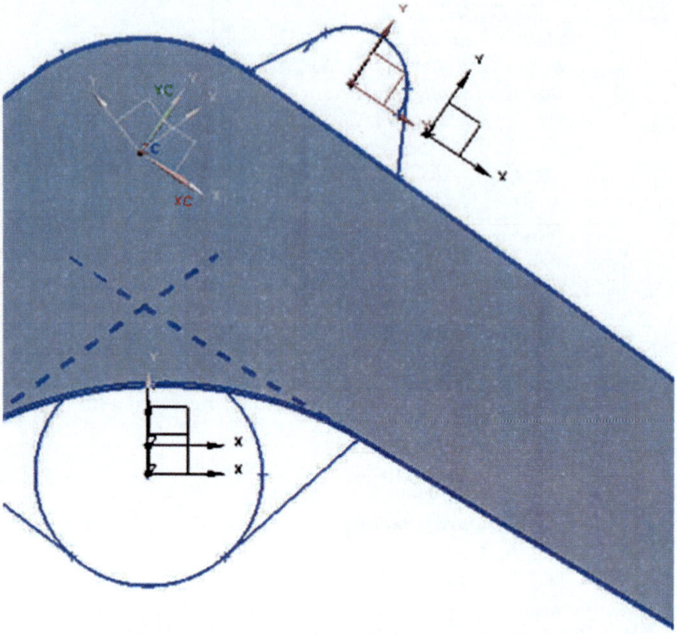

Fig. 30 Hinge joint profiles construction

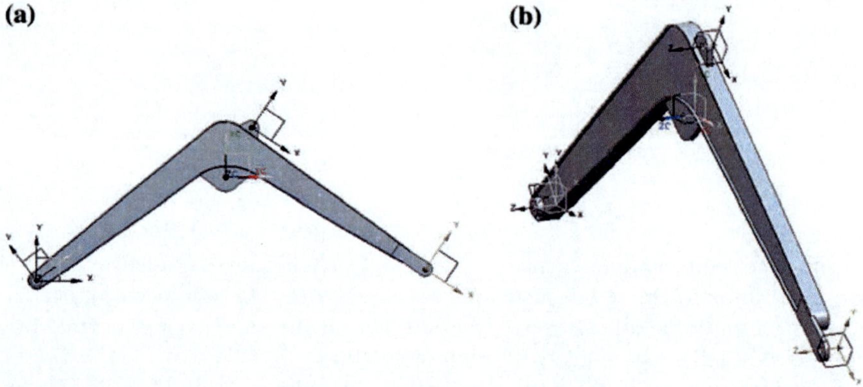

Fig. 31 Solid body features

12.3 Stick CAD Modeling

The programming and part creation procedures followed for the stick are very similar to the one adopted for the boom. Most of the functions developed are reused directly or, in some instances, with minor modifications to address stick-specific modeling needs.

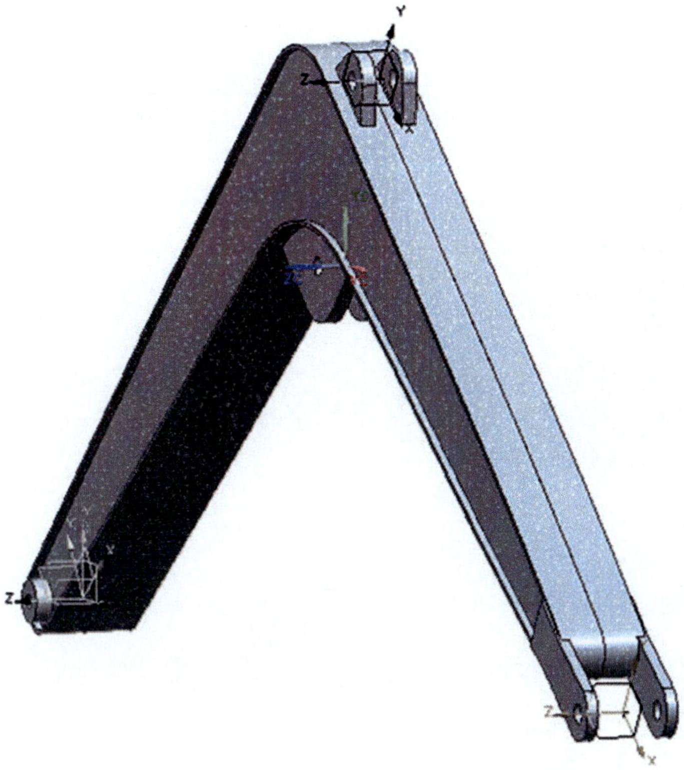

Fig. 32 Final CAD model of an excavator boom

As was done for the boom, the modeling process for the stick started by creating a new part file called Stick.prt using the same function. Data were imported from the intermediate files using similar procedures. User-defined CSYSs were created at the joint locations and some critical locations necessary for the creation of sketches.

Generally, the stick construction procedure is easier than that of the boom because of the parallelism of the stick CSYS features (see Fig. 33).

The arcs of the stick at the joints J2, J3, and L9 were constructed first based on the data imported from the neutral text files. Profiles defining the edges of the stick were created by joining these arcs with the maximum middle point straight tangent lines as shown in Fig. 34.

After performing some line modification operations, such as trimming and joining, the created closed loop curves are projected onto two different planes positioned parallel to the middle *x*–*y* plane.

Figure 35 shows the cleaned and projected stick profile curves. The green and blue curves will be used as guides to create a sheet body by a sweeping operation

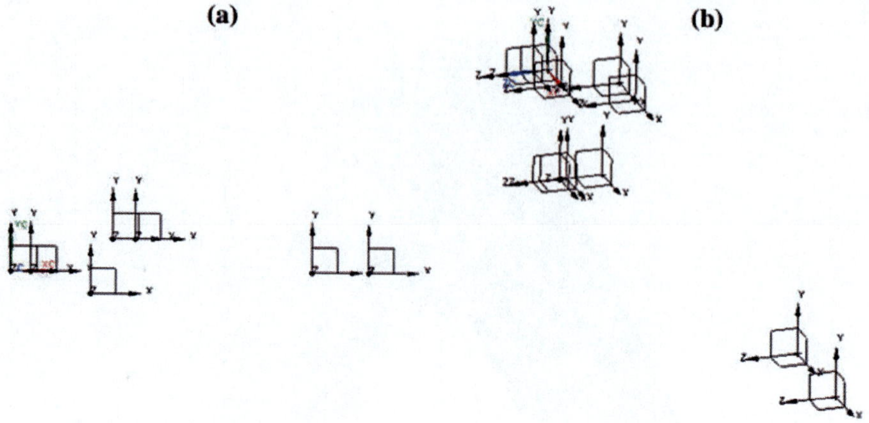

Fig. 33 Stick coordinate system features

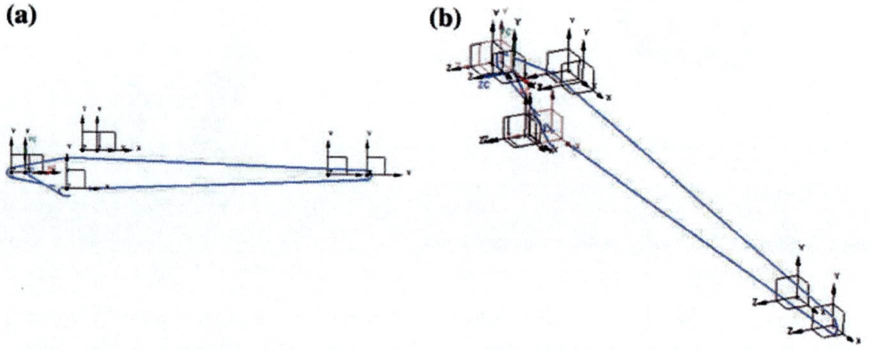

Fig. 34 Feature evolvement

while the red curve will be used as a boundary when creating a bounded plane. The pink closed curve will be extruded to create the hinge solid for joint J9.

A line element parallel to the z-axis was created and used for the sweeping operation. The following figures show the sweeping tool and the resulting sheet body features (Fig. 36).

The left side wall is created by using the other projected curve as a boundary in the bounding plane operation. These planes are thickened with appropriate thickness values and with consideration for necessary inference tolerances. Joints are created using similar procedure as used in the boom modeling. The final left half of the boom is shown in Fig. 37.

The final complete stick solid body feature is created by merging individual extruded and thickened solid features into a single part and mirroring this part

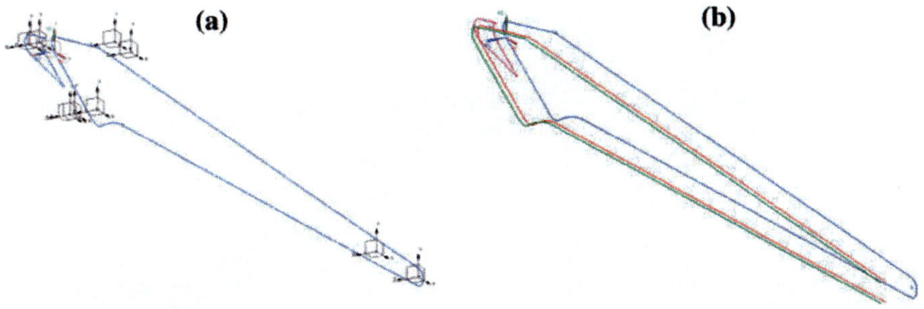

Fig. 35 Embodiment features

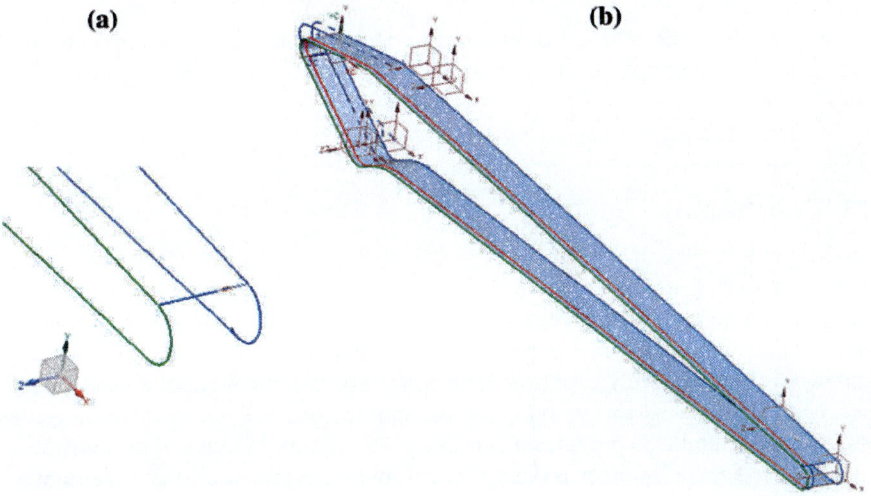

Fig. 36 Stick sheet body features

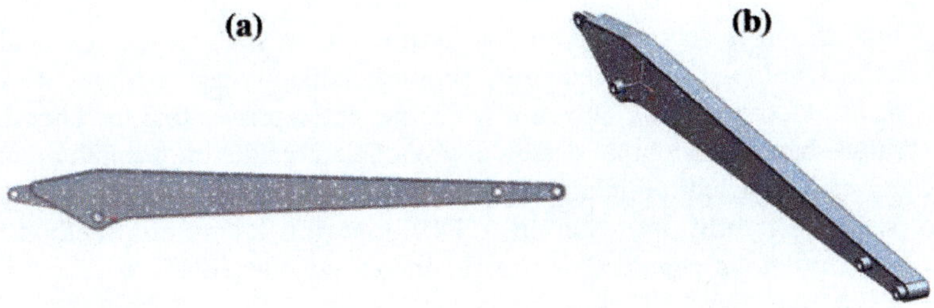

Fig. 37 Stick solid body feature

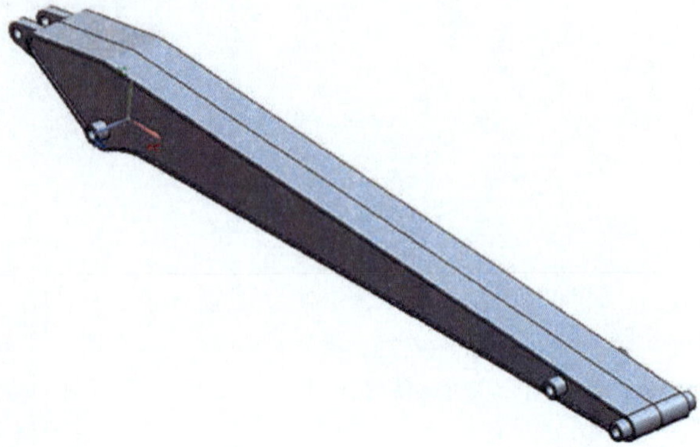

Fig. 38 Final CAD model of an excavator arm stick

about the x–y plane. The mirrored and its parent solid are converted again into single solid by merging them with similar command to get the final model shown by Fig. 38.

13 Conclusions

The proposed generative feature-based conceptualization method for product development is a promising response to addressing the issues found in the current information-fragmented and experience-based practice. The method discussed in this work can effectively capture engineering rules and facilitate the design cycle processes through insightful configuration optimization and embodiment development. The method also provides much-needed flexibility in terms of customization and standardization of other products which involve frequent changes. Reusability of the developed functions has provided evidence that, unlike traditional modeling methods, the knowledge in the design stages can always be embedded, harnessed for a new product, and be reused when developing future generations of similar products. However, it is worth noting that in the authors' opinion, regardless how intelligent a design system may be in the future, human design expertise is always required; the developed system can only support decision making more effectively with some productivity tools. The case study presented proves that the knowledge-driven feature-based conceptual design approach can handle traditionally complex challenges such as machine linkage optimization problems.

The proposed hybrid optimization-ANN dimensional synthesis method introduced in the previous chapter has greatly increased the reliability of calculated solutions. Training the ANN with larger size of data collected from the existing products in the market is believed to produce solutions reflective of common design intents and common industrial trends. The hybrid ANN-Optimization method has been proven to provide satisfactory results by generating close initial solutions of the linkage dimensions). The optimization procedure that was also introduced in the previous chapter was employed to calculate the final linkage dimensions) that satisfy the customer's specification feature requirements. This hybrid method can assist in generating optimal solutions in an integrated design environment.

Together with the previous chapter, this feature-based smart mechanism design method provides a novel approach to solving similar problems in the product design domain, i.e., a hybrid linkage dimension synthesis method. The general procedure can be followed for the conceptual design of other mechanisms, as all the process modules will remain valid. The only exception would be the details of design calculations and optimization criteria since they are usually very specific to the product under discussion. Regardless of the product being designed, the procedure is scalable.

14 Future Works

Formal definitions of generic conceptual design features need to be investigated such that embedded engineering rules, constraints, data representations, and behaviors can be modeled and managed generically in an object-oriented approach and systematically implemented.

Programming and engineering analysis tools such as Visual C ++ and MATLAB can be integrated with feature-based tools such as Siemens NX so that the analysis procedure can be part of the integrated conceptual design system. Their input and output as well as the constraints can be automatically managed according to the formal definition of concept problems.

The conceptual design process discussed was based only on the mechanical design aspect of the case study. The data structures and communication mechanisms can be similarly used in designing other aspects of the product. For example, the conceptual level design of hydraulic circuit subsystem in the case study can also be modeled and solved under the proposed data and information management scheme by implementing its own conceptualization contents.

Acknowledgments The presented research work was partially supported by Canada Natural Sciences and Engineering Research Council of Canada (NSERC) discovery grant (No. 355454-09) and the University of Alberta GRF (G121140079) grant.

References

1. Basak H, Gulesin M (2004) A feature based parametric design program and expert system for design. Math Comput Appl 9:359–370
2. Bronsvoort WF, Bidarra R, Van DM (2010) The increasing role of semantics in object modeling. Comput Aided Des Appl 7:431–440
3. Danjou S, Lupa N, Koehler P (2008) Approach for automated product modeling using knowledge-based design features. Comput Aided Des Appl 5:622–629

4. Durupt A, Remy S, Ducellier G (2010a) KBRE: a knowledge based reverse engineering for mechanical components. Comput Aided Des Appl 7:279–289
5. Durupt A, Remy S, Ducellier G (2010b) Knowledge based reverse engineering: an approach for reverse engineering of a mechanical part. J Comput Inf Sci Eng 10:044501
6. Larsen S, Jensen CG (2009) Converting topology optimization results into parametric CAD models. Comput Aided Des Appl 6:407–418
7. Li M, Zhang YF, Fuh JYH (2010) Retrieving reusable 3D CAD models using knowledge-driven dependency graph partitioning. Comput Aided Des Appl 7:417–430
8. Lourenco D, Oliveira P, Noort A (2006) Constraint solving for direct manipulation of features. Artif Intell Eng Des Anal Manuf 20:369–382
9. Ma YS, Britton GA, Tor SB (2007) Associative assembly design features: concept, implementation and application. Int J Adv Manuf Technol 32:434–444
10. Ma YS, Chen G, Thimm G (2008) Change propagation algorithm in a unified feature modeling scheme. Comput Ind 59:110–118
11. Ma YS, Tang S, Au CK (2009) Collaborative feature-based design via operations with a fine-grain product database. Comput Ind 60:381–391
12. Mantyla M, Nau D, Shah J (1996) Challenges in feature-based manufacturing research. Commun ACM 39:77–85
13. Myung S, Han S (2001) Knowledge-based parametric design of mechanical products based on configuration design method. Expert Syst Appl 21:99–107
14. Ong SK, Shen Y (2009) A mixed reality environment for collaborative product design and development. CIRP Annals: Manuf Technol 58:139–142
15. Pratt MJ, Anderson BD (2001) A shape modelling applications programming interface for the STEP standard. Comput Aided Des 33:531–543
16. Riou A, Mascle C (2009) Assisting designer using feature modeling for lifecycle. Comput Aided Des 41:1034–1049
17. Shigley JE, Misheke CR (2001) Mechanical engineering design. McGraw-Hill Higher Education, Inc, New York
18. Singh DK, Jebaraj C (2008) Feature-based design for process planning of the forging process. Int J Prod Res 46:675–701
19. Solazzi L (2010) Design of aluminium boom and arm for an excavator. J Terramech 47:201–207
20. Ter Hofstede AHM, Lippe E, Van DW (1997) Applications of a categorical framework for conceptual data modeling. Acta Inform 34:927–963
21. Thakur A, Banerjee AG, Gupta SK (2009) A survey of CAD model simplification techniques for physics-based simulation applications. Comput Aided Des 41:65–80
22. Van Der Meiden HA, Bronsvoort WF (2009) Modeling families of objects: review and research directions. Comput Aided Des Appl 6:291–306
23. Verdes D, Stan S, Manic M (2009) Kinematics analysis, workspace, design and control of 3-RPS and TRIGLIDE medical parallel robots. 2nd conference on human system interactions. Catania, Italy
24. Wang H, Zhou X, Qiu Y (2009) Feature-based multi-objective optimization algorithm for model partitioning. Int J Adv Manuf Technol 43:830–840
25. Wang Q, Li J, Wu B (2010) Live parametric design modifications in CAD-linked virtual environment. Int J Adv Manuf Technol 50:859–869
26. Wong LM, Wang GG (2003) Development of an automatic design and optimization system for industrial silencers. J Manuf Syst 22:327–339
27. Ye X, Liu H, Chen L (2008) Reverse innovative design: an integrated product design methodology. Comput Aided Des 40:812–827
28. Yoon J, Manurung A (2010) Development of an intuitive user interface for a hydraulic backhoe. Autom Constr 19:779–790

A Smart Knowledge Capturing Method in Enhanced Generative Design Cycles

G. P. Gujarathi and Y.-S. Ma

1 Introduction

Product development cycles involve numerous CAD and CAE interaction iterations to accommodate design evolvement, changes, and verifications. Since many common design considerations, constraints, and iterations are applied to the cycles for the same or a set of product configurations, it is useful to develop coded and generative programs to automate certain tedious and repetitive processes in order to minimize the time to market and the engineer's routine efforts. Such programs will also assure that a common design procedure is followed throughout the product lifecycle without redundancy or errors. The research question addressed is how the interactions can be structurally modeled, and the engineering knowledge captured and reused.

This chapter proposes a new method to capture and reuse engineering knowledge throughout CAD and CAE interactions with a generative approach. A design development process model was developed to record, interpret, and reuse embedded procedural knowledge with one-time initial interactive effort, and to facilitate CAD and CAE application program code creation. Further, CAD and CAE model updating processes are automated as much as possible by using a common data model (CDM). The proposed method is generic and systematic with combined use of parametric and procedural knowledge. Design parameters and analysis conditions can be managed and customized for specific design purposes and changes. It offers a design automation solution for those products with relatively predictable configurations and constraints.

G. P. Gujarathi
Suncor Energy Inc, Calgary, AB, Canada
e-mail: ggujarathi@suncor.com

Y.-S. Ma (✉)
University of Alberta, Edmonton, AB T6G 2G8, Canada
e-mail: yongsheng.ma@ualberta.ca

Y.-S. Ma (ed.), *Semantic Modeling and Interoperability in Product and Process Engineering*, Springer Series in Advanced Manufacturing,
DOI: 10.1007/978-1-4471-5073-2_11, © Springer-Verlag London 2013

Conventional design processes for product development use theoretical and analytical engineering calculation-based techniques. This traditional methodology usually involves iterative steps that are mostly performed interactively with the help of computer-aided design (CAD) and computer engineering analysis (CAE) tools and through a "trial and error" approach while trying to fit the generalized design practice to a specific problem. Currently, CAD modeling is well established with mature parametric design capability; design intent can be reflected by a set of design patterns defined with features. In turn, features are defined via parameters of dimensional values and various constraints [11]. Concurrently, engineering analysis and verification in conventional analytical processes are being complemented by CAE tools to which the modern computational methods, such as finite element analysis (FEA), are applied. Modern "easily accessible" and high power computing resources have enabled mass availability of popular numerical solutions for highly complex linear and nonlinear equations. Some commercial tools are Ansys, LS Dyna, and NX Nastran. CAE analyses provide good benchmarking measures to ensure that the design meets engineering requirements, although CAE results are not presumed to be so perfectly accurate as to predict real application scenarios. CAD technology coupled with CAE offers an effective cyclic product design approach with higher design flexibility and complexity than the traditional approach.

In advanced industrial practice, CAD and CAE processes are interwoven during product development cycles, as product development involves multiple iterations throughout the evolvement and change management processes. Performing CAD/CAE operations interactively requires a considerable amount of repetitive and tedious effort, and hence represents a drain on various kinds of resources.

The major limitations associated with such a loosely coupled and interactive approach consist of the following:

1. There is no systematic design process method to guide and manage the design evolvement that can effectively interpret the product model evolvement. For example, in the early stages of conceptual design, there is insufficient information on specific and localized "sensitive" issues; only later and gradually are the design contents enriched and stabilized with the product specifications and constraints. A design process model that can support the design contents, knowledge, constraints, and procedural logics from the beginning is in high demand.
2. While the traditional analytical design approach seems to be preliminary and rough in contrast to the demands of high material efficiency, product reliability, and design accuracy in the modern market, CAD and CAE interaction in design engineering presents a major hurdle for many companies. They struggle to increase productivity, but interactive operations for CAD and CAE activities are both complicated and time consuming. While CAD is more popularly used due to the necessity of creating the product geometry for downstream manufacturing processes, CAE analysis is considered a "luxury" due to the effort involved in setting up models and the computing time devoted to it. With the

cyclic nature of engineering design, the cost for fully CAE-supported product development is high, and only companies in high value-added industries, such as aerospace and the military, can afford it.

3. In current practice, it is quite difficult to implement certain design rules uniformly. Automation of design processes is a desired goal for many companies. Research in the effective integration of various computer application tools with effective interoperability and low technical skill requirements is highly demanded by industrial users.
4. There is a lack of optimization for design features and parameters. Most optimization has been done on the theoretical level, such as the thickness of the shell in a pressure vessel design; it has been too time consuming in interactive modeling and analysis to apply optimization for detailed design parameters and localized features, such as the stress concentration at some sensitive corners for material distribution considerations.

The goal of this research is to develop a practical methodology and an adaptive process model to accommodate numerous changes and "if-then" scenario explorations where the processes can be automated as much as possible. The proposed approach is to combine the parametric design techniques with knowledge-embedded computer programming. Since each design process follows a set of specified steps, a computerized reusable design development procedure can be used to develop different products with similar rules. To do this, it is useful to create a program to automatically generate CAD models and carry out CAE analysis processes. Then, considering the integration between CAD and CAE activities, a research question arises: how might those CAD and CAE interactions be structurally modeled and the engineering knowledge captured and reused automatically?

2 Research Background

Tomiyama et al. [17] have run a systematic review of various current design theories and methodologies for product development. Their study identified the insufficiencies of those methodologies, including lack of consideration for increasingly complex operational and geometric requirements, multidisciplinary collaboration, management of complex product development processes and information integration of various advanced technologies for computer-oriented design methods.

The feature concept [9, 12, 16], which is used to model template types of objects with semantic meanings in engineering design and manufacturing models, has been developed to associate geometric dimensions and other design-related parameters. When one of the feature parameters changes, it also requires modifications of other parameters and feature elements driven by the associated constraints accordingly. Hoffman and Kim [6] discussed the issues of under- and over-constrained models in CAD, and suggested an algorithm to compute the valid ranges of parameters and the numerous required constraints.

Historically, CAD and CAE data models have been developed separately in different software packages and defined with completely different data structures. To transfer product geometry definitions from a CAD to a CAE application, the geometry has to be further processed, e.g., converted to a mid-plane model, or simplified. Su and Wakelam [15] worked on creating an intelligent hybrid system to integrate various CAD, CAE, and CAM tools in a design process using a blend of rule-based systems. Artificial neural networks (ANNs) and a genetic algorithm (GA) were applied using a parametric approach for model generation as well as a rule-based approach to control the design logics.

Considerable research has been done to integrate geometric modeling and computer-aided analysis. Zeng et al. [19] suggested the use of a knowledge-based finite element (FE) modeling method to reduce design time, and suggested CAD-FEA integration at the knowledge level. They also stressed the importance of automation in the idealization of CAD and mesh generation. However, Johansson [7] identified some of the issues that their systems encounter, i.e., the compatibility with the available commercial CAD and CAE software tools, and difficulties in developing complicated models. Bossak [1] and Xu et al. [18] explored the feasibility of complete product development with integration of various computer-based technologies in order to create consistent and associated models. Cao et al. [2] developed middleware to transform CAD models into acceptable CAE mesh models, which is a process referred to as high end digital prototyping (HEDP). It can simplify and de-feature CAD models for FEA meshing, but the integration is one way and lacks recursive loop support.

One of the tasks involved in the design development process is managing semantic relations between various parameters while maintaining associativity between parameters and design features. Pinfold and Chapman [10] proposed a "design analysis response tool" (DART) to use knowledge-based engineering to automate the FE model creation process. The major objective behind this effort was to reduce the time associated with creating and modifying an FE model.

More recently, feature technology has enabled a parametric design approach [3, 4, 13] where the physical models of design geometry are coupled with associated dynamic behaviors. Each feature type is modeled to contain some generic and modular design semantic concepts, which reflect those commonly accepted patterns used in the engineering field. Geometric and nongeometric data structures are included to define various features, and they are parametrically associated among themselves across different aspects of a product model. Researchers have utilized knowledge based and unified feature information databases for information sharing, consistency, and control among different models [4].

Thus far, previous research has not considered the creation of the product model contents or changes to it over the dynamic cycles of evolvement. There is a notable gap in capturing accumulated engineering knowledge in the design processes and reusing the knowledge-rich process information in the iterative cycles of product evolvement via computer-executable codes. Unfortunately, the authors could not find any conclusive research that covers this focused research direction. Therefore, it can be concluded that the proposed research has strong relevance to enrich the theoretic domain of advanced engineering informatics automation and its effective applications.

3 Product Design Process Cycles

By definition, "Design is the act of formalizing an idea or concept into tangible information which is distinct from making or building" [8]. Any design process is a repetitive process that incorporates specific working and decision-making steps. The activities or sub-processes involved are commonly interlocked due to the evolvement of engineering details and to the availability of related information and decision results.

Product development involves many stages, from initial requirement collection to product functions and market information searches, generation of various solutions, calculations, CAD modeling, drawing generation, and evaluation. Finalizing the product requires several evaluation stages. If the results are not compliant with requirements, certain steps of the process need to be repeated to get better results. In order to save cost and reduce development time, these design loops must be effective and should be as efficient as possible.

3.1 Parametric Design Process

Parametric modeling allows for manipulation of model data on a microparametric level; an automated modeling process can be effectively used to propagate changes made in parameters to the specific area of the design object. The major advantages of parametric modeling are systematic control of the engineering design intent and the quick propagation of changes according to new input conditions, i.e., design changes. Parametric techniques can be employed within a number of software tools. Most available computer modeling and analysis software packages provide application programming interfaces (APIs). Model templates in the form of external executable files or functions can be called upon in the application sessions to generate product models or to run coded analysis tasks. This approach is widely known as the *generative approach* [2, 6, 15, 18]. With the help of APIs, automatic creation of computer models is much more convenient; in fact, automation of the design process has been increasingly adopted. The use of engineering knowledge embedded in application programs simplifies the automatic creation of design models via design features and the parameters reflecting the design intent, and offers better control over the product development processes.

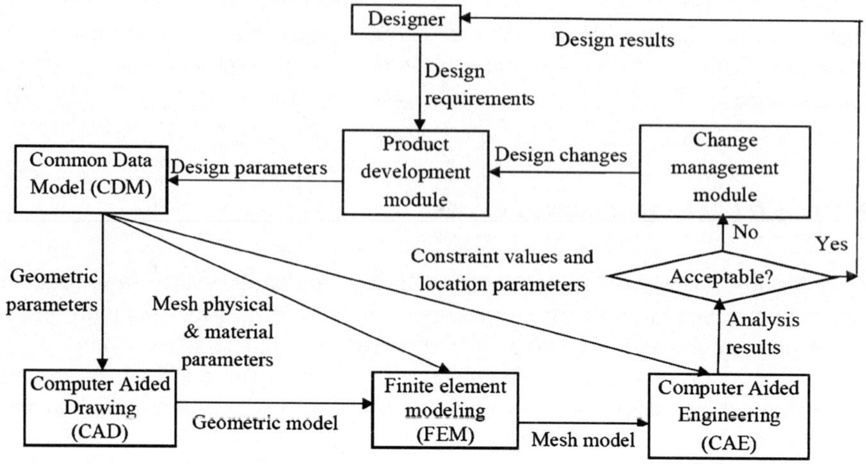

Fig. 1 Information flow diagram among various modules in the proposed design system [5]

3.2 *Design Information Flow and Sharing in a Design Cycle*

Figure 1 shows the modules of a computer-based design system proposed by the authors. The development of each process module depends upon the relative progress of succeeding and preceding modules. Figure 1 also illustrates the information flow between the modules. This proposed design process is semi-automated with the help of generative CAD and CAE programs using a centralized data repository called a CDM [5]. In order to make the entire process as flexible as possible, it has been designed to keep the information associated with every module in a neutral format for data sustainability.

4 Purpose of the Research

The purpose of this research is to achieve the partial automation of CAD modeling and analysis processes by recording predefined procedures, then creating reusable program templates through programming coupled with parametric modeling for changes in design conditions. Theoretically, it is preferable to use a single software tool to maintain the associativity between modeling and analysis. The diversity of applications makes a complete and coherent product development solution too complicated to be handled conveniently. This is because of the differences of model definitions associated with different software tools; there are always inconsistencies in mapping the transformation of models from one to another. Most commercial software tools cannot uniformly support all the engineering areas, and thus the designer has to work with two or more software tools concurrently and collaboratively. To support effective software functionality and sustainability, it is essential to develop a unified data structure [5]. User-defined information and knowledge should be represented in a neutral, reusable, and scalable data structure. In addition, for a generative product development system with multiple applications or modules, a user-defined procedure should be interfaced with all the software APIs, such as those related to the modeling and analysis software tools.

5 Proposed Design Process Model

5.1 Cyclic Design in Conceptual and Detailed Stages

Although an engineering design process is a continuous evolvement of changes, in most engineering design projects, design activities and deliverable models can be divided into stages, such as the conceptual design stage and the detailed design stage. The conceptual design stage creates a workable solution that verifies the physics principles, solving the problem of design feasibility. The detailed design stage completely defines the product with all manufacturing components and assembly details; phase by phase, they are fully optimized according to different considerations such as assembly, outsourcing, cost, and so on. Hence, as to the contents of the deliverables for each stage or phase, the interested entities and their related constraints which are created and managed via the integrated and synchronized design model, are quite different from those in other stages or phases. The characteristic measures include the level of intelligent representation, geometrical completeness, refinement of justification and documentation, etc. When a product model is represented in the conceptual stage, for example, a pressure vessel with nozzles is represented as a thin shell surface model with a constant thickness assumed and little consideration for the manufacturing issues and the final shape factors. Such a simplified evaluation model is referred to as abstract conceptual model. The abstract conceptual product model can also have sub-models including a CAD abstract model, a preliminary process model, a mechanical engineering analysis model, and so on. Similarly, when a product reaches the production stage, its model represents the final design and manufacturing drawings with the complete representation before outsourcing and in-house manufacturing. The deliverable models at that time are then collectively referred to as the detailed product model, which has the complete definitions for the product, comprehensive engineering calculations, and fully developed 3D solid model geometry as well as the exhaustive 3D CAE analysis procedures and results. To handle the evolvement processes from conceptual to detailed design stages and their corresponding deliverables for other stages, different computer system models can be used; such models include application software tools, routines, representation schemes, data structures, and analysis functions or algorithms.

5.2 CAD and CAE CDM

As discussed above, integrating CAD and CAE data models is helpful in unifying the product's key parameters and their related constraints, so that the design processes and built-in engineering knowledge can be captured and represented coherently in a programmable manner and reused in both CAD and CAE models. That is why a CDM has been suggested, as shown in Fig. 1; the details of CDM structure and implementation have been introduced by Gujarathi and Ma [5]. Essentially, CDM can be understood as a central data structure that stores all the user-defined parameters and their values related to all stages of product development, as well as the explicit references and constraints among them. CDM allows for the neutral and coherent integration of data with explicit relations that can be easily interfaced with any of the functional software tools, such as CAD, FEM, and CAE packages. Due to the centralized and shared nature of the CDM, the product modeling in each of the software tools deals with only a specific view of the total information model, and the user programming for the specific product model becomes much more manageable.

During the design processes, CAD modeling serves the purpose of providing the product's initial geometry, its visualization model, and geometric inputs for the FE meshes. The additional information required for generating the mesh, such as physical and material properties of the design object, are taken as parametric input from the CDM. The FE mesh is then used in an analysis environment to apply the required loads and constraints and to carry out the analysis calculations. Similarly, to perform the analysis, additional associated information can also be taken from the CDM. Thus, parametric input from a neutral CDM allows for the consolidation of user-defined product models with specific information for various software tools. In a template structure, associations of specific data in the CDM to the modeling or analysis tools can be conveniently achieved by data and file-handling functions.

5.3 Integrated CAD and CAE Processes Via the CDM

The proposed design process is semi-automated with the help of generative CAD and CAE programs using the CDM (the centralized data repository) [5]. The design methodology integrates two different design cycles to increase the effectiveness and efficiency of product development.

The proposed process model begins with design parameterization. All the designer's requirements and product operational conditions are taken as input for the program to automatically calculate all the parameters associated with the design. To achieve this, the program uses embedded engineering design knowledge and engineering calculations along with necessary professional regulatory codes and industrial standards.

Once an initial set of parameters is developed and stored in the CDM, the next stage is the construction of an abstract conceptual model to confirm the basic physics principles via calculations and structures of the design object. The abstract CAD model illustrates a general idea of the product and serves as the geometric input for FEA. Both the CAD model and FE mesh model are generated automatically with software APIs coupled with parametric modeling. The CDM created previously is used as the parametric input to generate these models. The parameters and the values of the newly generated conceptual models are appended to the previous CDM in a structural form.

The next step is to generate the initial finite element mesh model from the previous conceptual CAD design. The conceptual numerical CAE analysis is then carried out. The results from the analysis provide the designer with the indicative structural and performance measures, and verification is done by the designer such that the design integrity is checked. The designer can select the necessary course of action from the available strategies for change management. Once the necessary changes have been made to the design, the program code recalculates all the parameters and stores a new version of the CDM, which is used to generate a new set of CAD models and the CAE models for analysis. The analysis results are again provided to the designer for verification. The cycle continues until the desired results are obtained. This kind of design cycle is referred to as the conceptual design interaction cycle.

Once the design process passes through the conceptual phase, it then enters the next stage, the detailed design cycle. A more comprehensive product model is created using the refined CDM from the conceptual design phase. The model created in this phase is to resemble the actual product definition as closely as possible. The detailed CAD model is then used to create a full 3D FE mesh model which is used to perform detailed CAE analysis. Such detailed computer models and analyses are also generated automatically through parametric modeling with CDM as a centralized parametric source. The development process goes through multiple iterations until satisfactory results are obtained. Once finalized, the final CAD models along with analysis data and final CDM records are documented to the user for further use and for knowledge reuse in future projects.

6 Knowledge Capturing and Reuse

6.1 Knowledge Capturing Processes

In order to fully utilize the potential of product modeling automation, the program development process needs to be recorded and further parameterized, so that it can be reused for similar design problems. Two forms of knowledge are involved in the development of computer-based design processes: (1) automated parametric knowledge captured in the form of accumulated parameters and their values that are specific to the product, as discussed below; and (2) procedural information or knowledge for design object handling and flow mechanism.

The automation of handling parametric data and knowledge has been proven feasible in a programming environment [3, 4]. However, since a design process involves a number of iterations, the creation of a coded program for model generation involves complicated logical reasoning and consumes a great deal of programming resources. This technical barrier leads to a reliance on traditional, manual, procedural, and interactive methods in industry. In order to achieve the automation of design procedures, it is desirable to develop an easy method to capture and reuse procedural knowledge in a computer interpretable and executable form.

During product design, many computer-based interactions use expert knowledge to achieve valid and efficient designs and to facilitate manufacturing processes. Such knowledge inputs are reflected by the features and their parameters. Engineering design parameters, which constitute part of the engineering knowledge, were taken care of by using the CDM [5]. For example, the initial stage of a pressure vessel design requires following engineering design rules, regulatory codes, and standard sizes. In addition, a detailed design procedure reflects equally critical information or knowledge such as those sequenced operations, options chosen, and parameter values used. The proposed idea is to program the above-mentioned CAD and CAE interactions with a one-time effort, and incorporate them into an automated procedure that serves as the captured procedural knowledge while following an accepted industrial design methodology. Once the program has been developed, it can be reused to calculate all the required design parameters automatically, and to store the parameters and their values in an external neutral file format, i.e., the CDM.

The modeling and analysis parameters can be further changed during the design process; however, the data structure incorporated and the procedural knowledge required is largely unchanged. With the proposed method, the procedural knowledge can be captured and reused as an executable program with the one-time interactive and GUI-based effort.

It is expected that during the design cycles, the design processes will require expert judgment and change management knowledge for design evolvement and modifications. For the current study, this is achieved by providing CAE analysis results directly to the user; an interface for the designer is created to accommodate the intended design changes automatically. The interface program recalls and validates embedded engineering formulae.

6.2 Knowledge Capturing and Reuse Mechanism

As discussed above, one of the most challenging tasks involved with automation of the design process is the coding effort for the generation of the CAD geometric model, FE mesh model, and CAE analysis model. With the increasing complexity in modeling and analysis, API-based programming becomes more and more lengthy and requires the user to have detailed knowledge of model development and programming. These obstacles can be overcome by using the journaling application in modern software tools coupled with parametric modeling.

In the proposed method, a CAD model for the design object is first developed interactively through the software tool graphical user interfaces (GUIs). At the same time, the entire modeling procedure is recorded into a journal file, which consists of program functions and arguments according to the software command interaction protocol applied. The journal file thus records all the steps taken. In fact, many software tools have developed such journaling applications, and the recorded commands can be embedded into a program code, which can easily be used to develop CAx models through API automatically, in a so-called generative approach.

6.3 Recording Interactions Between CAD and CAE

In the first iteration of the design process, every product modeling and analysis operation needs to be carried out manually through the software GUIs while recording every step taken by the designer using the journaling application. For example, with Siemens NX software, the journal file contains application functions and can be generated automatically in the form of a required programming language such as C++. In other words, the journal file contents are command operation records, generated in the background, and follow the exact steps and logic used by the designer during interactive processes.

Although the recorded journal file can be directly used to reproduce the created scenario, this capability is only for strictly re-running the user-computer interactive steps. To make the journal file more flexible, it has to be post-processed manually, so that every property of the model and analysis settings can be imported from an external data repository, i.e., the CDM for the design. In such a way, the journaling application provides a ready structure of programming functions involved with the modeling process in the required logic flow coupled with variable parametric modeling. Figure 2 shows the program development process for automatic generation of the computer models and analysis.

In order to create the program codes, a specific set of steps during interactive modeling has to be designed and tested before recording the journal file. To develop a well-defined process model, the user has to know the basic logic and algorithms required for constraint solving and management. It is essential for the designer to predetermine the engineering modeling and design development processes. The first step in creating a journal file is making a reproducible "fresh start point" by initialization of the application recording session. For example, all the previous data in the form of history must be erased, so that no previous data get recorded into the journal file. The starting setting conditions for each journal file must be well documented and assured. The designer should use those available

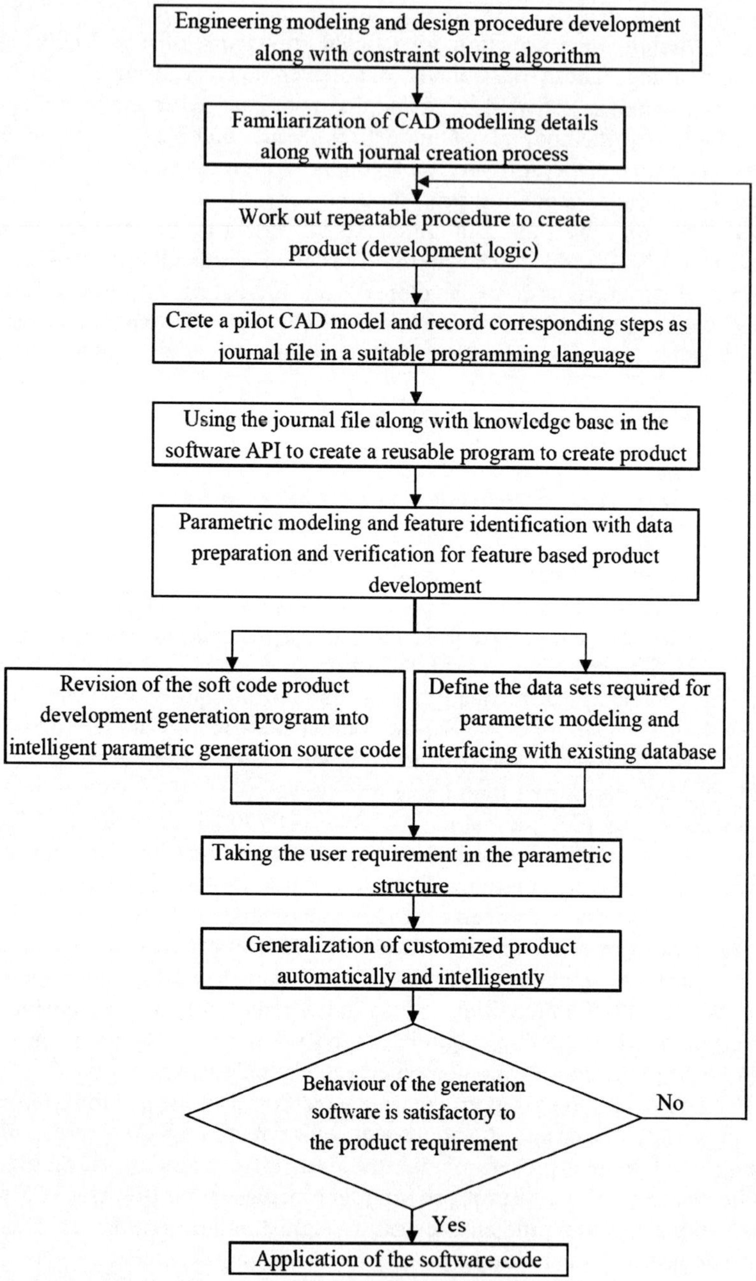

Fig. 2 Modeling and analysis program development flow using a pre-recorded journal filejournal file

functions compatible with the journal file recording application as much as possible. The designer should also be familiar with the nature of the journal file, and know how to separate the useful portion for creating the reusable program in the next step from the rest of the file.

A journal file provides a list of all the process steps and data in an orderly manner. By incorporating the journal operations into a compilation structure and organizing them in programmable logics, a program can be developed with an appropriate reusable data structure.

When developing the model through GUIs, the designer needs to be sure about the corresponding associations between the parametric input requirements of feature-related modeling and the feeding data from the external data file (CDM) as the new input for parametric modeling [5]. By associating journaling with parametric modeling, the program code generated becomes flexible enough to follow changes made into the external data file as the input data set. Each of the program codes generated is then added to the proposed design process structure. It is clear that using the external data file to tune the procedural knowledge embedded journal program (and hence to manipulate the design models and analysis procedure) can create a flexible, reusable, automatic, and customizable product development procedure.

6.4 Conversion of Journal Files Into a Reusable Program

By integrating the recorded journal files into a programming environment and associating each command operation with the corresponding API library functions, the journal file can be used to develop a compiled program. This program then can be used to generate design models and run analysis routines automatically. The relevant API functions are incorporated into the program structure through the specific individual header files. This guarantees that the program follows the same logic as the actual manual process and also ensures that the same model development process is followed each time.

Since the program exactly reflects the steps taken by the designer during the pilot interactive procedure, any specialized step required to create the model is automatically embedded into the program, thus preserving the expert knowledge associated with the process while making it reusable. By capturing the designer's knowledge and intent associated with the procedure, this compiled program enables generalization of the customized product automatically and intelligently.

A list of input and output parameters of different types involved with model development can be derived from the journal file. In order to integrate the CDM with the reusable program developed, those input and output parameters need to be converted into a data structure that can be compiled. Therefore, various data type extraction and variable changes are required, in line with the CDM definitions [5], as outlined here:

1. *Recording journal files step by step*. Users usually go through a single recording procedure to create a "whole" journal file if the modeling task is not complicated. In fact, a complex interactive design or analysis session can be recorded portion by portion, and then the journal files can be merged. The only condition that the user has to be aware of is that the start and stop points of the journal files correspond to the exact interactive operation state in the CAD or CAE session. If the user makes a mistake, the recording can be stopped. After cancelling the unwanted operation in the application session, the user can then edit the recorded journal file by deleting the unwanted command. The user can then continue the recording by using a new journal file. The previous journal files can later be merged into a final complete journal file by simple text editing.
2. *Creating a user-defined program environment for reusable code*. The CAD and CAE software tools used are Siemens NX v6 and NX Nastran. To code the NX for the integrated external program, the user must first create a user-defined program entry section in the main program development environment (Visual C++) which can be compiled (see Fig. 3). A programming template provided by the NX Open module is used. The recorded journal file contents are then copied into the user-defined program section.
3. *Replacing hard-coded inputs with variables*. A journal file records all the macro commands, which are the operation steps that the software takes while modeling the product. During the process, all the expressions and associated functions in the form of "hard numbers" are recorded, as shown in Fig. 4. To make use of the CDM parameters and their values, those hard-coded inputs must be replaced with variables that are to be imported into the generated model from CDM data files as expressions. Note that the variable names defined in the reusable program, which are to be interpreted in the NX API functions as modeling expressions, have to be associated with the corresponding variables to be imported from the CDM.

```
extern void main( char argc, char *argv[] )
{
    /* Initialize the API environment */
       if( UF_CALL(UF_initialize()) )
       {
              /* Failed to initialize */
              return;
       }
    /*Application code starts*/
       Session *theSession = Session::GetSession();
       ...
       ...
    /* Terminate the API environment */
    UF_CALL(UF_terminate());
}
```

Fig. 3 Creating a program environment for reusable code

Journal

```
/* Hard coded functions with variables*/
  expression2417->
  SetRightHandSide("15.25");
  expression2418->
  SetRightHandSide("26.00");
  expression2419->
  SetRightHandSide("-29.0");
  ...

/* Perpetual trial and error based recorded functions*/
  Expression *expression2153;
  expression2153 = associativeArcBuilder22->Radius();
  expression2153->SetRightHandSide("94");
  ...
  Expression *expression2170;
  expression2170 = associativeArcBuilder22->Radius();
  expression2170->SetRightHandSide("100.5");
  ...
```

Reusable code

```
/* Hard coded functions with variables*/
  expression2417->SetRightHandSide
  ("Saddle dimension A/2");
  expression2418->SetRightHandSide
  ("Saddle location 1-(Saddle dimension C/2)");
  expression2419->SetRightHandSide
  ("-Saddle dimension B");
  ...

/* Perpetual trial and error based recorded functions*/
  Expression *expression2170;
  expression2170 = associativeArcBuilder22->Radius();
   workPart->Expressions()->EditWithUnits
  (expression2170, unit1, "Gas nozzle diameter+Gas nozzle thickness");
  ...
```

Fig. 4 Replacing hard-coded functions with variable input

4. *Deleting unnecessary data and functions.* As the journal file is developed during the process of creating a model through a GUI, it records graphical operations as well, such as modeling view changes. These operations are not necessary for the reusable modeling program, and should be deleted. The journal file also records the intermediate and temporary functions and variables generated, as shown in Fig. 5. Since such information is not required in the reusable program code, it needs to be removed.
5. *Grouping similar sections and functions together.* A journal file records the model development process step by step, and thus keeps track of each individual process. Hence, a journal file often contains repetitive functions in the form of step-by-step incremental commands. In order to simplify the reusable code and enhance the efficiency of execution, all the repetitive functions and even sections are restructured as logic loops in the compiled program and sorted out together. Some commands have to be cleaned up, as shown in Fig. 6.

```
/* Function for modeling view orientation*/
  ModelingView *modelingView1;
  modelingView1 = workPart->ModelingViews()->WorkView();
  modelingView1->SetRenderingStyle
  (View::RenderingStyleTypeStaticWireframe);
  modelingView4 = workPart->ModelingViews()->WorkView();
  Point3d scaleAboutPoint1(39.8975562618742, 21.9294807031849, 0.0);
  Point3d viewCenter1(-39.8975562618742, -21.9294807031849, 0.0);
  modelingView1->ZoomAboutPoint(0.8, scaleAboutPoint1, viewCenter1);
  ...

/* Intermediate expressions*/
  workPart->Expressions()->SystemRename(expression84, "p25__x");
  workPart->Expressions()->SystemRename(expression85, "p26__y");
  workPart->Expressions()->SystemRename(expression86, "p27__z");
  ...
```

Fig. 5 Deleting unnecessary data and function

Journal

```
  Section *section14;
  section14 = joinCurvesBuilder1->Section();
  section14->SetDistanceTolerance(0.001);
  ...
  Section *section15;
  section15 = joinCurvesBuilder1->Section();
  section15->SetChainingTolerance(0.00095);
  ...
  Section *section16;
  section16 = joinCurvesBuilder1->Section();
  section16->SetAngleTolerance(0.5);
  ...
```

Reusable code

```
  Section *section14;
  section14 = joinCurvesBuilder1->Section();
  section14->SetDistanceTolerance(0.001);
  ...
  section14->SetChainingTolerance(0.00095);
  ...
  section14->SetAngleTolerance(0.5);
  ...
```

Fig. 6 Grouping data sections and functions together

The incremental expressions are also examined, and repetitive steps are replaced by combined equivalent cycles; again, if possible, always replace "static" numbers with variables.

6. *Associating the program code with appropriate ".dll" files*. In the NX environment, when the journal file is recorded, it includes lists of all the header files required. In order to compile the program code, the header files need to be included; the corresponding ".dll" files (dynamically linked library files) are also to be included in the program's compiling and debugging directory.

7. *Maintaining associative relations among design parameters*. The associative relationship among design features can be maintained by developing a semantic map with a well defined and semantically interconnected design parameter scheme. The parameters are interconnected with engineering design concepts and constraints. With well-defined parameters, constraints and interdependencies between various features can be easily verified, checked, and maintained.

7 Case Study

The proposed method is applied for the design of separator vessels, which are used in many processing industry plants. There are various types of separators, including horizontal, vertical, and spherical. Separators are designed and manufactured taking into account the specific advantages and limitations of different configurations. The basic criteria for the selection of configuration are optimized lifecycle cost, operational safety, and accordance with design codes and standards for the required design domain. Separators are usually customized according to the specific operational requirements, and thus every separator requires particular design, but the overall design procedure and engineering considerations are similar. Thus, it is convenient to automate the design processes with a common programmed procedure.

7.1 The Case Set: Horizontal Separators

Figure 7 shows the basic layout of a horizontal separator [14]. The mixed flow of gas and liquid enters the separator and impacts on the inlet diverter, causing a sudden change in momentum and hence the preliminary separation. The larger droplets of liquid then separate over the gravity settling section and fall to the bottom liquid collection section. The liquid is given enough retention time for the dissolved gases to escape from the liquid gathered at the bottom and to collect in the vapor space above.

A separator vessel is designed according to gas capacity constraints. Each type of separator has its own merits and drawbacks. The major advantage of horizontal separators over vertical ones is that they are compact and less expensive for the required gas and liquid flow rate, and are more effective in cases with high gas–liquid ratio and foaming crude. Because of the large liquid–gas interface in a horizontal separator, there are more opportunities for gases to escape and more time for the liquid droplets to settle down. However, horizontal separators cannot handle solid sediments as well as vertical separators can. Vertical separators offer more liquid surge capacity than similar horizontal vessels for a steady-state flow rate. Under normal conditions, horizontal separators are better for oil–gas

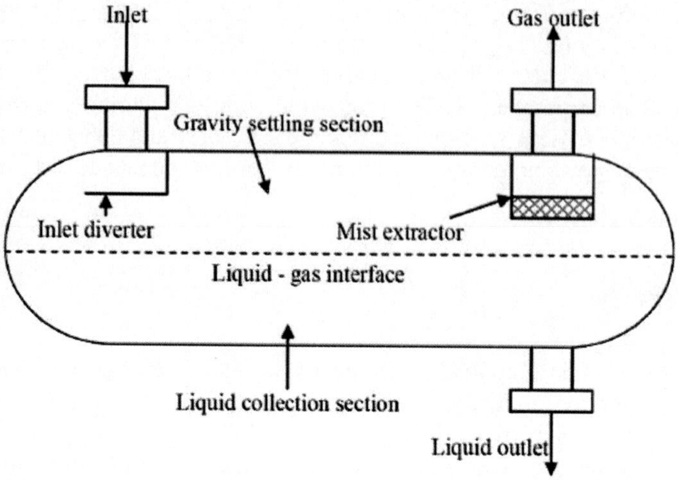

Fig. 7 Layout of a horizontal separator

separation of high gas–oil ratios. They are also better for handling problems with emulsions or foam.

Separator vessels are designed according to the American Society of Mechanical Engineers' boiler and pressure vessel code (ASME Code), section VIII. Pressure vessels are designed to withstand the loadings exerted by internal and external pressures, weight of the vessel, reaction of support, and impacts. Temperature, pressure, and feed composition and its mass flow rate are considered in the selection of type and the design of the vessel, and to come up with the dimensions of the vessel. For this particular study, the only loads considered are internal pressure and temperature. In general, vessel size is decided depending upon the flow rate requirement. For horizontal vessels, the support dimensions are standard and are based on the vessel diameter. During the initial product development stage using engineering design formulae and industrial standards, all the required parameters associated with the separator are calculated. By programming in C++, the above-mentioned design calculations for the separator vessel design are implemented as the embedded engineering knowledge in a program. In a typical application scenario, the designer is provided with operating conditions for the separator vessel in the field. The first task for the designer is to select the type of vessel required; in this case it is either a vertical or a horizontal vessel. The input requirements include flow rate, fluid properties, and working conditions such as pressure and temperature.

In this study, designs for vertical and horizontal separators are developed according to the same set of operating conditions and requirements. The embedded calculation programs have been written generically to handle different designs. The design input requirements used by the designer for designing the separator are given in Table 1.

Table 1 Design requirements for two phase separator

Input parameter definition	Unit	Value
Liquid droplet size in fluid to be separated	μm	140
Liquid flow capacity	BPD	2,000
Gas flow capacity	MMscfd	10
Retention time	min	3
Operating pressure	psi	1,000
Operating temperature	°F	60
Specific gravity of oil	–	40
Specific gravity of gas	–	0.6
Material strength	psi	20,000
Joint efficiency	–	0.9

7.2 Application in CAD Environment for the Data Structure

In the separator case study, the flow of the design process implemented by the program is set up to follow the generally accepted design procedure (ASME pressure vessel design code, section VIII) to solve an engineering problem. First, all the input values are converted from field units to the unit category used for calculations; in this case all the calculations are made with SI units. Next, the relevant nongeometric parameters, associated with operation and the fluid flow, are calculated. These parameters include gas flow rate, liquid flow rate, maximum allowable working pressure, and allowable stress values. They constitute the engineering information essential to calculating the geometric parameters required to build the CAD model for the required type of vessel, and further the FEM and CAE models. All the standard sizes for the components such as vessel diameter and nozzle sizes are stored in external files and can be modified, as per the requirements, by the user. This provides a flexible option for the user to design the vessel using the available standardized elements or components in the inventory. The calculation steps for the vertical and horizontal separator vessels are the same; the difference is in the design of the support, considering the different requirements. For a vertical vessel, a supporting skirt is used, for which design equations are coupled with standard sizes, whereas for a horizontal vessel, saddles are used and standardized for the given vessel diameter. Some of the assumptions made for the design process are: (1) the vessel is designed for the loading of internal pressure; (2) the vessel is designed with essential working components; (3) the pressure and temperature are assumed to be uniform; and (4) design is only for static loading.

After the engineering calculations, the next step of the design process involves a complete conceptual design model. In this step, based on the preliminary calculations, a mid-plane conceptual CAD model in a thin shell is constructed. This is a simplified model of the vessel utilizing the parameters determined from the engineering calculation program, as discussed above, and forms the basic structure of the vessel with the operating essentials. This model is created for the initial

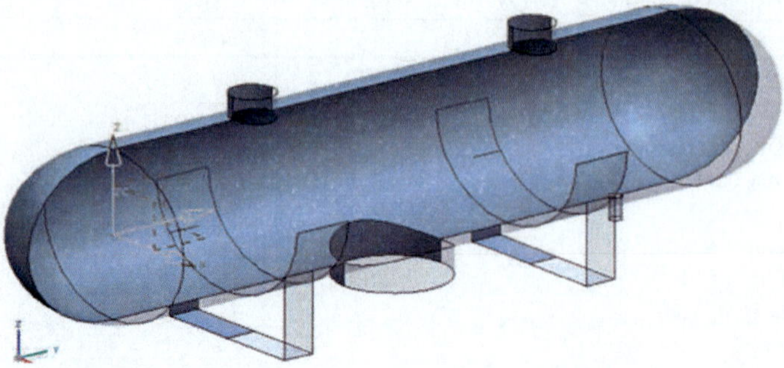

Fig. 8 A separator's conceptual mid-plane CAD model

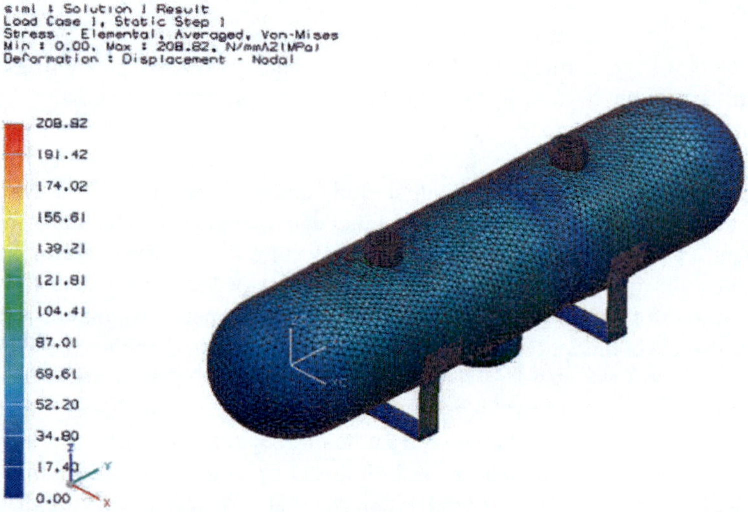

Fig. 9 Von Mises stress analysis results via Nastran with the conceptual mid-plane model (MPa)

CAE verification of the design vessel and the development of the basic vessel shell structure. The FE model for this stage contains only 2D elements; the simulation time (or computing time for solving the FE model) is thus much less than that for an equivalent 3D solid model. The use of a mid-plane analysis cycle reduces the conceptual design time, considering the numerous iterations of design changes required in the early design stages. Figures 8 and 9 show the mid-plane CAD model and the CAE result. They also provide the concept design for the overall structure and can be used to create an early estimation of manufacturing expenses based on material requirements and manufacturability.

To make any changes regarding the design during iteration, the designer is given design-specific choices such as geometric changes, functional changes, or direct changes to specific parameters. Again, depending upon the requirements and the corresponding changes made by the user, the program code regenerates the new parameter values and a new CDM. Older versions of the CDM are stored separately in order to maintain design records for previous iterations with the help of a file-handling mechanism.

7.3 Detailed Design for the Sample Separator Design

Once the conceptual mid-plane model is deemed acceptable, a 3D CAD model is generated and the corresponding 3D FE analysis is performed using the detailed CDM to perform the final analysis iterations. Similar to the mid-plane design process cycle, all the models are automatically generated using external API functions. Design changes are handled in a similar manner. In addition, at this time the user can still return to the mid-plane conceptual stage to change the basic structure and start the process again. The design process goes through a number of iterations until the results obtained are within an allowable stress limit. Figures 10 and 11 show the final CAD model and CAE result. The final result, based on the 3D FE model, shows the maximum stress and deformation values along with their locations. Just like the mid-plane conceptual phase, CDM files in each design iteration phase are maintained for records. Once the design model and the CAE results are finalized, the manufacturing drawings generated from the 3D CAD solid models and the final CDM data can be handed over to the downstream manufacturing engineer for further use.

7.4 FEA Details

The current finite element method used for CAE is based on a division of the design model into smaller elements and a calculation of the loads associated

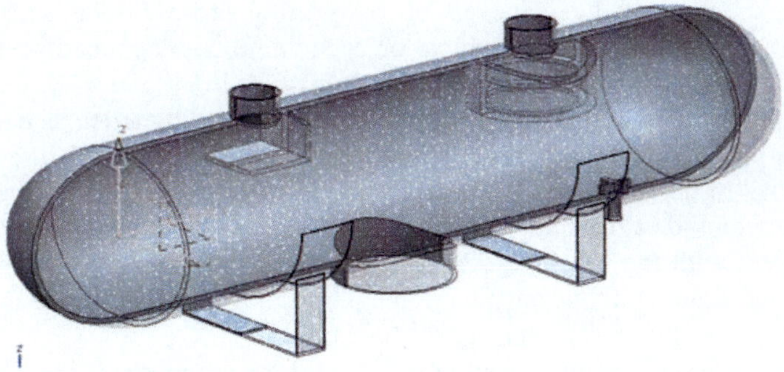

Fig. 10 Detailed 3D solid CAD model

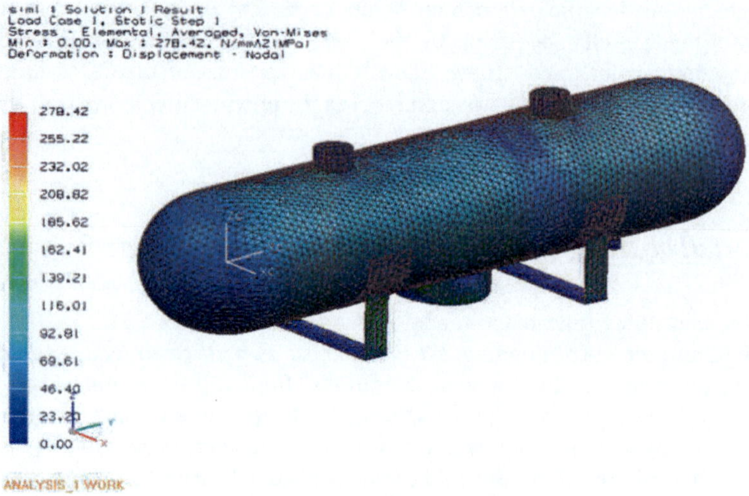

Fig. 11 Von Mises stress analysis results from detailed 3D model via Nastran (MPa)

separately for each element, followed by a combination of the various results for the entire model to get final results. Design of the separator vessel was done under the loading of internal pressure and temperature. All the loads on the structure are static and the material is assumed to be homogeneous and isotropic.

The first step of the analysis requires meshing the available design CAD model into suitable elements. One of the important factors for meshing is to select an optimum mesh size. The mesh element size has to be small enough to produce reliable results, but should be big enough, so that it will take minimum computational time. For any given vessel design, the smallest features of the model are the various thickness associated with the nozzles and vessel body. In the case of mid-plane model mesh, each part of the model is meshed separately to assign specific mesh and physical properties. Well-defined objects during the CAD modeling process make it easier to map those geometry elements with the corresponding mesh properties. Mapping of information between CAD and FE software modules can be achieved by associating those similar sets of input parameters corresponding to individual objects within both applications through a neutral and centralized data structure (the CDM file) [5].

In the preliminary design stage with the mid-plane mesh model, as each design object has no physical thickness associated in the form of geometry, the thickness attribute has to be added in as a parametric physical property. The associativity between various design features in the mid-plane model is maintained by connecting them with the help of grid elements. The connecting mesh elements are massless and rigid, and thus only serve the purpose of transmitting the loads; they do not affect the structural integrity of the model.

In the detailed design stage using the 3D solid model, mesh size was chosen to obtain uniform meshing for those small features, which would ensure their mesh

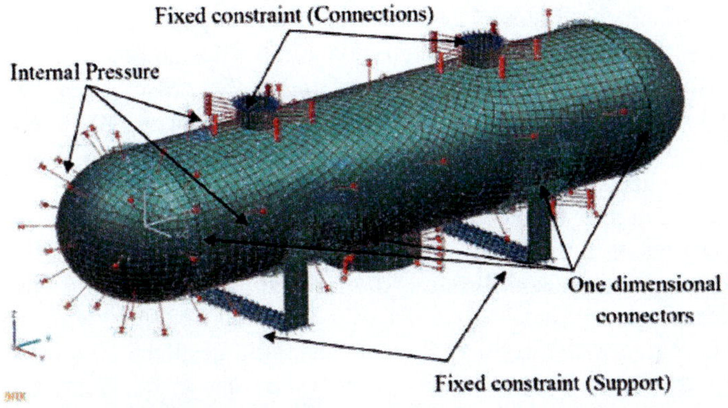

Fig. 12 Load and constraint model

grids are compliant with the other larger features. In the mesh modeling process for the current case study, the mesh creation is made automatic by using NX Open coupled with parametric modeling; however, various physical and material properties need to be assigned manually. The element type used for meshing is tetrahedral, which achieves a uniform distribution around openings and corners as well as controlling the maximum available nodes. Next, the mesh model is applied with specific loads and constraints at the required locations automatically by running the reusable program compiled with NX Open API and parametric modeling functions using the CDM. Figure 12 shows the constraint and load map for a horizontal separator analysis model. Internal pressure is applied on the main vessel body as uniformly distributed load directed outward from the vessel surface. The temperature is also applied to the vessel body and is assumed to be uniformly distributed over the entire surface. A vessel support base at the ground end is given fixed constraints to represent fastening. The nozzle ends are constrained for axial movement along the pipe continuation in respective dimensions. The model then goes through static analysis calculations and results in the form of maximum Von Mises stress and maximum deformation are obtained (see Fig. 11).

The designer observes the results to judge the validation and the expected ranges of parameters based on the constraint settings. The loading conditions can be modified, and the analysis can be repeated to re-evaluate the solution. In the case of interpolated continuously changing loads and constraints, an iterative solver should be used to get a more refined and accurate solution.

7.5 Expansion of the Case Study to Vertical Separators

To test the general applicability of the proposed method to different product configurations, the same design development structure is used to design two different types of vessels with some conceptual design differences and considerable geometric differences. Figure 13a shows the basic layout of a vertical separator along with various computer models involved in the design process. For a vertical separator, the mixed liquid and gas flow enters the vessel from the side and impacts the inlet diverter. The initial gross separation happens at the inlet diverter. The liquid droplets fall down by gravity in the collection section. There are a few internal components in the collection section, such as the vortex breaker for the outlet; aside from that, this section is just a storage space for liquid designed as per the required retention time. The design should provide enough retention time, so that liquid should reach equilibrium and dissolved gases should rise upward to the vapor space.

While rising up with the gases, the heavier liquid droplets fall to the collection section by gravity only; those droplets having a very small diameter (less than 100 μm, approximately) are usually carried by the gases until reaching the mist extractor. When gas passes through the mist extractor, these small droplets collide with the mist extractor coalescing section and fall downward to the collection section owing to lost momentum. Vertical separators are used for a flow with low or medium gas–liquid ratio. A vertical separator fitted with a false cone at the bottom can be effectively used to separate heavier solid particles such as sand and other sediments. Thus, in case of a flow with a small amount of sediment, residue can be effectively handled by the vertical separator. In the case of a low oil–gas ratio, vertical vessels are more effective. They can also be used as gas scrubbers where only fluid mists need to be removed from the gas, which demands extra surge capacity.

For the design of a vertical separator, a liquid capacity constraint is used because of the assumed higher liquid-to-gas ratio. Locations of all the nozzle openings for a vertical separator are usually fixed and are subject to change only because of major installation problems. A support skirt for the vertical vessel is custom designed for each vessel depending upon the weight load and momentum. In normal practice, external wind pressure or momentum has to be considered for the design, as the whole structure is considered to be a "column." Regarding nozzle locations for vertical vessels, locations for manholes and gas outlets are fixed. For fluid and liquid nozzle opening locations, only height is fixed; peripheral location depends upon ease of installation. The design process for the vertical separator follows the same procedure as the horizontal separator. Figure 13b shows the mid-plane thin shell model as well as the 3D solid detailed model. The CAE analysis results based on the mid-plane mesh model are given in Fig. 13(c), while the 3D results are shown in Fig. 13d.

8 Advantages and Limitations

The proposed design method is an innovation beyond the conventional design process with CAD and CAE tools used separately. The product development time, especially the effort associated with cyclic CAD modeling and CAE analysis, can

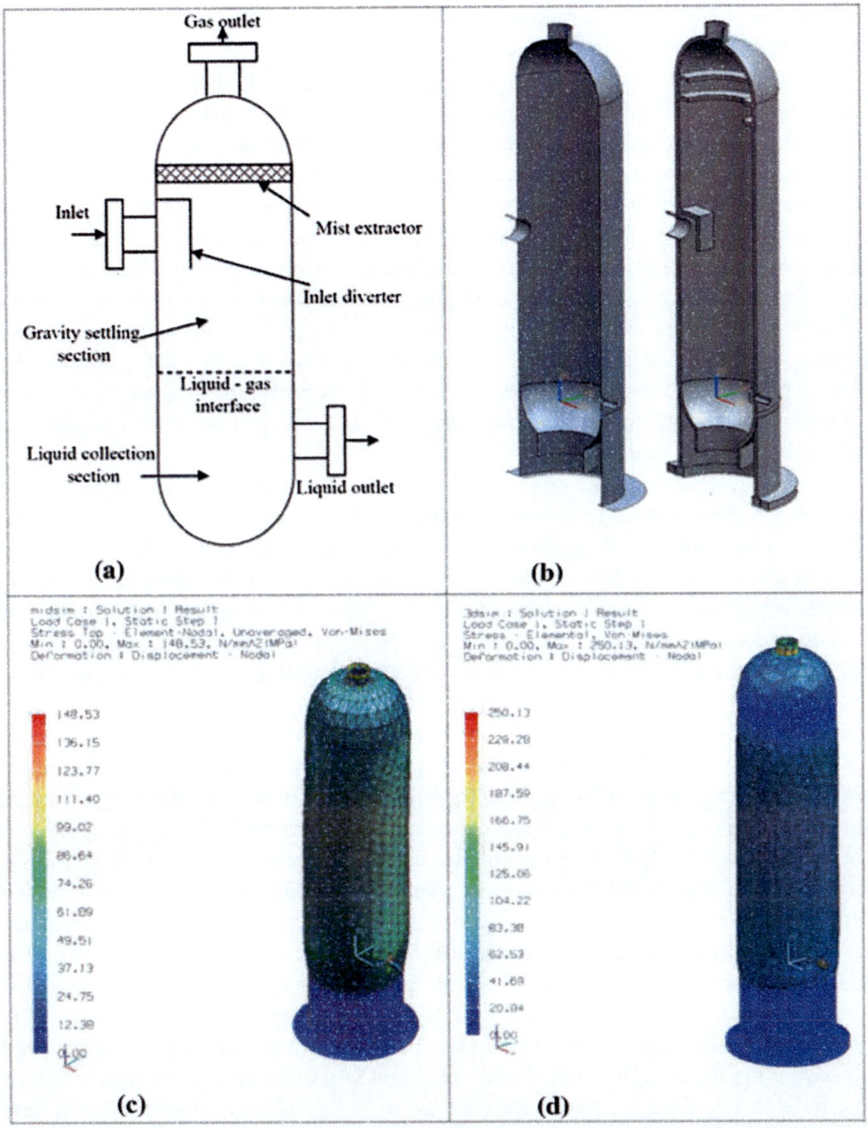

Fig. 13 Example vertical separator design and analysis results. **a** Layout of a vertical separator. **b** Mid-plane 3D detailed CAD models. **c** Mid-plane CAE analysis. **d** 3D detailed CAE anaysis

be greatly reduced. For example, running automated design and analysis for one cycle with the given case takes about 25 min after the development of the reusable programs, which include modules for CAD construction and CAE analysis. In contrast, the interactive approach takes 6 h for a skilled user to run the same procedure. However, the automated approach requires CDM construction and code preparation, which takes a one-time effort of 12 h initially. If the cycles of CAD and CAE activities require several rounds, this one-time effort can easily be justified, as subsequent cycles will save quite a bit of time. Potential time saved can be likened to how much a computerized numerical control (CNC) metal cutting method can save in comparison to manual machining methods.

Since the proposed process makes better use of integrated CAD and CAE software capabilities, it helps reduce the modeling and analysis time significantly via the use of parametric and procedural knowledge captured in much shorter program development time, while still maintaining the traditional design procedure. The journaling application for program development clearly reduces the need for programming expertise as well as the development time associated with the generative CAD/CAE approach. The method proposed here also provides the flexibility to handle complex modeling and analysis problems. The method makes it easier than before to capture modeling and analysis procedural knowledge as well as to make it reusable.

Parametric CAD models represent a smart product representation at a point of time. Procedural models reflect the product construction operations over a period of time. They are complementary and cannot replace each other. The proposed method leverages both the knowledge representations and processing mechanisms. Hence, it makes a unique contribution to this aspect of design methodology.

The use of a CDM makes it easier to transfer and control the specific information associated with various models as well as coupling the computer modeling process with an engineering knowledge-embedded program. Parameter-based modeling and analysis facilitates changes associated with specific design features.

The limitation of this method comes from the restriction of the journal file: not all the software capabilities can be fully utilized with a journaling application, since it is not capable of recording every manual operation. Though the program created is controllable and changeable to a certain degree, it lacks ideal flexibility. However, the traditional limitation of using the journal files, i.e., repeating the exact scenario as recorded, has been significantly relaxed by incorporating parametric modeling capabilities. In the new method, so long as the design and analysis procedures are the same, and the parameters given are valid inputs, different products can be designed with the same generated program code. The parameter values can be checked upfront before being used in the automated procedure via a constraint-checking algorithm. Only those sets of parameter values that have satisfied all the constraints are listed in the CDM as the acceptable data sets to be used.

As the journal file itself has limited capabilities for data and file handling, after converting it into a reusable program, the program has to be associated with an external data repository (the CDM) and file management system to make it flexible enough to handle changes. This task requires programming skill. To adopt the process, engineers will need to go through certain initial training and must understand software API programming.

9 Conclusion and Future Work

An innovative design process integration mechanism has been developed to efficiently create a reusable program structure obtained by editing journal files where the procedural engineering knowledge is embedded and by incorporating parametric modeling knowledge. With this novel mechanism, this chapter has explored a systematic automation method for knowledge capture and reuse via CAD and CAE programming. This method is able to improve the efficiency of the overall product development process. The use of a common parametric data model, which semantically connects CAD and CAE models, enables the designer to have better control over the process. By changing the design contents and parameter values, an automatic design process with minimal human interface is proposed and a demonstration case developed. The adaptive nature of the proposed method enables faster development of those products that have relatively clear development strategies and processes. The proposed method offers parametric flexibility for different design applications. The method enables the automation of design modeling and analysis processes more systematically, effectively, and conveniently.

Ideally, in the proposed method, the knowledge can be captured in the form of modular structures coupled with optimization algorithms and decision-making procedures and matrices. This element of the method has to be explored in future work due to limitations on available research resources. Future work is also necessary for refining the data and knowledge management capability with more generic program code development.

References

1. Bossak MA (1998) Simulation based design. J Mater Proc Technol 76:8–11
2. Cao BW, Chen JJ, Huang ZG, Zheng Y (2009) CAD/CAE integration framework with layered software architecture. In: Proceedings of 11th IEEE international conference on computer-aided design and computer graphics. Huangshan, China
3. Chen G, Ma YS, Thimm G, Tang SH (2004) Unified feature modeling scheme for the integration of CAD and CAx. Computer-Aided Des Appl 1:595–601
4. Chen G, Ma YS, Thimm G, Tang SH (2006) Associations in a unified feature modeling scheme. J Comput Inf Sci Eng 6:114–126
5. Gujarathi GP, Ma YS (2011) Parametric CAD/CAE integration using a common data model. J Manuf Syst 30:117–186
6. Hoffmann CM, Kim KJ (2001) Towards valid parametric CAD models. Computer-Aided Des 33:81–90
7. Johansson J (2009) Manufacturability analysis using integrated KBE, CAD and FEM. In: Proceedings of the ASME international design engineering technology conference, DETC 2008, vol 5, pp 191–200
8. Mital A, Desai A, Subramanian A, Mital A (2008) Product development a structured approach to consumer product development design and manufacture. Elsevier, Oxford
9. Monedero J (2000) Parametric design: a review and some experiences. Autom Constr 9:369–377

10. Pinfold M, Chapman C (2001) The application of KBE techniques to the FE model creation of an automotive body structure. Comput Ind 44:1–10
11. Radhakrishnan P, Subramanyan S, Raju V (2008) CAD/CAM/CIM. New Age International, Daryaganj, Delhi
12. Shah JJ, Ntylä MM (1995) Parametric and feature-based CAD/CAM: concepts, techniques, and applications. Wiley-Interscience, New York
13. Shephard MS, Beall MW, O'Bara RM, Webster BE (2004) Toward simulation-based design. Finite Elem Anal Des 40:1575–1598
14. Stewart M, Arnold K (2009) Gas–liquid and liquid–liquid separators. Gulf Professional Publishing, Oxford
15. Su D, Wakelam M (1998) Intelligent hybrid system for integration in design and manufacture. J Mater Proc Technol 76:23–28
16. Susca L, Mandorli F, Rizzi C, Cugini U (2000) Racing car design using knowledge aided engineering. Artif Intell Eng Des, Anal Manuf 14:235–249
17. Tomiyama T, Gu P, Jin Y, Lutters D, Kind CH, Kimura F (2009) Design methodologies: industrial and educational applications. CIRP Annals Manuf Technol 58:543–565
18. Xu X, Weiss U, Gao G (2002) The integration of CAD/CAM/CAE based on multi-model technology in the development of cylinder head. J Autom Technol 3:47–52
19. Zeng S, Peak RS, Xiao A, Sitaraman S (2008) ZAP: a knowledge-based FEA modeling method for highly coupled variable topology multi-body problems. Eng Comput 24:359–381

Index

Y.-S. Ma (ed.), *Semantic Modeling and Interoperability in Product and Process Engineering*, Springer Series in Advanced Manufacturing,
DOI: 10.1007/978-1-4471-5073-2, © Springer-Verlag London 2013

D

E

F

G